BIOTECHNOLOGY IN AGRICULTURE SERIES

General Editor: Gabrielle J. Persley, Science Adviser, The Agricultural Research Group, Environmental and Sustainable Development, The World Bank, Washington DC, USA.

For a number of years, biotechnology has held out the prospect for major advances in agricultural production, but only recently have the results of this new revolution started to reach application in the field. The potential for further rapid developments is, however, immense.

The aim of this book series is to review advances and current knowledge in key areas of biotechnology as applied to crop and animal production, forestry and food science. Some titles focus on individual crop species or groups of species, others on specific goals such as plant protection or animal health, with yet others addressing particular methodologies such as tissue culture, transformation or immunoassay. In some cases, relevant molecular and cell biology and genetics are also covered. Issues of relevance to both industrialized and developing countries are addressed, and social, economic and legal implications are also considered. Most titles are written for research workers in the biological sciences and agriculture, but some will also be useful as textbooks for senior-level students in these disciplines.

BIOTECHNOLOGY IN AGRICULTURE SERIES

Titles Available:

1: Beyond Mendel's Garden: Biotechnology in the Service of World Agriculture
G.J. Persley
2: Agricultural Biotechnology: Opportunities for International Development
Edited by G.J. Persley
3: The Molecular and Cellular Biology of the Potato
Edited by M.E. Vayda and W.D. Park (out of print)
4: Advanced Methods in Plant Breeding and Biotechnology
Edited by D.R. Murray
5: Barley: Genetics, Biochemistry, Molecular Biology and Biotechnology
Edited by P.R. Shewry
6: Rice Biotechnology
Edited by G.S. Khush and G.H. Toenniessen
7: Plant Genetic Manipulation for Crop Protection
Edited by A. Gatehouse, V. Hilder and D. Boulter
8: Biotechnology of Perennial Fruit Crops
Edited by F.A. Hammerschlag and R.E. Litz
9: Bioconversion of Forest and Agricultural Plant Residues
Edited by J.N. Saddler
10: Peas: Genetics, Molecular Biology and Biotechnology
Edited by R. Casey and D.R. Davies
11: Laboratory Production of Cattle Embryos
I. Gordon
12: The Molecular and Cellular Biology of the Potato, 2nd Edition
Edited by W.R. Belknap, M.E. Vayda and W.D. Park

The Molecular and Cellular Biology of the Potato

Second Edition

W.R. Belknap
United States Department of Agriculture
Agricultural Research Service
Albany, California
USA

M.E. Vayda
Department of Biochemistry, Microbiology and Molecular Biology
University of Maine
USA

W.D. Park
Department of Biochemistry and Biophysics
Texas A&M University
USA

CAB INTERNATIONAL

CAB INTERNATIONAL
Wallingford
Oxon OX10 8DE
UK

Tel: Wallingford (01491) 832111
Telex: 847964 (COMAGG G)
E-mail: cabi@cabi.org
Fax: (01491) 833508

A catalogue record for this book is available from the British Library.

ISBN 0 85198 900 4

Printed and bound in the UK by Biddles Ltd, Guildford

Contents

List of Contributors

William R. Belknap (13, 17), *USDA/ARS, Western Regional Research Center, Albany, CA, USA*

Merideth W. Bonierbale (1), *Cornell University, Ithaca, NY 14583, USA*

Jaap S. Bosscher (9), *Department of Genetics, University of Groningen, Haren, The Netherlands*

B.B. Brodie (1), *Cornell University, Ithaca, NY 14583, USA*

M.M. Burrell (4), *Advanced Technologies (Cambridge) Limited, Cambridge, UK*

A.J. Conner (18), *New Zealand Institute for Crop and Food Research Ltd., Christchurch, New Zealand*

Dennis Corsini (17), *University of Idaho Research and Extension Center, USDA/ARS, Aberdeen, ID, USA*

H.V. Davies (6), *Department of Cellular and Environmental Physiology, Scottish Crop Research Institute, Invergowrie, UK*

I.B. Dry (11), *CSIRO Division of Horticulture, Adelaide 5001, Australia*

Klaus Düring (16), *Federal Center for Breeding of Cultivated Plants, Institute for Breeding Methods in Vegetables, Quedlinburg, Germany*

Ali El-Kharbotly (2), *Department of Plant Breeding, Wageningen Agricultural University, The Netherlands*

Elmer E. Ewing (1), *Cornell University, Ithaca, NY 14583, USA*

Will J. Feenstra (9), *Department of Genetics, University of Groningen, Haren, The Netherlands*

Ulf-Ingo Flügge (5), *Institut für Botanik, Würzburg, Germany*

Wolf Bernd Frommer (5), *Institut für Genbiologische Forschung, Berlin, Germany*

Joan E. Garbarino (13), *USDA/ARS, Western Regional Research Center, Albany, CA, USA*

L.W. Godfrey (15), *Department of Plant Molecular Physiology, University of Hawaii, Honolulu, HI, USA*

David J. Hannapel (12), *Department of Horticulture, Iowa State University, Ames, IA, USA*

Bas te Lintel Hekkert (2), *Center for Plant Breeding and Reproduction Research, Wageningen, The Netherlands*

Hans Heldt (5), *Institut für Botanik, Göttingen, Germany*

Carrie Hiser (10), *Michigan State University, DOE Plant Research Laboratory, East Lansing, MI, USA*

Jeanne M.E. Jacobs (2), *Center for Plant Breeding and Reproduction Research, Wageningen, The Netherlands*

Evert Jacobsen (2), *Department of Plant Breeding, Wageningen Agricultural University, The Netherlands*

Jens Kossmann (5), *Institut für Genbiologische Forschung, Berlin, Germany*

Lee McIntosh (10), *Michigan State University, Biochemistry Department, East Lansing, MI, USA*

James K. Morelli (14), *Department of Biochemistry, University of Maine, Orono, ME, USA*

Bernd Müller-Röber (5), *Institut für Genbiologische Forschung, Berlin, Germany*

Paul A. Nakata (3), *Department of Biochemistry and Biophysics, Washington State University, Pullman, WA 99164, USA*

Thomas W. Okita (3), *Institute of Biological Chemistry, Washington State University, Pullman, WA 99164, USA*

Joseph J. Pavek (17), *University of Idaho Research and Extension Center, USDA/ARS, Aberdeen, ID, USA*

Andy Pereira (2), *Center for Plant Breeding and Reproduction Research, Wageningen, The Netherlands*

Omaira Pineda (1), *Cornell University, Ithaca, NY 14583, USA*

Robert L. Plaisted (1), *Cornell University, Ithaca, NY 14583, USA*

Jorg Riesmeier (5), *Institut für Genbiologische Forschung, Berlin, Germany*

S.P. Robinson (11), *CSIRO Division of Horticulture, Adelaide 5001, Australia*

David R. Rockhold (17), *USDA/ARS, Western Regional Research Center, Albany CA, USA*

Gordon W. Snyder (17), *USDA/ARS, Western Regional Research Center, Albany CA, USA*

Uwe Sonnewald (5), *Institut für Genbiologische Forschung, Berlin, Germany*

Joe Sowokinos (7), *Department of Horticultural Science, University of Minnesota, St Paul, MN 55108, USA*

James P. Spychalla (8), *Spychalla Farms, Inc., Antigo, WI, USA*

Willem J. Stiekema (2), *Center for Plant Breeding and Reproduction Research, Wageningen, The Netherlands*

S.M. Sun (15), *Department of Plant Molecular Physiology, University of Hawaii, Honolulu, HI, USA*

P.W. Thygesen (11), *CSIRO Division of Horticulture, Adelaide 5001, Australia*

Ward M. Tingey (1), *Cornell University, Ithaca, NY 14583, USA*

H.M. Tu, (15), *Department of Plant Molecular Physiology, University of Hawaii, Honolulu, HI, USA*

Jan Van den Berg (1), *Cornell University, Ithaca, NY, 14583, USA*

Feike R. van der Leij (9), *Department of Biotechnology, Swiss Federal Institute of Technology, Zürich, Switzerland*

Michael E. Vayda (14), *Department of Biochemistry, University of Maine, Orono, ME, USA*

R. Viola (6), *Department of Cellular and Environmental Physiology, Scottish Crop Research Institute, Invergowrie, UK*

Lothar Willmitzer (5), *Institut für Genbiologische Forschung, Berlin, Germany*

G. Craig Yencho (1), *Cornell University, Ithaca, NY 14583, USA*

Preface

The four years since the publication of the first edition in this series have seen marked advances in our basic understanding of potato molecular genetics and the application of this technology to improvement of the crop. Considerable progress has been made in such areas as gene mapping and expression, analysis of disease and pest resistance loci, characterization and manipulation of carbohydrate metabolism, as well as safety and performance issues associated with the field release of transgenic clones. These advances reflect the development of important new molecular tools which, when integrated with traditional selection and breeding techniques, are expected to lead to the introduction of transgenic potatoes into the marketplace within five years.

The chapters in this volume review material presented at the Third International Symposium on the Molecular Biology of the Potato (Santa Cruz, CA 1993). While this volume is not intended as a comprehensive survey, we have attempted to incorporate many of the more important developments since the publication of the previous edition.

The editors wish to thank the authors of the individual chapters for their valuable contributions.

William R. Belknap
Michael E. Vayda
William D. Park

Chapter One:

Potato Improvement by Traditional Breeding and Opportunities for New Technologies

Robert L. Plaisted, Merideth W. Bonierbale, G. Craig Yencho, Omaira Pineda, Ward M. Tingey, Jan Van den Berg, Elmer E. Ewing and B.B. Brodie
Cornell University, Ithaca, NY, USA.

Potato breeding programs worldwide are undergoing a period of rapid change. Like most endeavors, in order to be successful, breeders must adapt and incorporate appropriate techniques as they become available. To date, potato cultivars have been developed using traditional means. However, the use of the term 'traditional' should not be interpreted as being static. As new tools or techniques become available, breeders have adopted those which give them an advantage. The advances in biotechnology, analytical procedures, instrumentation and computation that have occurred during the last decade promise to speed greatly the development of new cultivars.

However, certain aspects of the breeding process will remain relatively constant, such as the time consuming and often expensive field evaluation and replicated yield trial procedures which are the foundation of a potato breeding program. In this chapter, we provide a general introduction to a traditional potato breeding program. Because it is difficult to judge progress if an appreciation for the past is lacking, we first provide an overview and brief history of potato breeding in North America. We then outline a typical potato breeding program and review the shifts in varieties that have occurred in recent years. Finally, new opportunities for advances in potato breeding are discussed, using examples from our own program.

Potato Breeding in North America: an Overview

There are currently 13 variety development programs in the US. Ten of these are state sponsored, two are USDA programs and one is in private industry. The 12 public programs have a capacity of screening an initial population of about half

The Molecular and Cellular Biology of the Potato, Second Edition
Edited by W.R. Belknap, M.E. Vayda and W.D. Park

Table 1.1. Number of years between cross and variety release for 35 US and Canadian varieties released between 1983 and 1992.

Years	No. of varieties
10	2
11	1
12	2
13	5
14	8
15	3
16	1
17	2
18	2
19	3
20	2
21	1
22	1
23	1
24	1

Mean = 16 years
Mode = median = 14 years

a million genotypes each year. In Canada the major breeding effortis at four sites which are wholly or largely national programs. In addition there is one provincial, and four quite small private programs. The public programs in Canada have a capacity of about 120,000 new seedlings each year.

In the past ten years, the public programs in the United States and Canada have produced 58 new varieties. In the previous decade, 49 varieties were named, and between 1932 and 1985, 201 varieties were named. Since World War II, about five varieties have been released each year with over 100,000 seedlings being grown to find one worth giving a name. Generally, about one in ten newly named varieties becomes established in the seed industry. Thus, breeders typically screen a million seedlings for each new variety.

If one looks at the number of years between the time a cross is made and a variety named, the range for the varieties released in the most recent decade is between 10 and 24 years (Table 1.1). The mode and the median length of time is 14 years. There have been occasions on which these numbers have been used to declare the traditional method to be too slow and too expensive to be continued. However, three points warrant emphasis:

1. The time taken to produce new genotypes is very brief and the least expensive part of the process. The bulk of the time is spent evaluating advanced selections and getting them into seed programs.

Table 1.2. Major potato varieties, US and Canada, 1991.

Variety	Acres	Year of release
Russet Burbank	89,950	1886
Superior	21,345	1961
Shepody	18,376	1980
Norchip	15,668	1968
Kennebec	15,577	1948
Atlantic	15,488	1976
Russet Norkotah	14,575	1987
All Varieties	259,118	

2. Traditional programs are producing new varieties as rapidly as the seed production system can accommodate them.
3. The traditional system is not static, but changes as new techniques arise and prove to be worthwhile.

Variety Development Since 1920

In 1991, seven varieties occupied 74% of the total US and Canadian seed acreage (Table 1.2). One of these, Russet Burbank, which represents 34% of the total, is a special case. Its predominance is based on special market properties, merchandizing and national shifts in regions of production. Of the remaining six, one was released in the late 1940s, two in 1960s, one in the 1970s, and two in the 1980s.

The shifts in importance of varieties in the US since 1920 resulted from shifts in production regions and the popularity of varieties. Until 1940, three varieties were grown on about two thirds of the US acreage. They were Bliss Triumph, Green Mountain and Irish Cobbler. All three originated between 1875 and 1885. In 1933, Katahdin and its sibling Chippewa were released and began to have an impact. By 1955, these two took over about one third of the US acreage, replacing Green Mountain, Cobbler and Triumph. In 1948 and 1949, Red Pontiac and Kennebec were released and about this time Russet Burbank acreage began to increase from the 5-10% it had occupied since the early part of the century. In 1965, Russet Burbank surpassed Katahdin as the leading variety and Kennebec was third in acreage. Together they occupied two thirds of US acreage. The current distribution is given in Table 1.2.

At the present, New York State (NY) does not represent a significant portion of US potato production, but at the start of this century, NY ranked first in US production, with about the same number of acres that presently are grown in Idaho. The changes in variety preference and utilization that have occurred in

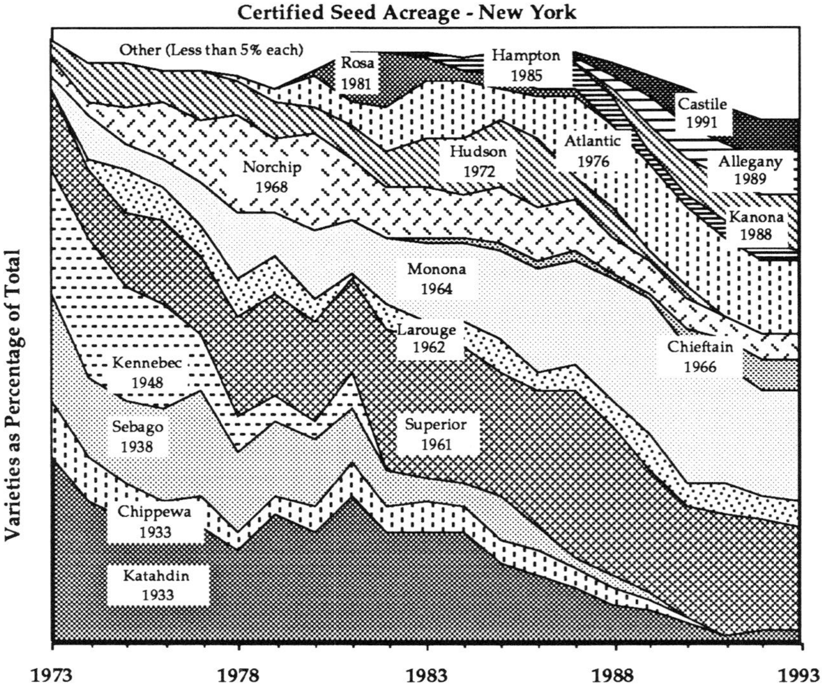

Figure 1.1. Shifts in importance of varieties produced in New York State during 1973-1993. Data compiled from *New York Certified Seed Potatoes* (Crop Directory, NYSIC, Ithaca, NY).

the NY potato industry, in many ways, reflect those that have occurred elsewhere in North America (with the exception of Russet Burbank, which is primarily grown in the Pacific Northwest). Seed production records for the 40 year period between 1917 and 1957 show that for the first 20 years the variety acreage in NY was very stable. Green Mountain, Russet Rural, Smooth Rural and Irish Cobbler were about the only varieties certified. Katahdin, Chippewa and Sebago rapidly replaced them in the early 1940s so that by 1957, the three older varieties were about gone and the early USDA varieties, Katahdin, Chippewa and Sebago were grown on about 70% of the acreage. Kennebec and Ontario made an initial impact. Overall the number of varieties was still small and the shifts were not extreme. It was an easy time for seed growers to decide which varieties to grow. Most of the changes occurred between 1935 and 1945. The four early varieties had lasted about 60 years.

The next 15 years, between 1957 and 1972, were quite stable with Katahdin, Chippewa and Sebago being the predominant varieties and with Kennebec and Superior gaining in acreage. The last 20 years (Fig. 1.1) has seen the disappearance in NY of Katahdin, Chippewa, Sebago and Kennebec. These were important for about 50 years. Superior, Monona and Atlantic have taken

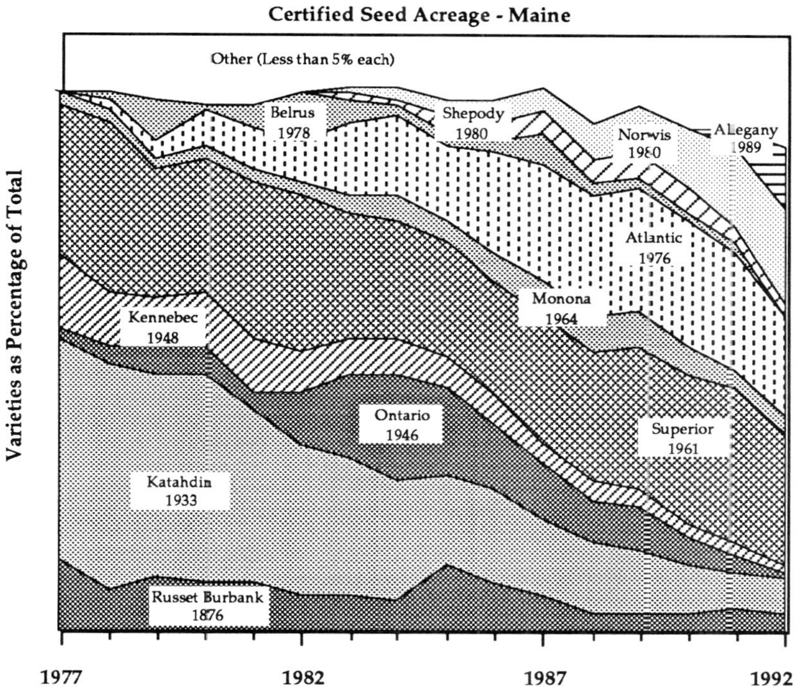

Figure 1.2. Shifts in importance of varieties produced in Maine during 1977-1992. Data compiled from *Maine Seed Potatoes* (Department of Agriculture, Food and Rural Resources, Augusta, ME).

over as the predominant varieties, but only comprise about half the total acreage. Unlike prior years, there is a considerably greater number of varieties and the shifts between varieties have been more rapid, making it more difficult for seed growers to decide which varieties to grow. A big factor has been the increase in importance of producing potatoes for chipping (making potato chips or crisps). In this case breeding for better chipping quality has had a marked effect. In the 1950s, Russet Rural was the variety of choice. This was replaced by Kennebec in the 1960s. Then in the 1970s, Monona and Norchip replaced Kennebec. In the 1980s, as storage quality became more important, Monona began to replace Norchip and the high solids of Atlantic made it a favorite for chipping directly from the field. Another impact of breeding can be illustrated by the infusion of a single gene for resistance to the golden nematode. In 1992, the resistant varieties, Atlantic, Hampton, Kanona, Allegany and Castile were grown on about one third of the total acreage.

Since NY seed acreage is not as large as it once was, it is useful to look at Maine acreage in the same way (Fig. 1.2). The situation is very similar. Katahdin, Kennebec and Ontario occupied about half the acreage up to 1985.

Superior and Atlantic have been strong throughout the period. Norwis and Allegany are gaining in importance. Thus, in both Maine and NY, and to some extent the rest of North America, the number of varieties being certified is becoming greater and the shifts between varieties is more rapid.

Breeding Objectives

Each of the North American breeding programs has a different blend of objectives dependent upon the needs of the industry it serves. The first division is based on market goals, usually tablestock or processing. The tablestock use is primarily for baking or boiling. The processing is primarily for french fries or chips (potato crisps). Another subdivision, not independent of the first, is tuber shape and color: the primary classifications are round white, round red, or long russet. There is a small, but growing market for yellow flesh and purple skin color.

Each program also has some disease and insect resistance features that are important to its clientele. The most common objective is resistance to common scab. Next would be resistance to soil-borne pests which cause early dying, such as wilts and nematodes. Breeding for resistance to the late blight fungus which at one time was the major breeding objective in North America, is now almost totally absent. Extreme resistance to the viruses potato virus X (PVX) and potato virus Y (PVY) are available in two dominant genes, but until recently the seed certification programs controlled these diseases. Now there is some interest in incorporating resistance to PVY into new varieties, but when the incidence of this disease declines, as it has in the past, this interest in PVY-resistant varieties will likely disappear again.

A Typical Program of Variety Development

Each breeding program has a system of selection adapted to the local situation, and frequently is altered as opportunities change. An outline that is fairly representative is as follows:

Year 1	Make crosses between selected parents. Extract seeds from the fruits.
Year 2	Sow seeds and transplant to pots or directly to the field. At harvest may select for tuber color.
Year 3	Plant seedling tubers in one to four hill plots. Select for tuber appearance. Save one tuber or entire hill.
Years 4 and 5	Plant plots of four to 20 hills in year four and 40 to 160 hills in year five for seed and selection purposes. Select for tuber appearance, some disease resistance, internal tuber defects and processing quality.

Years 6-8	Conduct small plot replicated yield trials with increasing replication and number of sites each year. Seed increase continues.
Years 9-11	Continue replicated plot yield trials and seed increases. Begin farm-scale evaluations. Introduce virus testing and *in vitro* multiplication.
Years 12 and 13	Continue replicated plot yield trials and seed increases. Initiate farm-scale tests using commercially grown foundation seed. Conduct special trials to evaluate all production features.
Year 14	Name the variety.

As soon as the first generation is grown in the field, the process of trying to find a desirable clone begins. The earliest generation selection has to be on traits that are highly consistent and quickly identified. Examples are skin color, tuber defects and some disease susceptibilities. As the number of clones is reduced and the amount of seed of each is increased, then selection can be applied for more variable traits which require replication, such as yield, quality features and disease resistance. In the earliest generations the selection rate may be as low as 5%, but as the generations proceed the rate increases. In the early yield trials it may be about 30%. Each generation the size of plot, number of replications and number of test sites increase.

The number of clones typically represented by each cohort at a specific stage of selection is as follows:

Year 1	> 500,000 true seeds
Year 2	45,000 as transplants
Year 3	40,000 as seedling hills
Year 4	1500 in unreplicated plots
Year 5	150 in larger unreplicated plots
Years 6-8	50 in year 6 to fewer than ten in year 8 in replicated plots
Years 9-11	Six in regional trials in year 9 to 1 in year 11
Years 12 and 13	None or one in preliminary grower evaluation and commercial seed production
Year 14	None or one to be released

Seed Production

Most breeding programs are located where they can produce virus-free seed and are associated with a seed production system. Of the 16 states which produce certified seed, nine have breeding programs and these produce 90% of the certified seed in the US. Depending on the program, seed production systems may be selected seed growers, private *in vitro* plantlet multipliers, or special foundation seed farms that are generally managed by a public agency. It is often necessary to use seed produced by the breeding program in the initial stages of grower evaluation, but as soon as possible the breeders' seed should be replaced by foundation or certified seed produced in the normal seed channels. It is critical at the time of release that an adequate supply of commercial seed be available. The success of the release will be damaged by too little or too much seed at this stage. The actual amount of seed is dependent on the location of release and the market need.

Advances in Potato Breeding

As indicated earlier, the use of the term traditional should not be interpreted as being static. As new tools or techniques become available, breeders adopt those which give them an advantage. Before discussing the more recent advances in potato breeding technologies, it is useful to consider techniques that have shaped potato variety development and to comment on genetic diversity.

Seedling selection

Between 1850 and 1900 there was a considerable activity of growing seeds from open pollinated berries by private individuals. Partly, this was in response to the idea that potato varieties 'ran out'. The decline was attributed to the repeated asexual propagation - a cycle of sexual propagation was necessary for rejuvenation - but it was probably due to the accumulation of viruses in the seed stock. Part of the incentive for this activity was the excitement of the variability present in the seedlings, and part of it was the financial gains that could be made in the initial years of seed increase. One of the problems was the shortage of varieties that would set berries in the field. A few selectors made controlled pollinations.

In 1910, the USDA initiated a formal breeding project. In the same period several states also initiated breeding projects, often as collaborators with the USDA project. These programs were based on seeds from controlled crosses, but a major limitation was the small number of clones that functioned as males. Then in 1932, the USDA released the variety Katahdin which in a few years became the most popular variety in the Northeast because of its yield, appearance and pest resistance properties.

Some of the techniques developed by this generation of breeders are still used today. These include conditions to induce plants to flower, techniques to collect and store pollen, and techniques to extract seeds.

Somatic variation

Somatic variation is the selection of naturally occurring variants called 'sports'. This is a technique that is not often applied in a routine way, rather it is an opportunistic one. The primary examples are the chance occurrence of a russet skin on a white-skinned variety (Russet Burbank), of dark red versions of lighter red-skinned varieties (Red Pontiac from Pontiac), and of clones with later maturity than the original release.

In the 1950s it was forecast that radiation would facilitate the use of somatic variation in potato breeding. Later, the somaclonal variation that can occur after plants are recovered from a callus phase in *in vitro* culture was promulgated by some as being a replacement for the hybridization method. Neither prophecy came to pass. However, somaclonal variation has proven to have merits in special situations, for example, for producing tetraploid forms of a diploid clone, or for promoting genetic interchange in interspecific crosses. In general, somatic variation introduces more undesirable traits than desirable ones (Karp, 1990).

Genetic diversity

The assessment of the extent of genetic diversity is a matter of speculation. Some argue that the preeminent acreage (34%) of the single variety, Russet Burbank is a sign of genetic restriction. However, in the first half of this century, three varieties occupied 67% of US acreage, and at mid-century Katahdin occupied half the acreage. Thus, in terms of the numbers of varieties being grown, the diversity is increasing (the number of varieties was greater a century ago, but many of these were synonyms resulting from line selections). With the notable exception of the introduction of a clone from South America, Rough Purple Chili, in 1851 by Chauncey Goodrich (Plaisted and Hoopes, 1989) presumably most of the varieties were derived from a few European introductions. Either Goodrich's seedling had exceptional value, or its widespread incorporation in pedigrees reflects the need for diversity in the clones of that time.

For all the diversity present in the unadapted tuber-bearing species of *Solanum* and in spite of the ease of securing this germplasm from the USDA repository in Sturgeon Bay, Wisconsin, breeders have been limited in their ability to utilize this diversity. Historically, it has taken about 25 years to incorporate a single gene from an unadapted source into a successful variety. Quantitative traits are an even greater challenge. In spite of this difficulty and the limited number of breeding programs, the use of exotic germplasm is increasing. There have been long-term programs to select for adaptation in the two cultivated species, *S. phureja* and *S. tuberosum* subspecies *andigena,* to capitalize on their quantitative traits and there are more recent projects to do the same for additional species. A survey in 1989 (Plaisted and Hoopes, 1989) revealed that 11 wild species were present in the pedigrees of 124 varieties released to that date. Thus it appears that as well as an increase in the number of varieties being grown, their genetic complexity is increasing.

Recent Advances in Potato Breeding

Haploids

Nearly forty years ago Hougas *et al.* (1958) discovered a routine procedure for developing haploids of tetraploid *S. tuberosum* clones and since then, the Wisconsin breeding program in particular has devoted much effort to expanding upon this research. Haploids have proven to be an invaluable tool in potato genetic research as a means of securing fertile hybrids with most of the wild diploid *Solanum* species, and have been proposed as a method for variety development.

Viroid, virus and glycoalkaloid detection

Up until the mid 70s, the potato spindle tuber viroid (PSTV) was a serious problem for potato breeders because it could be transmitted through the true seeds and diagnosis was very difficult. Indicator plants were of limited value. However, soon after gel electrophoresis and nucleic acid hybridization techniques became routine, PSTV became a rare problem.

The experience with Lenape in 1970 (Sinden and Webb, 1972) and the subsequent GRAS (Generally Regarded As Safe) regulations of the FDA caused breeders to make the test for glycoalkaloids a routine pre-release procedure.

Additional significant developments which have occurred in the past decade include the development of enzyme-linked immunosorbent assays (ELISA) and the use of tissue culture. Visual roguing for viruses and off-types continues to be a key part of the seed multiplication phase of a breeding program. Every breeding program now uses ELISA to test routinely some phase of seed multiplication, or at least to verify the health of questionable plants. *In vitro* multiplication has become a key component of most foundation seed programs, and is a routine component of most breeding programs. In some cases it may be only for maintenance of certain clones, but for others it is the form for release to foundation seed multiplication and has become a routine part of the multiplication cycle. *In vitro* multiplication has been used more routinely for assurance of freedom from disease for introduction into the seed trade than for acceleration of seed production.

A technique which utilizes both *in vitro* multiplication and ELISA tests is the eradication of viruses by means of heat treatment coupled with chemotherapy. Until now breeders have had only the technique of using very small meristem tips as a method to remove viruses. Before that, a clone that was totally infected with a serious virus usually had to be discarded.

Endosperm balance number and fusion hybrids

The concept of endosperm balance number (EBN) that has been developed by Johnston and Hanneman (1980) has in a great measure eliminated the mystery in making crosses between species. The characterization of species on the basis of EBN as well as chromosome number eases the use of wild species. It provides insight into crossing schemes that overcome barriers to direct hybridization. To a considerable extent it has alleviated the need for making fusion hybrids. This technique of making fusion hybrids is one more recently developed tool that can be used to tap into the diversity present in a few species that have been difficult to hybridize.

Genetic transformation

The use of transformations to correct weaknesses in existing varieties represents a technique which will be adopted as both utility and cost-effectiveness are proven. This is a very conservative method of crop improvement in that it can be used to change a single trait. In this way it gives breeders an equivalent to the backcross method of self-pollinated crops. Potato varieties, being heterozygous hybrids, cannot be recovered through repeated backcrossing. If transformed varieties become accepted in the marketplace, and if the cost of transformation is reasonable, then certain weaknesses of otherwise desirable varieties can be corrected. Or from another perspective, if transformations of a specific nature become quite routine, then breeders can give that trait a lower priority in their breeding and selection procedure. Marketability, effectiveness, reliability and cost will determine the breeder's adoption of this technique. It is likely that breeders will not be the ones to do the transformations so proprietary rights of each contributor will need to be settled.

Restriction fragment length polymorphisms and related technologies

Another technique which shows promise of becoming a part of traditional potato breeding is the use of restriction fragment length polymorphism (RFLP) and random amplified polymorphic DNA (RAPD) markers to identify clones with desirable traits. The development of a molecular genetic map of potato by Bonierbale *et al.* (1988, 1993) (Fig. 1.3) and Gebhardt *et al.* (1989) has greatly facilitated these efforts. Our experiences indicate that cost will be a major factor in the adoption of RFLP and RAPD procedures in routine screening, although it is likely that cost will diminish rapidly as sequence data, automation and primers become more available.

RFLP markers can be used to increase efficiency of selection for quantitative traits because they offer rapid and precise identification of the regions of the chromosomes required for the expression of a given trait. In the near term, two types of traits seem to be appropriate candidates for molecular genetic markers. One will be for traits subject to extreme variation due to environment or inability to obtain a routine and uniform infestation by a pathogen or insect. If the trait is easily identified at a few loci by RFLP techniques, this character

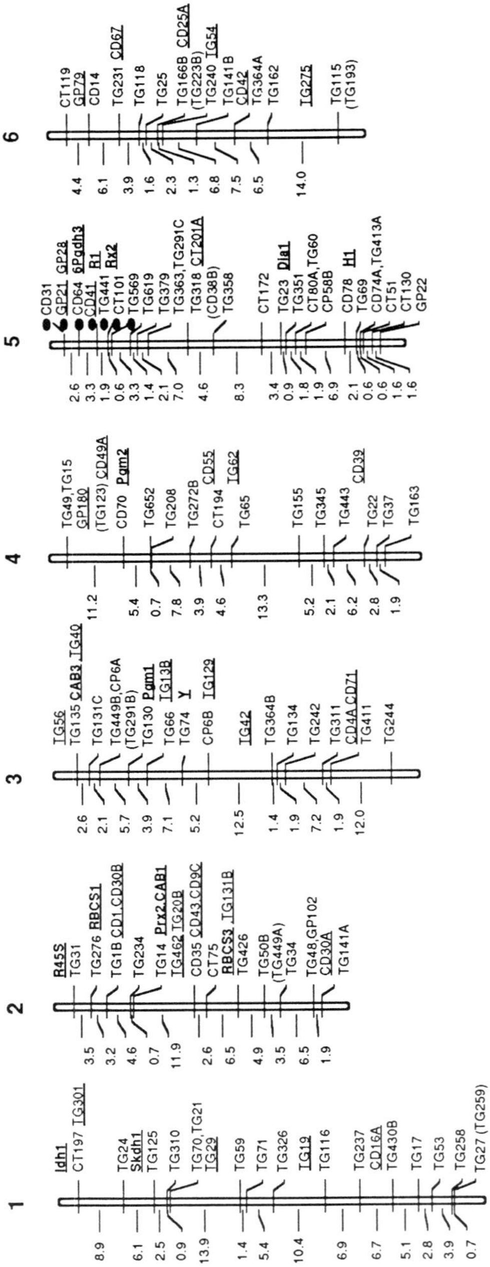
1
Idh1
CT197 TG301
TG24
Skdh1
TG125
TG310
TG70,TG21
TG29
TG59
TG71
TG326
TG19
TG116
TG237
CD16A
TG430B
TG17
TG53
TG258
TG27 (TG259)
8.9
6.1
2.5
0.9
13.9
1.4
5.4
10.4
6.9
6.7
5.1
2.8
3.9
0.7
2
R45S
TG31
TG276 RBCS1
TG1B CD1,CD30B
TG234
TG14 Prx2,CAB1
TG462 TG20B
CD35 CD43,CD9C
CT75
RBCS3 TG131B
TG426
TG50B
(TG449A)
TG34
TG48,GP102
CD30A
TG141A
3.5
3.2
4.6
0.7
11.9
2.6
6.5
4.9
3.5
6.5
1.9
3
TG56
TG135 CAB3 TG40
TG131C
TG449B,CP6A
(TG291B)
TG130 Pgm1
TG66 TG13B
TG74 Y
CP6B TG129
TG42
TG364B
TG134
TG242
TG311
CD4A CD71
TG411
TG244
2.6
2.1
5.7
3.9
7.1
5.2
12.5
1.4
1.9
7.2
1.9
12.0
4
TG49,TG15
GP180
(TG123) CD49A
CD70 Pgm2
TG652
TG208
TG272B CD55
CT194 TG62
TG65
TG155
TG345
TG443 CD39
TG22
TG37
TG163
11.2
5.4
0.7
7.8
3.9
4.6
13.3
5.2
2.1
6.2
2.8
1.9
5
CD31
GP21 GP28
CD64 6Pgdh3
CD41 R1
TG441 Rx2
CT101
TG569
TG619
TG379
TG363,TG291C
TG318 CT201A
(CD38B)
TG358
CT172
TG23 Dia1
TG351
CT80A,TG60
CP58B
CD78 H1
TG69
CD74A,TG413A
CT51
CT130
GP22
2.6
3.3
1.9
0.6
3.3
1.4
2.1
7.0
4.6
8.3
3.4
0.9
1.8
1.9
6.9
2.1
0.6
0.6
1.6
1.6
6
CT119
GP79
CD14
TG231 CD67
TG118
TG25
TG166B CD25A
(TG223B)
TG240 TG54
TG141B
CD42
TG364A
TG162
TG275
TG115
(TG193)
4.4
6.1
3.9
1.6
2.3
1.3
6.8
7.5
6.5
14.0

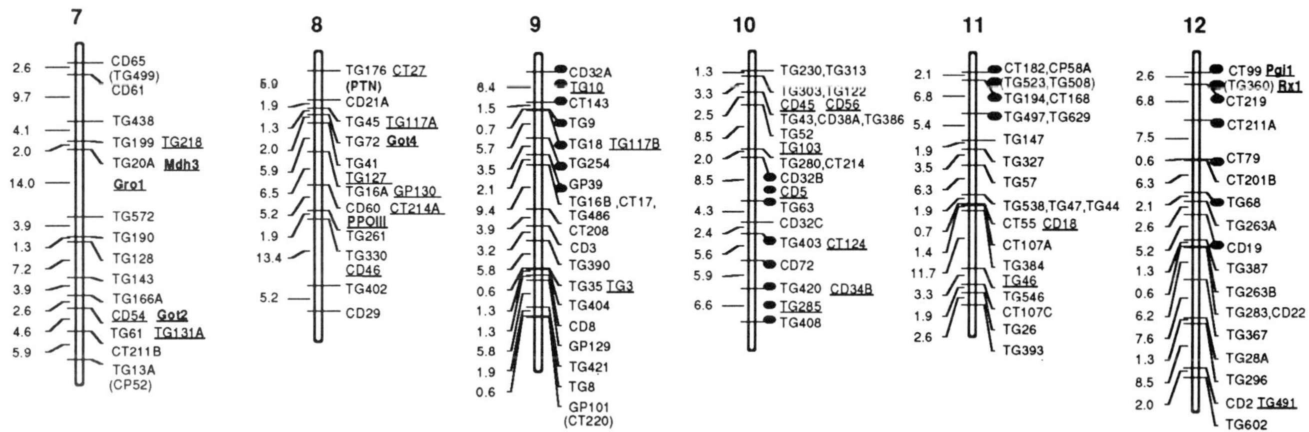

Figure 1.3. The molecular linkage map of the potato is based on genetic mapping in diploid derivatives and relatives of *S. tuberosum*. The RFLP markers were based on genomic or cDNA probes from tomato and potato. The markers indicated by dots on chromosomes 5, 9, 10, 11 and 12 indicate intrachromosmal inversions between the tomato and potato genomes. The linkage order among other markers is conserved which allows the prediction of the map position of more than 1000 markers from the tomato RFLP map in potato (adapted from Bonierbale *et al.*, 1993).

could be a good candidate. A second case is one in which screening depends on the use of a pest that for quarantine or safety reasons cannot be introduced.

RFLPs and resistance to nematodes

RFLPs could be employed in screening for resistance to races of the potato cyst nematodes that do not yet occur in the US. Currently, the Cornell potato breeding program is dependent upon screening at the International Potato Center (CIP) for the races of *Globodera pallida*. It would be more efficient to use molecular markers to screen segregating generations and use the pathogen test at CIP to verify the resistance of our selections.

We are beginning to make use of RFLP markers to identify potato clones which have the *H1* gene for resistance to race Ro1 of the cyst nematode. At present, this technique is not as cost-effective as exposure to the nematodes for routine screening but the costs will certainly drop as the techniques are further refined. The *H1* gene was recently located on chromosome 5 (Pineda *et al*., 1993). The resistance of clones identified through the use of RFLP markers tightly linked to the *H1* gene has been confirmed by means of exposure to the nematodes. The RFLP probe may become the standard method for determining whether resistant varieties are being grown on regulated fields because, for this test, speed and accuracy are more important than cost.

Another use for RFLP markers will be to identify parents that are duplex, triplex or quadriplex for simply inherited traits. Currently, in our cyst nematode program we use test results from our progenies to determine which parents are duplex. A cross between a duplex and a simplex will yield only 1/12 susceptible offspring. If there are only a few escapes, it will be difficult to identify triplex and quadriplex genotypes by means of test crosses. RFLP genotyping methods may prove to be more efficient.

RFLPs and resistance to insects

The wild relatives of potato are a valuable source of genes for resistance to insect pests (Flanders *et al*., 1992; Tingey and Yencho, 1994). However, few have been used in breeding programs. In addition to using RFLPs to identify nematode resistant clones we have been exploring the utility of using RFLP markers to assist in the identification and introgression of quantitative trait loci (QTL) that confer resistance to a number of insect pests in the wild diploid potato *S. berthaultii*. Extensive field and laboratory screening of two reciprocal hybrid *S. tuberosum* x *S. berthaultii* backcross populations of potato mapped with RFLPs (Fig. 1.4) spaced an average 10 cM apart have facilitated progress toward the mapping of QTL for traits associated with resistance to the Colorado potato beetle (Bonierbale *et al*., 1994; Yencho, 1993).

We have used knowledge of the biochemistry of the glandular trichomes (see Tingey, 1991 for review) to facilitate identification of QTL associated with resistance to the Colorado potato beetle. The density of the trichomes can be

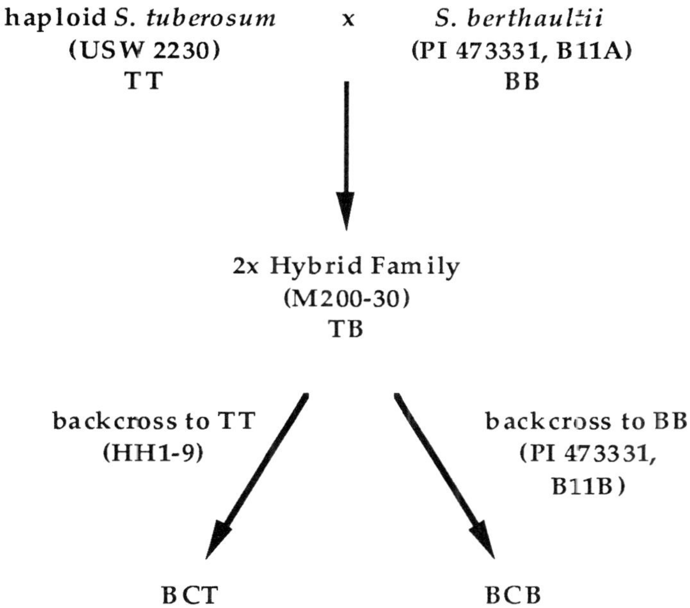

Figure 1.4. Pedigree of the diploid mapping populations. The clones involved in the crosses were: the dihaploid USW2230 ($2x = 2n = 24$), which was derived from *S. tuberosum* cv. Saco; *S. berthaultii*, PI 473331, clones B11A and B11B, which were selected for their trichome properties and crossability; and HH1-9, a clone selected for its male fertility. (Adapted from Bonierbale *et al.*, 1994.)

quantified relatively easily, and biochemical assays which measure the enzymatic browning activity (MEBA) (Avé *et al.*, 1986) and concentration of polyphenol oxidase (Kowalski *et al.*, 1993) in the type A trichomes, and the amount of sugar ester produced by the type B trichomes (Goffreda *et al.*, 1990) have been developed. We have used these traits to identify QTL associated with resistance in a population of potato that has been mapped with RFLP markers and is segregating for the trichome traits (Fig. 1.5) because they are associated with resistance to the Colorado potato beetle, and they are less variable and easier to obtain than direct measures of resistance. To date, we have identified QTL with highly significant LOD scores (LOD>2) for type A and B trichome density, polyphenol oxidase (PPO), MEBA and the concentration of sugar ester produced by the type B trichomes (Bonierbale *et al.*, 1994). In contrast, we have had more difficulty detecting significant QTL for direct measures of Colorado potato beetle resistance (Yencho, 1993). However, there is strong (but not complete) correlation between highly significant QTL (LOD>2) identified for the trichome

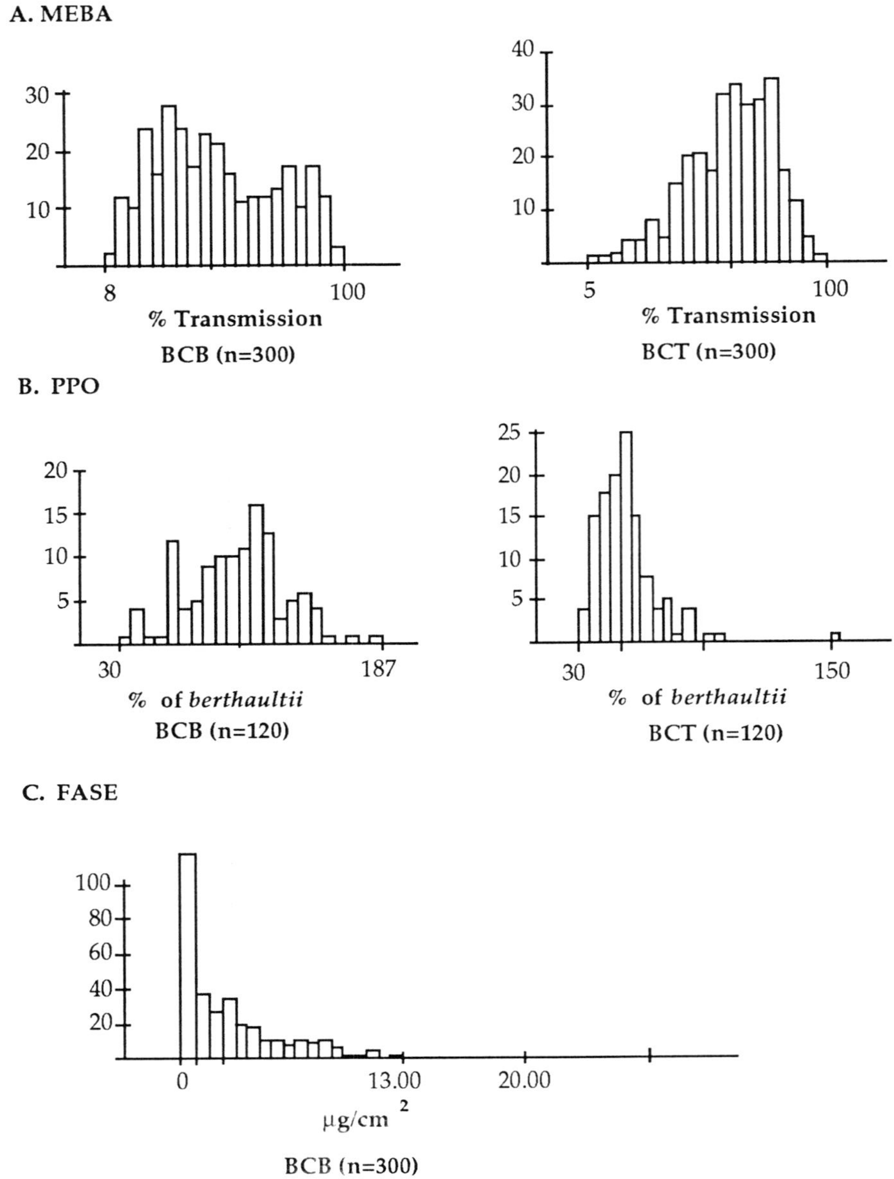

Figure 1.5. Frequency distributions of trichome phenotypes for individuals of BCB and BCT. (A) Modified enzymatic browning reaction (MEBA) measuring phenolic oxidation products of type A trichomes. (B) Polyphenol oxidase (PPO) concentration of type A trichomes as percentages of *S. berthaultii.* (C) Fatty acid sucrose ester (FASE) production of type B trichomes per cm^2 of leaf area. (Adapted from Bonierbale *et al.*, 1994)

properties and borderline significant (LOD ca. 1) QTL for resistance to the Colorado potato beetle (Yencho, 1993; Bonierbale *et al.*, 1994).

Since the trichome and Colorado potato beetle QTL identified do not correlate completely, and since there are other chemicals present in *S. berthaultii* that impart resistance that may or may not be associated with the trichomes, our future efforts to identify QTL for resistance will focus on reducing the variability of our bioassay techniques in order to identify the QTL that are associated with direct measures of resistance to adults and larvae of this insect. We are also beginning to apply our knowledge of QTL required for the expression of resistance to advance our diploid mapping populations and a population of tetraploid *S. tuberosum* x *S. berthaultii* potatoes in an attempt to implement an RFLP-assisted breeding program.

RFLPs for tuberization and earliness

The RFLP-mapped reciprocal backcrosses between *S. tuberosum* and *S. berthaultii* have also been used to identify QTL associated with earliness of tuberization. To date, we have identified ten QTL on eight chromosomes. Earliness of tuberization is favored by an *S. tuberosum* allele at seven QTL, and by an *S. berthaultii* allele at three QTL. Three of the seven *S. tuberosum* alleles and two of the three *S. berthaultii* alleles were at least partly dominant, and five of the seven *S. tuberosum* earliness alleles were detected by following RFLPs segregating from the recurrent *S. tuberosum* parent.

Interestingly, the allele with the greatest effect on earliness was among the five alleles segregating from the recurrent *S. tuberosum* parent, not from the hybrid parent (Van den Berg *et al.*, unpublished data). This demonstrates the importance of following segregating alleles from the original parent in an outbreeding species like potato. The detection of many loci, most of which had small effects on earliness, also appears to be in agreement with an hypothesis based on heritability studies with other *Solanum* species (Mendoza and Haynes, 1977).

Use of RFLPs to detect linkages between traits

Another intriguing aspect of our studies has been the documentation of the association between insect resistance and late maturity. *S. berthaultii* like most wild species, does not tuberize under long photoperiods. Breeders have had difficulties in separating the presence of droplets on the insect defensive type B trichomes from late tuberization in progenies of this species, although the type A trichomes could readily be combined with acceptable tuberization (Kalazich and Plaisted, 1991). The explanation for this became clear when QTL detected for trichome-mediated insect resistance were compared with QTL detected for earliness of tuberization (Van den Berg *et al.*, unpublished data). On chromosomes 1, 4 and 6, *S. berthaultii* alleles promoting early tuberization appear to be tightly linked to *S. berthaultii* alleles for the type A trichome action. However, the story for type B trichomes is quite different. The *S. berthaultii* allele that had the most effect on type B trichomes is tightly linked to an allele that in certain genetic combinations gave late maturity (e.g. *TG441* on chromosome 5, Fig. 1.3). The problem for the breeder who wants to utilize type B trichomes from *S. berthaultii* is not as impossible as this tight linkage might suggest because clones are late only if an *S. berthaultii* allele is also present at *TG413.A*. This means that in our crosses the problem of linkage between B trichomes and lateness at *TG441* would be overcome if a crossover permitted a *S. tuberosum* allele at *TG413A* when a *S. berthaultii* allele is present at *TG441*, about 35 cM distant. Selection of parents and screening of progeny using RFLP markers may permit us to select parents that would complement each other with earliness and trichome alleles in the offspring of the above-mentioned crosses.

Pleiotropic loci for earliness of tuberization, tuber dormancy, branching and leaf size

We have also uncovered pleiotropic effects between QTL associated with earliness of tuberization, tuber dormancy, leaf size, and branching using RFLPs (van den Berg *et al.*, unpublished data). Under short photoperiods tuberization is promoted while leaf area increases (Lorenzen and Ewing, 1990) and branching decreases (Gregory, 1956). QTL analysis of related traits provides the opportunity to study the differences and similarities in genetic control of related traits. At least three of the QTL detected for branching and two of the QTL for leaf size were pleiotropic or tightly linked with loci affecting tuberization. The relationships between the traits at those loci were as might be expected based on theory - early tuberization was associated with reduced branching and longer leaves - and these associations were not detected in conventional correlation analyses.

In addition, physiological studies indicate that hormones controlling tuber dormancy are similar to the ones controlling tuber initiation (Hemberg, 1985). At least three of the QTL detected for tuber dormancy were pleiotropic with, or tightly linked to, QTL for earliness of tuberization. Again at these three QTL the relationship between tuberization and dormancy was as expected; early tuberization was associated with long tuber dormancy. Thus, the comparisons of QTL detected for related traits could prove very useful in unraveling the different physiological components of quantitative traits such as potato tuberization.

Conclusion

The number of variety development programs in North America is gradually decreasing, but the number of seedlings being evaluated stays about the same. Breeders have always looked for new ways to do their job better or more efficiently. As new tools become available to breeders they are tested and adopted or rejected based on their merits, just as varieties are tested and adopted or rejected. The present time is particularly exciting because there are so many new tools to be considered. Even if some do not improve our ability to produce new varieties, they will broaden the base of germplasm we work with and will certainly improve our understanding of breeding behavior.

Predictions about the future of potato variety development are obscured by some of the significant changes of the present. A major factor is the consideration of breeders' rights for potato varieties in the US to match that of other crop varieties in the US and of potatoes in Canada and Europe. Another factor to be determined is the extent to which alien germplasm will need to be identified in specific varieties or strains in order to be marketed. This could lead to marketing potatoes by variety name as is currently done for apples. This could have a positive effect on consumer choices. Variety identity might promote diversity, but it could also restrict the opportunity for introduction of new tablestock varieties.

References

Avé, D.A., Eannetta, N.T. and Tingey, W.M. (1986) A modified enzymatic browning assay for potato glandular trichomes. *American Potato Journal* 63:553-558.

Bonierbale, M.W., Plaisted, R.L. and Tanksley, S.D. (1988) RFLP maps of potato and tomato based on a common set of clones reveal modes of chromosomal evolution. *Genetics* 120:1095-1103.

Bonierbale, M.R., Plaisted, R.L. and Tanksley, S.D. (1993) Molecular map of the potato *(Solanum tuberosum)* 2n=48. In: O'Brien, S. (ed.), *Genetic Maps*. Cold Spring Harbor Laboratory Press, Cold Spring Harbor, NY, pp. 6.88-6.90.

Bonierbale, M.R., Plaisted, R.L., Pineda, O. and Tanksley, S.D. (1994) QTL analysis of trichome-mediated resistance in potato. *Theoretical and Applied Genetics* 87:973-987.

Flanders, K.L., Hawkes, J.G., Radcliffe, E.B. and Lauer, F.I. (1992) Insect resistance in potatoes: sources, evolutionary relationships, morphological and chemical defenses, and ecogeographical associations. *Euphytica* 61:83-111.

Gebhardt, C., Ritter, E., Debener, T., Schachtschabel, U., Walkemier, B., Uhrig, H. and Salamini, F. (1989) RFLP analysis and linkage mapping in *Solanum tuberosum*. *Theoretical and Applied Genetics* 78:65-75

Goffreda, J.C., Steffens, J.C. and Mutschler, M.A. (1990) Association of epicuticular sugars with aphid resistance in hybrids with wild tomato. *Journal of Applied Horticultural Science* 115:161-165.

Gregory, L.E. (1956) Some factors for tuberization in the potato. *Annals of Botany* 43:281-288.

Hemberg, T. (1985) Potato rest. In: Li, P.H. (ed.), *Potato Physiology*. Academic Press, Orlando, FL, pp. 354-388.

Hougas, R.W., Peloquin, S.J. and Ross, R.W. (1958) Haploids of the common potato. *The Journal of Heredity* 49:103-106.

Johnston, S.A. and Hanneman, R.E. (1980) Support of the endosperm balance number hypothesis utilizing some tuber-bearing *Solanum* species. *American Potato Journal* 57:7-14.

Kalazich, J.C. and Plaisted, R.L. (1991) Association between trichome characters and agronomic traits in *Solanum tuberosum* (L.) x *S. berthaultii* (Hawkes) hybrids. *American Potato Journal* 68:833-847.

Karp, A. (1990) Somaclonal variation in potato. In: Bajaj, Y.P.S. (ed.), *Biotechnology in Agriculture and Forestry II. Somaclonal Variation in Crop Improvement*, Springer-Verlag, Berlin, pp. 379-399.

Kowalski, S.P., Plaisted, R.L. and Steffens, J.C. (1993) Immunodetection of polyphenol oxidase in glandular trichomes of *S. berthaultii, S. tuberosum* and their hybrids. *American Potato Journal* 70:185-200.

Lorenzen, J.H. and Ewing, E.E. (1990) Changes in tuberization and assimilate partitioning in potato *(Solanum tuberosum)* during the first 18 days of photoperiod treatment. *Annals of Botany* 66:457-464; note errata in *Annals of Botany* 67:191.

Mendoza, H.A. and Haynes, F.L. (1977) Inheritance of tuber initiation in tuber bearing *Solanum* as influenced by photoperiod. *American Potato Journal* 54:243-252.

Pineda, O., Bonierbale, M.W., Plaisted, R.L., Brodie, B.B. and Tanksley, S.D. (1993) Identification of RFLP markers linked to the *H1* gene conferring resistance to the potato cyst nematode *Globodera rostochiensis*. *Genome* 36:152-156.

Plaisted, R.L. and Hoopes, R.W. (1989) The past record and future prospects for the use of exotic potato germplasm. *American Potato Journal* 66:603-627.

Sinden, S.L. and Webb, R.E. (1972) Effect of variety and location on the glycoalkaloid content of potatoes. *American Potato Journal* 49:334-338.

Tingey, W.M. (1991) Potato glandular trichomes: defensive activity against insect attack. In: Hedin, P.A. (ed.), *Naturally Occurring Pest Bioregulators*. American Chemical Society Symposium Series 449, pp. 126-135.

Tingey, W.M. and Yencho, G.C. (1994) Insect resistance in potato: a decade of progress. In: Zehnder, G.W., Jansson, R.K., Powelson, M.L. and Raman, K.V. (eds), *Advances in Potato Pest Biology and Management*. APS Press, Minneapolis, MN, pp. 405-425.

Yencho, G.C. (1993) Behavioral, biochemical and molecular genetic aspects of resistance to the Colorado potato beetle in *Solanum berthaultii* Hawkes. PhD Dissertation, Cornell University, 156 pp.

Chapter Two:

Ac-Ds Transposons Mapped Near Disease Resistance Loci for Targeted Tagging in Potato

Jeanne M.E. Jacobs, Bas te Lintel Hekkert, Ali El-Kharbotly†, Evert Jacobsen†, Willem J. Stiekema and Andy Pereira

Center for Plant Breeding and Reproduction Research, Wageningen, The Netherlands. †Department of Plant Breeding, Wageningen Agricultural University, The Netherlands.

The maize Ac-Ds transposable element system has been shown to be functional in potato. We present evidence that in potato, similar to other plants, Ac transposes to closely linked sites. Therefore, a transposable element closely linked to the gene of interest can be used for efficient tagging and isolation of that gene. In order to develop a transposon tagging system in potato for the cloning of genes, we have mapped the chromosomal position of T-DNA containing Ac or Ds elements and of transposed Ac elements from a large set of diploid transformants. We have isolated potato genomic DNA flanking the transposed Ac elements or T-DNA inserts, and used these probes for RFLP mapping in relation to other molecular markers of known position. About 30 such transposon positions are reported here. Some T-DNA::Ds as well as T-DNA::Ac and Ac elements have been located close to resistance loci *R1* (*Phytophthora infestans*), *H1* and *Gro1* (both *Globodera rostochiensis* and Gpa, *G. pallida*). Targeted transposon tagging of these resistance loci can now be initiated after making the appropriate crosses.

Isolation of Genes by Transposon Tagging

Transposon tagging is a method for the isolation of genes which have an easily screenable mutant phenotype. This method of gene isolation has been the most successful technique in plant species like maize and *Antirrhinum* which contain well characterized endogenous transposons (see Walbot, 1992). Most of the mutants obtained by tagging in maize are by targeted tagging of specific loci, using transposons closely linked to the locus. A large number of F1 individuals heterozygous for the specific target locus (*A*/*a*) are screened for the mutant-transposon tagged phenotype (*A* -> *a*). In recent years the efficiency/frequency

The Molecular and Cellular Biology of the Potato, Second Edition
Edited by W.R. Belknap, M.E. Vayda and W.D. Park

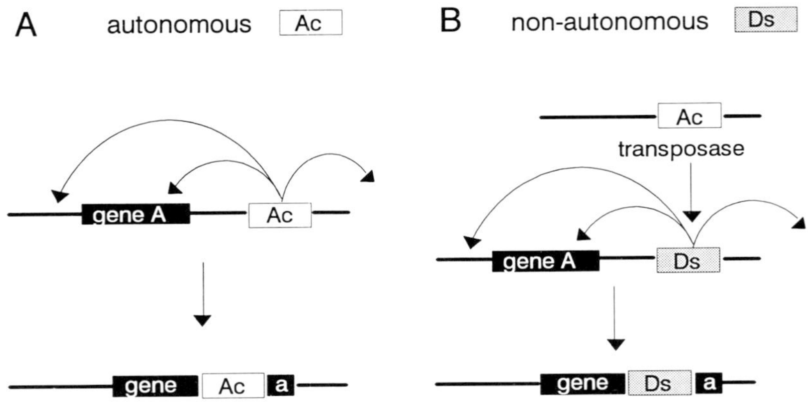

Figure 2.1. The two component maize Ac-Ds transposable element system consists of (A) the autonomous component, Ac, capable of transposing by itself and (B) the non-autonomous component, Ds (internally deleted derivative of Ac), which needs the transposase product of Ac in order to transpose

of obtaining mutants has been increased tremendously using closely linked transposons. For example the frequency of obtaining the TS2 Ac tagged mutant (DeLong *et al.*, 1993) was 1/2000 using an Ac located 2 cM from the TS2 locus (Walbot, 1992). With such high frequencies it is also possible to efficiently tag and isolate genes which have difficult phenotypes for screening e.g. quantitative traits. The introduction of maize transposons into heterologous plants has very recently resulted in a number of transposon tagged genes from *Arabidopsis* (Aarts *et al.*, 1993; Bancroft *et al.*, 1993) and *Petunia* (Chuck *et al.*, 1993). All these tagged mutants have one common character, they were isolated by a random transposon tagging approach. In potato both the maize transposon systems Ac-Ds (Knapp *et al.*, 1988; Pereira *et al.*, 1991) and En-I (Frey *et al.*, 1989) have been shown to function. Hence, transposon tagging offers an opportunity for the isolation of genes in potato. The use of the heterologous Ac-Ds two element system (Fig. 2.1) has the advantage of being able to manipulate the transposons *in vitro* before transformation into plants and thus to use a transposon tagging system tailored for efficiency. The critical step is to obtain transposons close to the target gene so that efficient tagging will be possible. We describe here our progress in mapping a number of T-DNA inserts containing Ac-Ds elements in the potato genome. The transposons mapped are from independent diploid fertile transformants. This strategy yields a restriction fragment length polymorphism (RFLP) map of potato in which the markers, the transposons, serve as the starting point for jumping to nearby locations on the chromosome.

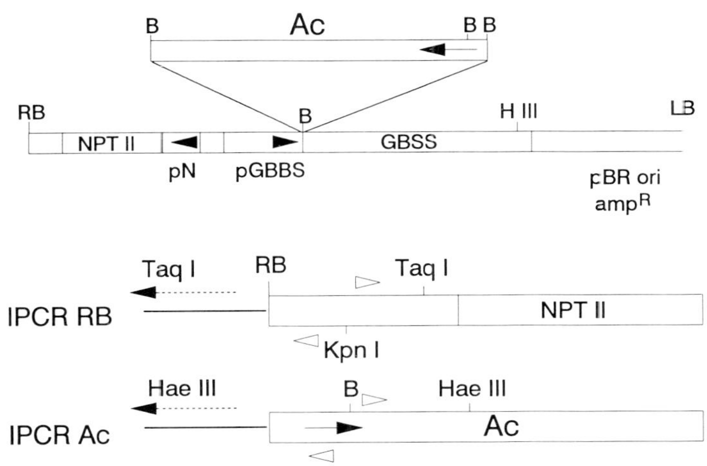

Figure 2.2. The T-DNA construct used for mapping transposed Ac elements is shown. Excision of the Ac element from the *GBSS* gene (Pereira *et al.*, 1991) provides a 'waxy' variegation phenotype. The *NPTII* gene is used for selection of transformants. Arrowheads indicate direction of transcription of *Ac*, *GBSS* and *NPTII* genes. For inverse polymerase chain reaction (IPCR) mediated rescue of flanking DNA from the T-DNA and Ac inserts in the genome, the enzymes used are indicated and the direction of primers used for synthesis shown as open arrowheads. The synthesized flanking DNA used as probes are shown as dashed arrows. In the case of the TDs construct (not shown) other sites are used. The pBR322 *ori* and ampicillin resistance (*amp*^{R}) present in the construct can provide for plasmid rescue instead of IPCR for the isolation of flanking DNA. Abbreviations: RB and LB = left and right borders of the T-DNA, B=*Bam*HI, HIII=*Hind*III.

Plant Material and Transposon Constructs

Diploid potato genotypes used for transformation were as described previously (Pereira *et al.*, 1991, 1992). The T-DNA::Ac construct used is shown in Fig. 2.2. In the Ac constructs the phenotypic 'waxy' variegation assay was used to detect excision (Pereira *et al.*, 1991). In a similar Ds construct (Pereira *et al.*, 1992) hygromycin resistance was used to monitor excision, and kanamycin resistance (*NPTII* gene) inserted in Ds served as the selectable marker for plant transformation. The Ac transposase T-DNA construct (not shown) used to transactivate the Ds element contained a methotrexate resistance gene as selectable marker for transformation.

Isolation of T-DNA and Transposable Element Flanking Sequences

We have produced about 400 different diploid potato transformants containing Ac/Ds T-DNA inserts. Inverse PCR (IPCR) (Triglia *et al.*, 1988) was used to isolate potato DNA flanking the Ac transposons and T-DNA inserts as described by Earp *et al.* (1990). The isolated IPCR fragments were cloned into the *Eco*RV site of pBluescript SK+ (Stratagene) and the identity of the clones was confirmed by hybridization to T-DNA/Ac border probes. Isolated inserts containing flanking DNA were used for radioactive labeling (Feinberg and Vogelstein, 1983) and hybridization to potato DNA. Most (95%) of the Ac probes were unique/low-copy as judged by hybridization to potato genomic DNA (data not shown) and could therefore be used for mapping when polymorphic in the mapping population. About 50% of the probes from T-DNA IPCRs were repetitive and could not be mapped. This suggests that Ac elements insert predominantly into low copy DNA which makes the mapping of transposed Ac/Ds elements more efficient than that of the T-DNAs.

RFLP Mapping and Linkage Analysis

T-DNA/Ac flanking genomic DNA probes were used for RFLP analysis (Fig. 2.3) in a diploid potato population (van Eck *et al.*, 1993). This mapping population was used for the construction of an integrated genetic map and containing both molecular (RFLP) and classical (morphological and isozymes) markers (Jacobs *et al.*, in preparation). A set of tomato genomic DNA probes with known chromosomal positions in potato (Bonierbale *et al.*, 1988) was used to enable comparison with already existing potato maps (Gebhardt *et al.*, 1991). Segregation of markers in either one of both parental lines can be used, because independent maps of both parents are present. Linkage analysis, mapping, and subsequent integration of the two parental maps into one was performed by the computer package JoinMap (Stam, 1993). The resulting map was drawn with the help of Drawmap (van Ooijen, 1994) The nomenclature of the probes is as follows.

TDsX: IPCR or plasmid rescue derived probe from T-DNA containing Ds insert; X = transformant number
TAcX: IPCR derived probe from T-DNA containing Ac insert; X = transformant number
AcX-Y: IPCR derived probe from transposed Ac originating from TAc; X = TAc number; Y = transposed Ac number

To date we have mapped six TAc on four different chromosomes, 17 Ac on seven chromosomes, and nine TDs on seven chromosomes. All chromosomes, with the exception of chromosome 1, contain at least one transposon marker.

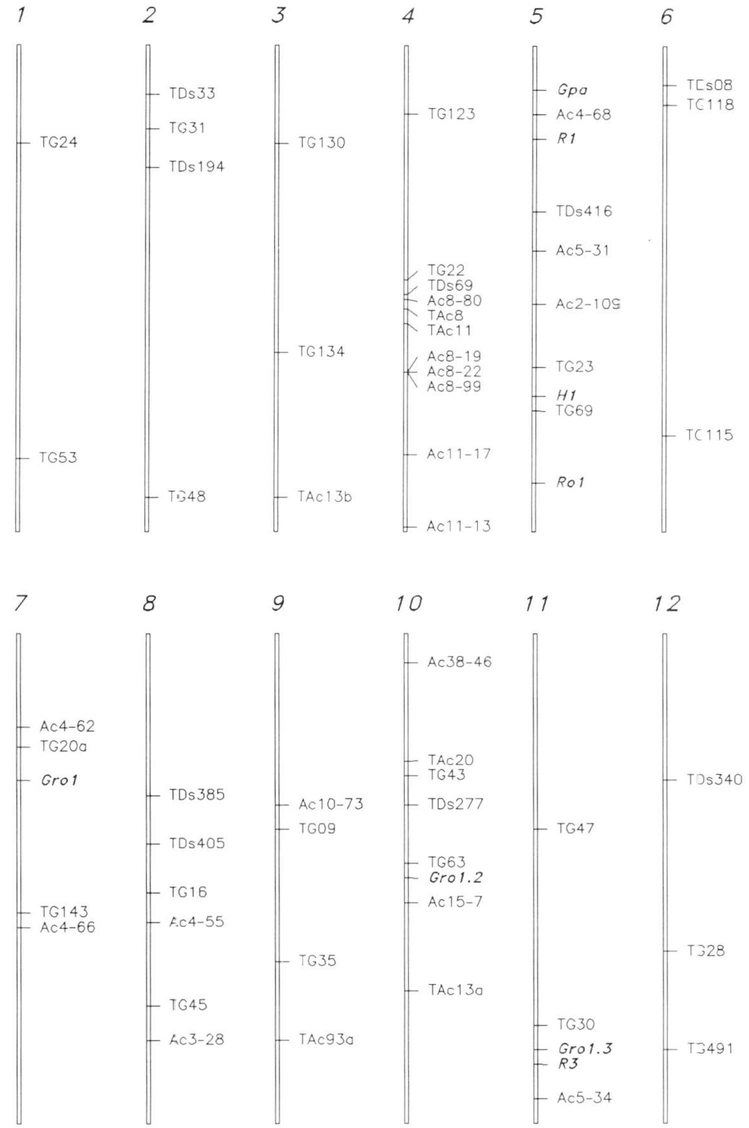

Figure 2.3. Genetic map of potato, showing the TAc, Ac and TDs markers. The map is a compilation of the independent maps of both parents in the mapping population. The markers are placed in a framework of two TG markers per chromosome to enable comparison with map positions in other potato maps. Resistance loci for potato cyst nematodes and *Phytophthora infestans* identified so far in several different potato populations are shown in italics.

Clustered and Scattered Ac Transposition

In order to characterize the transpositional behavior of Ac elements in potato we have used a set of transformants with actively transposing Ac elements. Potato DNA flanking (i) transposed Ac elements and (ii) the T-DNA inserts which gave rise to the transposed Ac elements, was isolated. In most cases the number of independent Ac inserts, as visualized in genomic Southerns of the transformants, correlated with the number of IPCR bands obtained. The map positions of a number of transposed Ac elements are shown in Fig. 2.3. The T-DNA insert labeled TAc11 on chromosome 4 gives rise to Ac transpositions to linked locations (located within 35 cM of the T-DNA) labeled Ac11-13 and Ac11-17. Similarly, Ac8-19, Ac8-22, Ac8-80 and Ac8-99 are within a cluster of 35 cM of TAc8. These results show that Ac does transpose to linked locations in potato. The frequency of transposition found between different transformants, using the same T-DNA construct, is highly variable based on somatic excision in tubers (Pereira *et al.*, 1991). Some TAc locations though do not give rise to many linked Ac transpositions as documented by TAc4 elements located on chromosomes 5, 7 and 8. TAc4, though unique, has not yet been mapped. Thus, depending on the original location of the T-DNA (position effect) the reinsertion pattern of Ac may be clustered or scattered. For efficient tagging it is necessary to have a transposon source that produces transposition events which are closely linked (like TAc11 or TAc8), and thus increase the frequency of tagging the nearby target locus. T-DNA loci like TAc4 are therefore not suitable as they give a set of scattered Ac transpositions. In maize as well as in several heterologous plant species it was found that reinsertion of an excised Ac element is mainly to closely linked sites. We now have evidence that this is also the case in potato.

Localization of T-DNA Inserts Containing Ds Transposons

For controlled transposon tagging in which Ds elements are mobilized by specific promoter-regulated Ac transposase sources it is necessary to use a Ds transposon located close to the target locus to be tagged. For this reason we have produced about 350 diploid potato transformants in a genotype containing the *R1* locus, which confers race-specific resistance to *Phytophthora infestans*. A set of transformants has been used to isolate the potato genomic DNA flanking the T-DNA borders by IPCR, and these probes have been used for mapping. This set of T-DNA::Ds inserts covers seven chromosomes (Fig. 2.3). Further mapping of the T-DNA::Ds elements is in progress, by obtaining flanking potato DNA sequences using the pBR322 *ori* sequences near the left border in the T-DNA construct for plasmid rescue (Dower *et al.*, 1988). This set of T-DNA::Ds lines can be used to tag a number of linked loci. First the T-DNA::Ds has to be brought near to a target locus present in another genotype by crossing and recombination. In most cases it is easier to use a Ds linked loosely to the gene of interest (20-40 cM) and then mobilize the Ds element to a position nearer to the target locus. The transposed Ds elements can be mapped by producing flanking probes by IPCR. The specific plant containing a very closely linked transposed Ds can then be used for further tagging.

Efficiency of Tagging

Closer linkage to a gene of interest results in higher frequency, and thus greater efficiency, of tagging of that gene. With an unlinked transposable element we expect the size of the population that has to be screened to find a mutant to be approximately 500,000 plants. This can be brought down to about 50,000 plants when the distance to the transposable element is 20 cM, and about 5,000 plants with a distance of 5 cM (Walbot, 1992). Especially with a phenotype that only becomes visible after laborious resistance testing, this reduction in plant number becomes very important. Therefore, the additional effort needed for the search of a line containing a transposable element closely linked to the gene of interest is largely compensated for at the stage of screening offspring for mutant (tagged) phenotypes. An additional promising strategy for increasing the efficiency of targeted tagging is the use of different promoters to regulate Ac transposase expression (Rommens *et al.*, 1992) to limit high transposition to specific developmental stages and/or tissues of the plant. Such transposase constructs will be introduced into TDs lines in order to control transposition.

Tagging Strategies

The mapping of sequences flanking transposable elements will be continued and will eventually provide a library of lines containing transposable elements at different positions equally spaced over the potato genome. Such a series of independent lines that have the genome saturated with transposable elements would make targeted tagging of any locus possible. The target genes we are focusing on are those conferring resistance to pathogens, which are present in wild species or their introgressions into potato. By means of RFLP mapping several resistance loci have been identified (Fig. 2.3). Monogenic resistances against the fungus *Phytophthora infestans* have been located on chromosome 5 (*R1*; Leonards-Schippers *et al.*, 1992) and 11 (*R3*; El-Kharbotly *et al.*, 1994). Several loci for resistance against potato cyst nematodes have been identified. These include the monogenic resistances directed against the species *Globodera rostochiensis* pathotype 1; *Gro1* from *Solanum spegazzinii* (Barone *et al.*, 1990), *H1* from *S. tuberosum* ssp. *andigena* (Gebhardt *et al.*, 1993; Pineda *et al.*, 1993) and *Ro1* from *S. vernei* (Jacobs *et al.*, unpublished results). Multigenic resistance against *G. rostochiensis* pathotype 1 from *S. spegazzinii* have been located on chromosomes 10 and 11 (*Gro1.2* and *Gro1.3*; Kreike *et al.*, 1994). A major locus for resistance against *Globodera pallida* pathotypes 2 and 3 from *S. spegazzinni* has been located on chromosome 5 (*Gpa*; Kreike *et al.*, 1994). Appropriate crosses to introduce transposon constructs close to our target genes (*R1*, *G. rostochiensis* and *G. pallida* resistances) have been performed. In the case of TDs transformants with T-DNA insertions close to the *R1* locus of *P. infestans* (chromosome 5) crosses with lines transformed with constructs having transposase activity have been performed. As the progeny of these crosses have the transposase activity *in trans*, further crosses will be necessary to obtain recombination to enable subsequent tagging.

Following the induction of transposable elements targeted to one of the resistance genes, screening for mutant phenotype(s), the identification of possibly tagged mutants, and the cloning of the resistance gene(s) from these tagged mutants will be carried out. A suitable (easy to perform with little space, rapid, and reproducible) resistance test will facilitate mutant screening. Especially in those cases where resistances from 'wild' sources are difficult to introgress into potato cultivars (e.g. reduced or no compatibility), transposon tagging should provide a useful tool for cloning of the resistance gene(s).

References

Aarts, M.G.M., Dirkse, W.G., Stiekema, W.J. and Pereira, A. (1993) Transposon tagging of a male sterility gene in *Arabidopsis*. *Nature* 363:715-717.

Bancroft, I., Jones, J.D. and Dean C. (1993) Heterologous transposon tagging of the DRL1 locus in *Arabidopsis*. *The Plant Cell* 5:631-638.

Barone, A., Ritter, E., Schachtschabel, U., Debener, T., Salamini, F. and Gebhardt, C. (1990) The localization of a dominant major resistance gene against the nematode *Globodera rostochiensis* in the potato genome using RFLP markers. *Molecular and General Genetics* 224:177-182.

Bonierbale, M.W., Plaisted, R.L. and Tanksley, S.D. (1988) RFLP maps based on a common set of clones reveal modes of chromosomal evolution in potato and tomato. *Genetics* 120:1095-1103.

Chuck, G., Robbins, T., Nijjar, C., Ralston, E., Courtney-Gutterson, N. and Dooner,H.K. (1993) Tagging and cloning of a Petunia flower color gene with the maize transposable element Activator. *The Plant Cell* 5:371-378.

DeLong, A., Calderon-Urrea, A. and Dellaporta, S.L. (1993) Sex determination gene TASSELSEED2 of maize encodes a short-chain alcohol dehydrogenase required for stage-specific floral organ abortion. *Cell* 74:757-768.

Dower, W.J., Miller, J.F. and Ragsdale, C.W. (1988) High efficiency transformation of *E.coli* by high voltage electroporation. *Nucleic Acids Research* 16:6127-6145.

El-Kharbotly, A., Leonards-Schippers, C., Huigen, D.J., Jacobsen, E., Pereira, A., Stiekema, W.J., Salamini, F. and Gebhardt, C. (1994) Segregation analysis and RFLP mapping of the R1 and R3 alleles conferring race specific resistance to *Phytophthora infestans* in progenies of dihaploid potato parents. *Molecular and General Genetics* 242:749-754.

Earp, D.J., Lowe, B. and Baker, B. (1990) Amplification of genomic sequences flanking transposable elements in host and heterologous plants:a tool for transposon tagging and genome characterization. *Nucleic Acids Research* 18:3271-3279.

Feinberg, A.P. and Vogelstein, B. (1983) A technique for radiolabelling DNA restriction endonuclease fragments to high specific activity. *Analytical Biochemistry* 132:6-13.

Frey, M., Tavantzis, S.M. and Saedler, H. (1989) The maize En-I/Spm element transposes in potato. *Molecular and General Genetics* 217:172-177.

Gebhardt, C., Ritter, E., Barone, A., Debener, T., Walkemeier, B., Schachtschabel, U., Kaufmann, H., Thompson, R.D., Bonierbale, M.W., Ganal, M.W., Tanksley, S.D. and Salamini, F. (1991) RFLP maps of potato and their alignment with the homoeologous tomato genome. *Theoretical and Applied Genetics* 83:49-57.

Gebhardt, C., Mugniery, D., Ritter, E., Salamini, F. and Bonnel, E. (1993) Identification of RFLP markers closely linked to the *H1* gene conferring resistance to *Globodera rostochiensis* in potato. *Theoretical and Applied Genetics* 85:541-544.

Knapp, S., Coupland, G., Uhrig, H., Starlinger, P. and Salamini, F. (1988) Transposition of the maize transposable element Ac in *Solanum tuberosum*. *Molecular and General Genetics* 201:513-518.

Kreike, C.M., de Koning, J.R.A., Vinke, J.H., van Ooijen, J.W. and Stiekema, W.J. (1994) Quantitatively inherited resistance to *Globodera pallida* is dominated by one major locus in *Solanum spegazzinii*. *Theoretical and Applied Genetics* 87:464-470.

Leonards-Schippers, C., Gieffers, W., Salamini, F. and Gebhardt, C. (1992) The R1 gene conferring race-specific resistance to *Phytophthora infestans* in potato is located on potato chromosome V. *Molecular and General Genetics* 233:278-283.

Pereira, A., Aarts, M., van Agtmaal, S., Stiekema, W.J. and Jacobsen, E. (1991) Waxy variegation in transgenic potato. *Maydica* 36:323-327.

Pereira, A., Jacobs, J.M.E., te Lintel Hekkert, B., Rutgers, E., Jacobsen, E. and Stiekema, W.J. (1992) Towards the isolation of resistance genes by transposon targeting in potato. *Netherlands Journal of Plant Pathology* 98:215-221.

Pineda, O., Bonierbale, M.W., Plaisted, R.L., Brodie, B.B. and Tanksley, S.D. (1993) Identification of RFLP markers linked to the *H1* gene conferring resistance to the potato cyst nematode *Globodera rostochiensis*. *Genome* 36:152-156.

Rommens, C.M.T., Van Haaren, M.J.J., Buchel, A.S., Mol, J.N.M., Van Tunen, A., Nijkamp, H.J.J. and Hille, J. (1992) Transactivation of Ds by Ac-transposase gene fusions in tobacco. *Molecular and General Genetics* 231:433-441.

Stam, P. (1993) Construction of integrated genetic linkage maps by means of a new computer package:JoinMap. *The Plant Journal* 3:739-744.

Triglia, T., Peterson, M.G. and Kemp, D.J. (1988) A procedure for *in vitro* amplification of DNA segments that lie outside the boundaries of known sequences. *Nucleic Acids Research* 16:8186.

van Eck, H.J., Jacobs, J.M.E., van Dijk, J., Stiekema, W.J. and Jacobsen, E. (1993) Identification and mapping of three flower colour loci of potato (*S.tuberosum* L.) by RFLP analysis. *Theoretical and Applied Genetics* 86:295-300.

van Ooijen, J.W. (1994) Drawmap:a computer program for drawing genetic linkage maps. *Journal of Heredity* 85:66.

Walbot, V. (1992) Strategies for mutagenesis and gene cloning using transposon tagging and T-DNA insertional mutagenesis. *Annual Review of Plant Physiology and Plant Molecular Biology* 43:49-82.

Chapter Three:

Studies to Enhance Starch Biosynthesis by Manipulation of ADP-Glucose Pyrophosphorylase Genes

Paul A. Nakata and Thomas W. Okita†

Department of Biochemistry and Biophysics, and †Institute of Biological Chemistry, Washington State University, Pullman, WA, USA

The study of photosynthesis and its enhancement is of extreme interest in the development of more efficient crop plants. Since relatively minor increments in the efficiency of photosynthesis can result in elevated crop yields much of the current research has been focused upon directly altering the components of the photosynthetic apparatus. Alternatively, a less direct approach of increasing the ability of plants to store photosynthate in their storage reserves may also be a feasible way of enhancing the photosynthetic capacity of plants (Thorne and Evans, 1964; Stark *et al.*, 1992). Several studies have shown that the rate and duration of photosynthesis in the leaf tissue (source) and the ability of the storage tissue (sink) to assimilate photosynthate are interrelated (Tollenaar, 1977; Nasyrov, 1978; Ewart, 1986). Under environmental conditions where photosynthesis is not limiting, overall plant productivity appears to be regulated primarily by sink strength, i.e. the capacity of the sink tissue to uptake and assimilate photosynthate from source leaves (Burt, 1964; King *et al.*, 1967; McArthur *et al.*, 1975). Since starch is the major storage form of photosynthate in the sink tissues of the potato, as well as many other crop plants, enhancement in its biosynthesis may stimulate photosynthesis and in turn increase overall crop yields. Currently, we are investigating the feasibility of altering sink strength through genetic manipulation of the enzyme, ADP-glucose pyrophosphorylase (AGPase), which is a key regulatory enzyme in the biosynthesis of starch in plants (Preiss, 1988) and glycogen in bacteria (Preiss, 1984). Such a molecular approach, however, first requires a general understanding of both the enzyme and gene structures, as well as the regulatory factors controlling gene expression. This chapter highlights our recent progress in these areas. For a more detailed discussion on the rationale of this approach the reader may refer to a number of recent reviews (Kleczkowski *et al.*, 1991; Preiss, 1991; Okita *et al.*, 1993).

The Molecular and Cellular Biology of the Potato, Second Edition
Edited by W.R. Belknap, M.E. Vayda and W.D. Park

The Role of AGPase in Starch Biosynthesis

An abundance of evidence indicates that the central enzyme controlling the sink capacity of starch-storing plants is AGPase (Preiss, 1988). A notable feature of this enzyme is its regulatory properties in which the activity of AGPase from both photosynthetic and most non-photosynthetic tissues are stimulated, *in vitro*, by 3-PGA and inhibited by Pi (Preiss, 1982, 1988). During diurnal growth, the ratio of these small effector molecules fluctuates in leaf tissue, in response to photosynthesis, and modulates the catalytic activity of AGPase. The functional significance of this allosteric regulation in controlling starch biosynthesis has been suggested by the studies of Steup *et al.* (1976), Heldt *et al.* (1977) and Neuhaus *et al.* (1989). Evidence obtained from both intact leaf and isolated chloroplast systems supports the hypothesis that the ratio of 3-PGA to Pi controls the activation of AGPase and hence the flow of carbon into leaf starch, *in vivo.*

Regulation of starch synthesis by AGPase activity is also believed to occur in non-photosynthetic tissues such as the potato tuber. Unlike transient starch metabolism in leaf tissue, deposition and mobilization in the non-photosynthetic storage organs are temporally separated to specific stages of plant development. In addition, the more autonomous chloroplast generates its own carbon and energy via CO2 fixation and photophosphorylation while the amyloplast is dependent on carbon imported from the source tissues. Several studies (Lucas and Madore, 1988; ap Rees and Entwistle, 1989) have suggested that the imported carbon arrives in the sink tissues in the form of sucrose which is subsequently processed into hexose phosphates. The generated hexose-phosphate(s) are then thought to be directly transported into the amyloplast where they are ultimately utilized as substrate (Glc-1-P) by AGPase (Keeling *et al.*, 1988; Tyson and ap Rees, 1988; Hill and Smith, 1991; Okita, 1992). This proposed pathway suggests a limited role for triose-P in amyloplastic starch biosynthesis and raises the question of whether 3-PGA serves as an *in vivo* activator of AGPase in these non-photosynthetic tissues. Enzymatic evidence has revealed little difference in effector sensitivity between the chloroplast and amyloplast derived enzymes (Sanwal *et al.*, 1967; Sowokinos and Preiss, 1982; Plaxton and Preiss, 1987). Whether the observed *in vitro* reponse of the non-photosynthetic enzymes (Sowokinos and Preiss, 1982) to the allosteric effectors operates in the storage organs, or simply reflects the evolutionary history of the structural genes, is not known. Nevertheless, the enzyme located in non-photosynthetic tissue is believed to operate in a highly inhibited environment and there is much interest in determining whether molecular modification of this enzyme could lead to increased rates of starch biosynthesis and, in turn, to increased plant productivity.

Subunit Structure of the Potato Tuber AGPase

To gain insight into the structure/function relationship of the tuber AGPase we extensively purified this enzyme by employing a modified protocol of Sowokinos and Preiss (1982) coupled with Mono-Q FPLC chromatography (Okita *et al.*, 1990) for both subunit structural analysis as well as antibody production. In single dimensional SDS-PAGE, the enzyme migrated as a single polypeptide band of about 50 kDa (Okita *et al.*, 1990). Analysis by 2-D PAGE, however, revealed the presence of two subunit types which could be distinguished by their slight difference in net charge and molecular weight (Okita *et al.*, 1990). Western blot analysis of the 2-D gel, using antiserum prepared against the purified tuber holoenzyme, showed cross-reactivity with both subunits. Antiserum raised against the spinach leaf small subunit cross-reacted solely with the tuber small subunit. Antiserum raised against the spinach leaf large subunit did not bind significantly to either potato tuber subunit. These immunological data coupled with evidence from the literature (Preiss *et al.*, 1989) suggested that the potato tuber enzyme was not a simple homotetramer as previously reported (Sowokinos and Preiss, 1982), but consisted of two immunologically distinct subunits like the AGPases from photosynthetic tissue.

Using the antibody raised against the purified potato tuber holoenzyme (Okita *et al.*, 1990), our laboratory (Anderson *et al.*, 1990; Nakata *et al.*, 1991), as well as others (Müller-Röber *et al.*, 1990; du Jardin and Berhin, 1991), has isolated cDNAs encoding each tuber subunit. A comparison of the small and large subunits revealed a 52% amino acid identity overall (Nakata *et al.*, 1991) which is somewhat higher than the 35% identity observed between the spinach leaf subunits (Preiss *et al.*, 1989). The potato tuber small subunit showed 84% identity to the rice endosperm small subunit (Anderson *et al.*, 1989), and 93% identity to the spinach leaf small subunit (Preiss *et al.*, 1989) at the primary level. The potato large subunit (Nakata *et al.*, 1991) exhibited 58% identity and 61% identity to the wheat (Olive *et al.*, 1989) and maize endosperm (Bhave *et al.*, 1990) large subunits, respectively. This homology study along with the previously discussed immunological data revealed two observations. First, there appears to be less identity between the two subunits from the same species than between similar subunits from the different plants. Second, the smaller subunits are more conserved among the various plants and the larger subunits more divergent. For a more in-depth analysis of the conserved motifs and their implied roles in catalytic and allosteric function of the plant and bacterial AGPase the reader may refer to recent discussions (Olive *et al.*, 1989; Nakata *et al.*, 1991; Smith-White and Preiss, 1992).

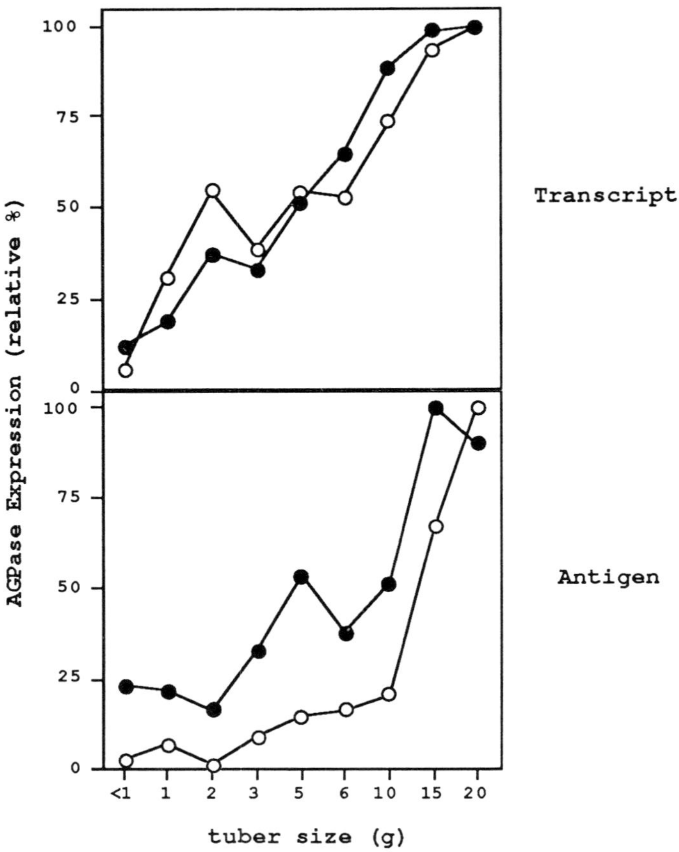

Figure 3.1. Temporal expression of AGPase in developing potato tubers. The filled circles represent the small subunit transcript and antigen levels while the open circles indicate the large subunit levels.

Temporal Expression of Tuber AGPase

In contrast to the abundant biochemical data available concerning the allosteric properties of the AGPase enzyme, little is known regarding other possible modes of regulating expression. As a step in this direction the temporal pattern of both the small and large subunits of the potato tuber AGPase was analyzed at the transcript and antigen levels (Fig. 3.1). Developmental Northern analysis revealed a 1.9 kb small subunit and a 2.1 kb large subunit transcript in which both mRNA populations exhibited a general trend of coordinately increasing levels with the developing tuber. Antigen levels also appeared closely to parallel the transcript accumulation profile, suggesting a role for the developmental regulation of AGPase expression in potato tubers as well as providing further support for our earlier work on the elucidation of the tuber's L_2S_2 subunit structure (Okita *et al.*, 1990). Similar temporal transcript accumulation profiles have also been previously reported in developing rice (Anderson *et al.*, 1991) and wheat seeds (Ainsworth *et al.*, 1993; Reeves *et al.*, 1986). Therefore, it appears that AGPase expression in storage organs may be controlled primarily at the level of transcription although regulation at the enzyme level also occurs. Results from Hnilo and Okita (1989) support the role of intracellular Pi levels having significant inhibitory role in regulating amyloplastic starch biosynthesis.

Spatial Expression of Potato AGPase

An abundance of both biochemical and genetic studies has established that the heterotetrameric subunit structure is essential for AGPase activity (Lin *et al.*, 1988; Müller-Röber *et al.*, 1992; Iglesias *et al.*, 1993). Contrary to these results, it has been reported that in potato leaf a large difference exists in the expression pattern of the genes encoding each subunit (Müller-Röber *et al.*, 1990). Subsequently, we have confirmed this report and have observed that the relative steady-state transcript levels of the small subunit are several-fold higher than the relative quantity of large subunit mRNA (Fig. 3.2). A similar phenomenon was recently reported revealing that the barley AGPase small subunit was also expressed in excess over the large subunit (Villand *et al.*, 1992). Although the molecular basis for the differential expression of the subunit genes remains to be resolved, we have identified an interesting feature which may reconcile this apparent difference between the steady-state transcript levels and the requirement for the presence of both subunits for enzyme activity. Although small subunit transcript levels are abundant in leaf tissue, corresponding antigen levels are much lower as compared to the other tissues (Fig. 3.2). It is likely that this apparent post-transcriptional event(s) regulates the amount of small subunit antigen resulting in corresponding levels of both subunits. Studies are currently in progress to distinguish the mode of post-transcriptional regulation. One cannot exclude the possibility that tissue-specific forms of the enzyme may also play a contributing role and preliminary analysis in this area will be discussed.

Influences of Sucrose on AGPase Expression

In addition to the temporal and spatial expression patterns, we also wanted to investigate potential signals which may influence AGPase expression. The preferential expression of the large subunit in the non-photosynthetic sink tissues of stolon and tuber (Fig. 3.2) suggested that regulation of this subunit may be related to tissue status, to the high levels of carbohydrate (i.e. sucrose)

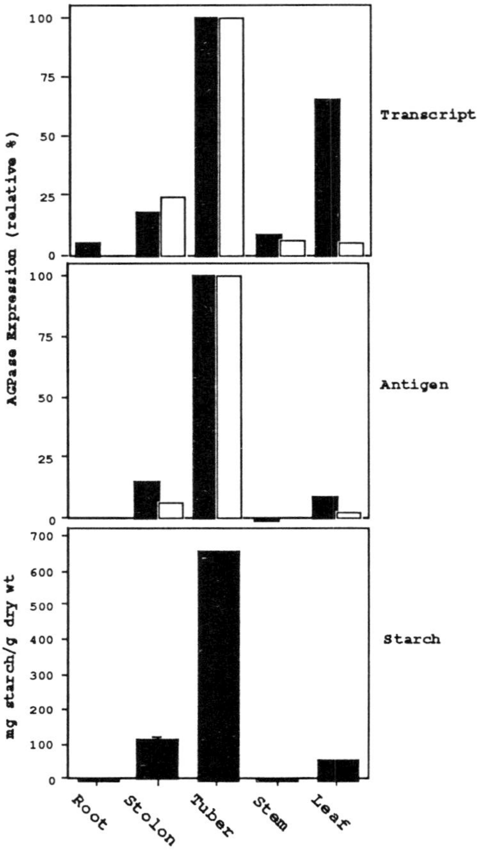

Figure 3.2

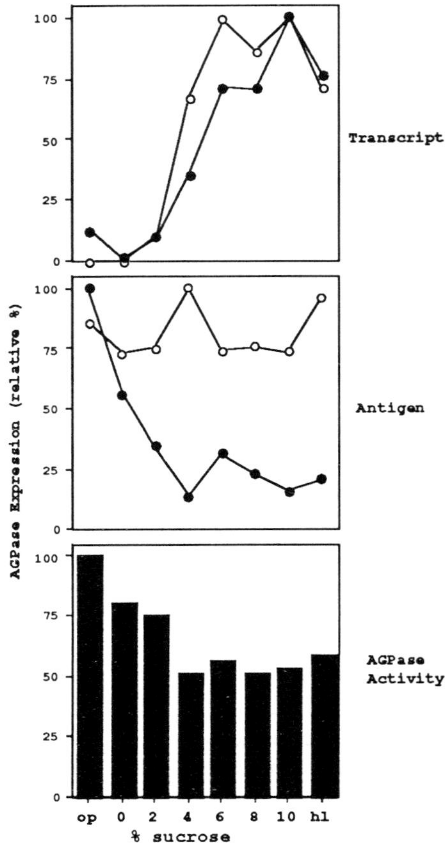

Figure 3.3

Figure 3.2. Spatial expression of AGPase. Small and large subunit transcript and antigen steady-state levels are represented by the filled bars and open bars, respectively.

Figure 3.3. Influence of sucrose on AGPase expression. Small subunit transcript and antigen accumulation patterns are shown as filled circles and large subunit profiles as open circles. Control leaves are designated 'op' and incubations in plain water under constant illumination 'hl'.

which are normally found within these tissues, or to a combination of both possibilities. Sucrose, in general, is often considered as the transportable product of photosynthesis and thus may be thought of as connecting photoassimilate production in the source tissues and photoassimilate utilization in the sink organs. Evidence has been accumulating that suggests a role for sucrose concentration as a regulator of plant genes involved in photosynthesis in leaf (Sheen, 1990) and starch biosynthetic (Visser *et al.*, 1991; Kossmann *et al.*,

1991; Müller-Röber *et al.*, 1990) and storage protein genes (Johnson and Ryan, 1990; Ohta *et al.*, 1991) in the reserve organ. In light of this potential role of sucrose in coordinating the expression of starch biosynthetic as well as other sink related genes, we decided to investigate an observation initially reported by Müller-Röber *et al.* (1990) which suggested a correlation between the sucrose-induced expression of the AGPase large subunit transcript and an increase in starch deposition. We analyzed RNA isolated from potato leaves which had been detached from the potato plant at the end of a dark cycle (op) or after they had been incubated on various concentrations of sucrose in the dark for 24 h or on plain water under constant illumination. Our Northern data suggest that the accumulation of not only the large subunit, but also the small subunit transcript may be similarly triggered by elevated levels of metabolizable carbohydrates, either supplied by exogeneous sucrose or via photosynthesis, in leaves that are not able to export the produced photosynthate (Fig. 3.3).

Extension of this analysis to the antigen level, however, revealed that each subunit in comparison with their transcript accumulation profile did not respond similarly to the exogeneous supply of sucrose (Fig. 3.3). This discrepancy in the relative abundance of transcripts and corresponding polypeptides does not appear to be due to a simple lag in antigen accumulation, as was observed in a similar study analyzing sporamin accumulation (Hattori *et al.*, 1991), since longer incubations of the water/light sample of up to seven days do not result in any increase in the antigen levels. Enzyme activity levels appears generally to follow the antigen accumulation profile and likewise do not show any induction in specific activity (Fig. 3.3). Müller-Röber *et al.* (1990) observed, however, a large increase in starch deposition under the conditions of elevated sucrose levels (6%) or constant illumination in plain water. Our analysis also revealed increases in starch accumulation under the latter condition (data not shown). Therefore our study suggests, in addition to a possible post-transcriptional event, that the observed increases in starch deposition with the addition of sucrose or under conditions of constant light is not a direct consequence of an enhancement in gene expression, but may simply reflect an increase in substrate availability and/or allosteric activation of the existing enzyme *in vivo*.

Small Subunit Expression in Both Photosynthetic and Non-Photosynthetic Tissue

Studies with barley (Villand *et al.*, 1992), wheat (Olive *et al.*, 1989) and rice (Krishnan *et al.*, 1986) have shown that each of these plants expresses tissue-specific forms of the enzyme based on tissue type, photosynthetic vs. non-photosynthetic. A major difference between the tissue-specific forms of the large subunit cDNAs appear to be contained primarily within the 5'-region in which they diverge in both sequence and length (Nakata *et al.*, 1991; Smith-White and

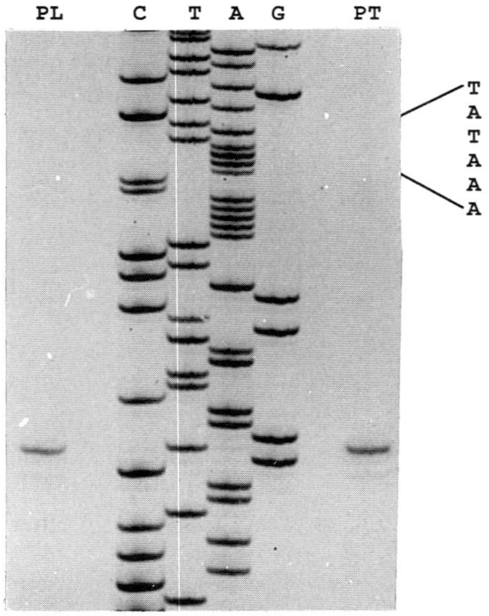

Figure 3.4. Primer extension analysis of the potato AGPase small subunit using leaf (PL) and tuber (PT) poly$(A)^+$ mRNA.

Preiss, 1992). Likewise, comparisons of the potato and rice small subunit sequences revealed an overall amino acid identity of 84%, but displayed a 91% identity when considering the putative mature proteins. This indicated that the main divergence between the small subunits would also most likely occur in the 5'-untranslated and putative transit peptide regions. Therefore, in an attempt to assess whether the same potato small subunit gene is expressed in the various tissues and/or if there are tissue-specific forms, Northern blot analysis was performed using probes derived from the 5'-untranslated and transit peptide regions of the tuber cDNA (data not shown). Tissue specific forms of the small subunit were not distinguishable by this method. Primer extension analysis was then conducted using poly (A)+ mRNA isolated from either tuber (non-photosynthetic) or leaf (photosynthetic) tissues. These two tissues exhibit the same size extension products indicating that the same gene may be expressed in both leaf and tuber (Fig. 3.4). Further evidence in support of the expression of same gene in both tissues was obtained upon isolation and characterization of leaf cDNAs encoding the small subunit. Several clones were partially sequenced and one full length leaf clone completely characterized. The full length leaf cDNA was virtually identical at the nucleotide level to the tuber sequence (data not shown). Overall, these experiments suggest that the same gene may be expressed in the different tissues but they do not rule out the possibility that multiple forms of the small subunit are expressed in a given tissue. The existence of multiple small subunit genes may simply be undetectable by the methods employed. This analysis does emphasize, however, that all plant small subunits have been under strong selective pressure to retain a particular primary structure regardless of whether it is from photosynthetic or non-photosynthetic tissue. Transgenic potato plants expressing ß-glucuronidase (GUS) fusions of the reporter gene driven by the small subunit promoter will allow further investigation as to whether there are tissue-specific forms of the small subunit in potato.

Isolation, Structure and Expression of the AGPase Small Subunit Gene

Genomic clones encoding the small subunit of AGPase have been isolated and the gene structure determined. In addition to the high degree of conservation of the primary level among the different plant small subunits, the genes encoding the rice endosperm (Anderson *et al.*, 1991) and potato small subunits also share a similarly complex structure (Okita *et al.*, 1993) as do other genes involved in starch metabolism (Callis *et al.*, 1987; Maas *et al.*, 1991). Transcriptional fusions of the 5'-promoter region of the potato small subunit to a GUS reporter gene were constructed to determine the effects of *cis* elements on expression and to investigate further whether there are multiple forms of the small subunit. Fluorometric analysis of a number of transformed plants displayed detectable reporter gene activity in all tissues assayed (Fig. 3.5). Relative GUS activity levels generally correlated with the small subunit Northern profile with the notable exception of leaf tissue where the observed methyl umbelliferyl-ß-D-glucuronide (MUG) activity was much lower than anticipated based on the spatial mRNA pattern. Currently, we are attempting to determine the reason for this unexpectedly low leaf activity.

Preliminary *cis* analysis revealed general regions which may be important in expression. The levels of GUS activity in the non-photosynthetic tissues generally decreased in a similar fashion upon deletion of the 5'-end. Stolon and tuber exhibited, on average, a 2-fold reduction in MUG activity upon exonuclease III digestion of the 3500 bp promoter region down to 1000 bp, a 7-fold decrease upon deletion down to 670 bp, and total elimination of GUS activity when digested down to 160 bp. In photosynthetic tissue, on the other hand, GUS activity levels remained fairly constant with the removal of the first 2500 bp, showed about a 2-fold reduction, on average, when digested down to 670 bp, and had low levels upon deletion down to 160 bp.

Summary

Plants accumulate starch primarily in leaf and storage organs. Leaf starch is transitory, being synthesized during the active periods of photosynthesis and catabolized to varying extents during the night. In storage tissue, on the other hand, synthesis and degradation occur at specific stages of plant ontology. In

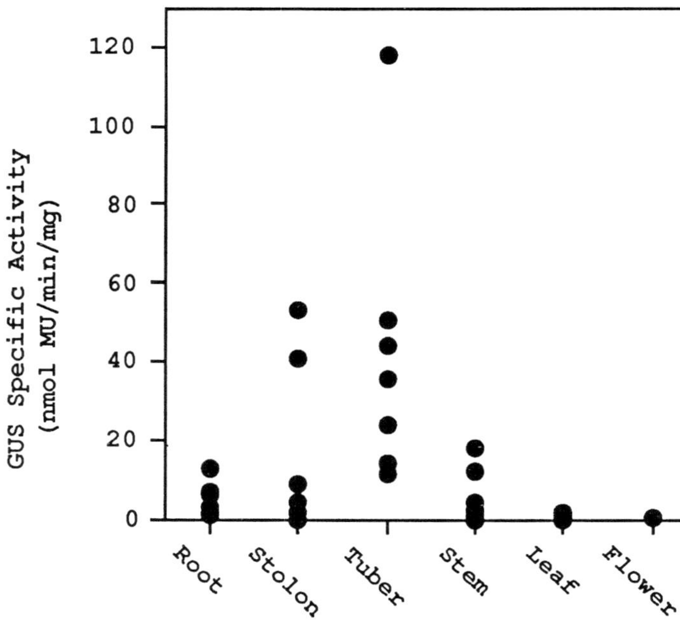

Figure 3.5. GUS activity in the different tissues of transgenic potato plants containing a chimeric GUS gene driven by the AGPase small subunit promoter.

both tissues the production of starch is metabolically controlled by the activity of AGPase . Our efforts to understand the control of AGPase expression are just beginning to provide information on the structure and regulation of the potato enzyme's subunits which may prove useful in attempts to enhance the conversion of photosynthate into starch.

Acknowledgements

This work was supported by a grant from the US Department of Energy (DE-FG06-87ER13699) and Project 0590, College of Agriculture and Home Economics, Washington State University. P.A.N. was supported by McKnight Foundation and Loyal Davis fellowships.

References

Ainsworth, C., Tarvis, M. and Clark, J. (1993) Isolation and analysis of a cDNA clone encoding the small subunit of ADPG-glucose pyrophosphorylase from wheat. *Plant Molecular Biology* 23:23-33.

Anderson, J.M., Hnilo, J., Larsen, R., Okita, T.W. and Morell, M. (1989) The encoded primary sequence of a rice seed ADP-glucose pyrophosphorylase subunit and its homology to a bacterial enzyme. *Journal of Biological Chemistry* 264:12238-12242.

Anderson, J.M., Okita, T.W. and Preiss, J. (1990) Enhancing carbon flow into starch: the role of ADPglucose pyrophosphorylase. In: Vayda, M.E. and Park, W.D. (eds), *Molecular and Cellular Biology of the Potato*. CAB International, Wallingford, UK, pp. 159-180.

Anderson, J.M., Larsen, R., Laudencia, D., Kim, W.T., Morrow, D., Okita, T.W. and Preiss, J. (1991) Molecular characterization of the gene encoding a rice endosperm-specific ADPglucose pyrophosphorylase subunit and its developmental pattern of transcription. *Gene* 97:199-205.

ap Rees, T. and Entwistle, G. (1989) Entry into the amyloplasts of carbon for starch synthesis. In: Boyer, C.D., Shannon, J.C. and Hardison, R.C. (eds), *Physiology, Biochemistry, and Genetics of Nongreen Plastids*. American Society of Plant Physiology, Rockville, Maryland, pp. 49-62.

Bhave, M.R., Lawrence, S., Barton, C. and Hannah, C. (1990) Identification and molecular characterization of *shrunken-2* cDNA clones of maize. *Plant Cell* 2:581-588.

Burt, R.L. (1964) Carbohydrate utilization as a factor in plant growth. *Australian Journal of Biological Science* 17:867-877.

Callis, J., Fromm, M. and Walbot, V. (1987) Introns increase gene expression in cultured maize cells. *Genes and Development* 1:1183-1200.

du Jardin, P. and Berhin, A. (1991) Isolation and sequence analysis of a cDNA clone encoding a subunit of the ADPglucose pyrophosphorylase of potato tuber amyloplast. *Plant Molecular Biology* 16:349-351.

Ewart, A.J. (1986) On assimilatory inhibition in plants. *Journal of the Linnean Society of Botany* 31:364-461.

Hattori, T., Fukumoto, H., Nakagawa, S. and Nakamura, K. (1991) Sucrose-induced expression of genes coding for the tuberous root stroage protein, sporamin, of sweet potato in leaves and petioles. *Plant Cell Physiology* 32:79-86.

Heldt, H.W., Chon, C.J., Maronde, D., Herold, A., Stankovic, Z.S., Walker, D., Kraminer, A., Kirk, M.R. and Heber, U. (1977) Role of orthophosphate and other factors in the regulation of starch formation in leaves and isolated chloroplasts. *Plant Physiology* 59:1146-1155.

Hill, L.M. and Smith, A.M. (1991) Evidence that glucose 6-phosphate is imported as the substrate for starch synthesis by the plastids of developing embryos. *Planta* 185:91-96.

Hnilo, J. and Okita, T.W. (1989) Mannose feeding and its effect on starch synthesis in developing potato tuber discs. *Plant and Cell Physiology* 30:1007-1010.

Iglesias, A.A., Barry, G.F., Meyer, C., Bloksberg, L., Nakata, P.A., Greene, T., Laughlin, M.J., Okita, T.W., Kishore, G.M. and Preiss, J. (1993) Expression of the potato tuber ADP-glucose pyrophosphorylase in *Escherichia coli*. *Journal of Biological Chemistry* 268:1081-1086.

Johnson, R. and Ryan, C. (1990) Wound-inducible potato protease inhibitor II genes; enhancement of expression by sucrose. *Plant Molecular Biology* 14:527-536.

Keeling, P.L., Wood, J.R., Tyson, R.H. and Bridges, I.G. (1988) Starch biosynthesis in developing wheat grain. *Plant Physiology* 87:311-319.

King, R.W., Wardlaw, I.F. and Evans, L.T. (1967) Effect of assimilate utilization on photosynthetic rate in wheat. *Planta* 77:261-276.

Kleczkowski, L.A., Villand, P., Lonneborg, A., Olsen, O.A. and Luthi, E. (1991) Plant ADP-glucose pyrophosphorylase - recent advances and biotechnological perspectives (a review). *Zeitschrift für Naturforschung C.- A Journal of Biosciences* 46:605-612.

Kossmann, J., Visser, R.G.F., Müller-Röber, B., Willmitzer, L. and Sonnewald, U. (1991) Cloning and expression analysis of a potato cDNA that encodes branching enzyme:evidence for co-expression of starch biosynthetic genes. *Molecular and General Genetics* 230:39-44.

Krishnan, H.B., Reeves, C.D. and Okita, T.W. (1986) ADPglucose pyrophosphorylase is encoded by different mRNA transcripts in leaf and endosperm of cereals. *Plant Physiology* 81:642-645.

Lin, T.-P., Caspar, T., Somerville, C.R. and Preiss, J. (1988) A starch deficient mutant of *Arabidopsis thaliana* with low ADPglucose pyrophosphorylase activity lacks one of the two subunits of the enzyme. *Plant Physiology* 99:1175-1181.

Lucas, W.J. and Madore, M.A. (1988) Recent advances in sugar transport. In: Preiss, J. (ed.), *The Biochemistry of Plants*. Academic Press, San Diego, CA, pp. 35-84.

Maas, C., Laufs, J., Grant, S., Korfhage, E.M. and Werr, W. (1991) The combination of a novel stimulatory element in the first exon of the maize *shrunken*-1 gene with the following intron 1 enhances reporter gene expression up to 1000-fold. *Plant Molecular Biology* 16:199-207.

McArthur, J.A., Hesketh, J.D. and Baker, D.N. (1975) Cotton. In: Evans, L.T. (ed.), *Crop Physiology: Some Case Histories*. Cambridge University Press, London, pp. 297-325.

Müller-Röber, B.T., Kossmann, J., Hannah, L.C., Willmitzer, L. and Sonnewald, U. (1990) One of two different ADP-glucose pyrophosphorylase genes from potato responds strongly to elevated levels of sucrose. *Molecular and General Genetics* 224:136-146.

Müller-Röber, B., Sonnewald, U. and Willmitzer, L. (1992) Inhibition of the ADP-glucose pyrophosphorylase in transgenic potatoes leads to sugar-storing tubers and influences tuber formation and expression of tuber storage protein genes. *EMBO Journal* 11:1229-1238.

Nakata, P.A., Greene, T.W., Anderson, J.M., Smith-White, B.J., Okita, T.W. and Preiss, J. (1991) Comparison of the primary sequences of the two potato tuber ADP-glucose pyrophosphorylase subunits. *Plant Molecular Biology* 17:1089-1093.

Nasyrov, Y.S. (1978) Genetic control of photosynthesis and improving of crop productivity. *Annual Review of Plant Physiology* 29:215-237.

Neuhaus, H.E., Kruckeberg, A.L., Feil, R., and Stitt, M. (1989) Reduced-activity mutants of phosphoglucose isomerase in the cytosol and chloroplast of *Clarkia xantiana* II. Study of the mechanisms which regulate photosynthate partitioning. *Planta* 178:110-112.

Ohta, S., Hattori, T. and Morikami, A. (1991) High-level expression of sweet potato sporamin gene promoter: ß-glucuronidase (GUS) fusion gene in the stems of transgenic tobacco plants is conferred by multiple cell type-specific regulatory elements. *Molecular and General Genetics* 225:369-378.

Okita, T.W. (1992) Is there an alternative pathway for starch synthesis? *Plant Physiology* 100:560-564.

Okita, T.W., Nakata, P.A., Anderson, J.M., Sowokinos, J., Morell, M. and Preiss, J. (1990) The subunit structure of potato tuber ADPglucose pyrophosphorylase. *Plant Physiology* 93:785-790.

Okita, T.W., Nakata, P., Ball, K., Smith-White, B. and Preiss, J. (1993) Enhancement of plant productivity by manipulation of ADPglucose pyrophosphorylase. In: Gustafson, J.P., Appels, R. and Raven, P. (eds), *Gene Conservation and Exploitation*. Plenum Press, New York, pp. 161-191.

Olive, M., Ellis, R.J. and Schuch, W.W. (1989) Isolation and nucleotoide sequences of cDNA clones encoding ADP-glucose pyrophosphorylase polypeptides from wheat leaf and endosperm. *Plant Molecular Biology* 12:525-538.

Plaxton, W.C. and Preiss, J. (1987) Purification and properties of nonproteolytic degraded ADPglucose pyrophosphorylase from maize endosperm. *Plant Physiology* 83:105-112.

Preiss, J. (1982) Regulation of the biosynthesis and degradation of starch. *Annual Review of Plant Physiology* 33:432-454.

Preiss, J. (1984) Bacterial glycogen synthesis and its regulation. *Annual Review of Microbiology* 38:419-458.

Preiss, J. (1988) Biosynthesis of starch and its regulation. In: Preiss, J. (ed.), *The Biochemistry of Plants*. Academic Press, New York, pp. 181-254.

Preiss, J. (1991) Biology and molecular biology of starch synthesis and regulation. In: Miflin, B.J. (ed.), *Oxford Surveys of Plant Molecular and Cellular Biology*, Vol. 7. Oxford University Press, Oxford, pp. 59-114

Preiss, J., Cress, D., Hutny, J., Morell, M., Bloom, M., Okita, T. and Anderson, J. (1989) Regulation of starch synthesis biochemical and genetic studies. In: Whitaker, J.R. and Sonnet, P.E. (eds), *Biocatalysis in Agricultural Biotechnology*. American Chemical Society, Washington, DC, pp. 84-92.

Reeves, C.D., Krishnan, H.B. and Okita, T.W. (1986) Gene expression in developing wheat endosperm. *Plant Physiology* 82:34-40.

Sanwal, G.G., Greenberg, E., Hardie, J., Cameron, E.C. and Preiss, J. (1967) Regulation of starch biosynthesis in plant leaves: activation and inhibition of ADPglucose pyrophosphorylase. *Plant Physiology* 43:417-427.

Sheen, J. (1990) Metabolic repression of transcription in higher plants. *Plant Cell* 2:1027-1038.

Smith-White, B.J. and Preiss, J. (1992) Comparison of proteins of ADP-glucose pyrophosphorylase from diverse sources. *Journal of Molecular Evolution* 34:449-464.

Sowokinos, J.R. and Preiss, J. (1982) Pyrophosphorylases in *Solanum tuberosum* III. Purification, physical and catalytic properties of ADPglucose pyrophosphorylase in potatoes. *Plant Physiology* 69:1459-1466.

Stark, D.M., Timmerman, K.P., Barry, G.F., Preiss, J. and Kishore, G.M. (1992) Regulation of the amount of starch in plant tissues by ADP glucose pyrophosphorylase. *Science* 258:287-292.

Steup, M., Peavy, D.G. and Gibbs, M. (1976) The regulation of starch metabolism by inorganic phosphate. *Biochemical and Biophysical Research Communications* 72:1554-1561.

Thorne, G.N. and Evans, A.F. (1964) Influence of tops and roots on net assimilation rate of sugar beet and spinach beet grafts between them. *Annals of Botany* 28:499-508.

Tollenaar, M. (1977) Sink-source relationships during reproductive development in maize: a review. *Maydica* 22:49-75.

Tyson, R.H. and ap Rees, T. (1988) Starch synthesis by isolated amyloplast from wheat endosperm. *Planta* 175:33-38.

Villand, P., Aalen, R., Olsen, O., Luthi, E., Lonneborg, A. and Kleczkowski, L.A., (1992) PCR amplification and sequences of cDNA clones for the small and large subunits of ADP-glucose pyrophosphorylase from barley tissues. *Plant Molecular Biology* 19:381-389.

Visser, R.G.F., Stolte, A. and Jacobsen, E. (1991) Expression of a chimaeric granule-bound starch synthase-GUS gene in transgenic potato plants. *Plant Molecular Biology* 17:691-699.

Chapter Four:

Control of Carbohydrate Metabolism in Potato Tubers

M.M. Burrell
Advanced Technologies (Cambridge) Limited, Cambridge, UK.

The substantial advances and refinements in the techniques of molecular biology have allowed the problems of crop productivity and quality to be approached in new ways. One of the factors which limits crop productivity is the efficiency with which the plant stores carbon. In the case of the potato this will be affected by the amount of carbon translocated to the tuber. Clearly the proportion of carbon translocated has a finite limit since ultimately if too much carbon is translocated there will be insufficient to develop leaves. This, in turn, would limit the production of more translocate. Similarly, the quality of the potato tuber will depend on the amount of carbon stored, the form in which it is stored and the amount used for energy production. The challenge is therefore to understand how the plant controls the balance between the competing uses for translocated carbon and how much this balance can be shifted in a particular direction to alter crop productivity.

To modify this balance requires the identification of mechanisms that control flux through a metabolic pathway and determination of whether certain steps in the pathway are more important than others under most environmental conditions. Classically plant biochemists have used three types of information to identify enzymes in control: the type of reaction and properties of the enzyme, the changes in metabolite levels when the environment is altered, and finally comparative biochemistry. Of these the first has tended to dominate because it is the easiest to investigate. Intuitively an enzyme which catalyzes an irreversible reaction can easily be seen as a good point of control since a different reaction is required to go in the opposite direction. Also if the enzyme activity is affected by allosteric regulators then it suggests that the reaction under consideration can be regulated. However, these arguments only indicate that an enzyme may regulate a reaction and do not indicate the role of the enzyme in the control of flux through a pathway. While enzymes catalyzing irreversible reactions are often favored as points of control there is no *a priori* reason why the regulation of an enzyme catalyzing a reversible reaction could not regulate the flux through a pathway. When the enzyme activity is insufficient to use the substrate at the

The Molecular and Cellular Biology of the Potato, Second Edition
Edited by W.R. Belknap, M.E. Vayda and W.D. Park

rate of supply then the enzyme will limit flux. Therefore particular environmental conditions may influence the rate of supply and use of intermediates and this could easily alter the importance of any one enzyme on the control of the overall pathway. Similarly, comparative biochemistry may be used as a pointer to indicate how control of flux is achieved. However, the balance of metabolism in different organisms is likely to be different and when the conditions change, the balance changes and different steps may assume more or less importance.

Kacser (1989) has, since the early, 1970s, developed a theory for metabolic control which evaluates how the control of flux in a pathway is shared between the component steps. For this theory to be applied it is necessary to cause a single change in a step of the pathway and measure the consequences. Genetic modification (GM) is a suitable mechanism to create this type of change because it can be precise, although usually the magnitude of the changes in enzyme activity obtained are larger than the ideal required to apply control theory. However, because one GM experiment will provide many progeny each with the amounts of enzyme altered to varying extents, the GM experiment provides both the means to assess control coefficients for each step and the means to test the applicability of them over a range of conditions. It should be noted that although genetic manipulation is a precise process there are always two changes that may occur when a gene is introduced into a plant. Firstly the introduced gene may be expressed and secondly the genome will be disrupted by the integration of the introduced gene. Thus, especially if control of metabolism is being considered, it is essential to demonstrate that any changes observed are due to the product of the introduced gene. Similarly, it is necessary to demonstrate that the introduced gene is not causing pleiotropic effects if a direct link between cause and effect is to be inferred.

Glycolysis

The synthesis of sucrose and starch and the use of these compounds to provide energy for growth are central to the development and quality of many crops. To produce energy by respiration starch and sucrose are converted to hexose phosphate. This is then metabolized to phosphoglyceraldehyde either by glycolysis (Fig. 4.1) or the pentose phosphate pathway (Fig. 4.2). Phosphofructokinase (PFK(ATP)) catalyzes the conversion of fructose-6-phosphate (Fru-6-P) to fructose-1,6,-bisphosphate and is generally regarded as the enzyme that regulates the entry of carbon into glycolysis and, therefore, the availability of carbon for respiration (Anderson and Beardall, 1991; Bryce and Hill, 1993). Evidence for this, however, is equivocal. For example, in castor bean endosperm (Kobr and Beevers, 1971), in grapes (Ruffner and Hawker, 1977) and in potato tubers (Dixon and ap Rees, 1980), the amount of Fru-6-P changed in the opposite direction to the flux. This would support an argument for control of flux at the step catalyzed by PFK(ATP) by regulation of the activity of this enzyme. However, no such change in Fru-6-P was found in carrot storage tissue

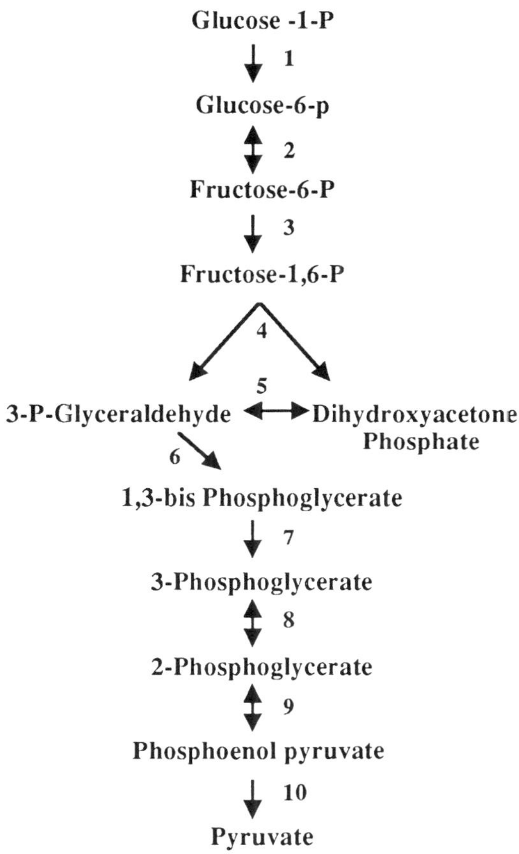

Figure 4.1. Glycolysis. (1) Phosphoglucomutase, (2) glucose phosphate isomerase, (3) phosphofructokinase, (4) aldolase, (5) triose phosphate isomerase, (6) glyceraldehyde phosphate dehydrogenase, (7) phosphoglycerate kinase, (8) phosphoglycerate mutase, (9) enolase, (10) pyruvate kinase.

(Faiz-ur-Rahman *et al.*, 1974), *Arum* spadix (ap Rees *et al.*, 1977), and cells of *Chenopodium* (Hatzfeld *et al.*, 1990).

Genetic manipulation of the amount of PFK(ATP) activity was used to study the effect of different amounts of PFK(ATP) on respiration and glycolysis in the potato tuber (Burrell *et al.*, 1993). The *Escherichia coli* PFKA was introduced and expressed in potato tubers and altered the enzyme activity up to 20-fold. The change in PFK(ATP) activity in the PFK(ATP) transgenic plants was attributed to the introduced gene and not to changes in expression of the

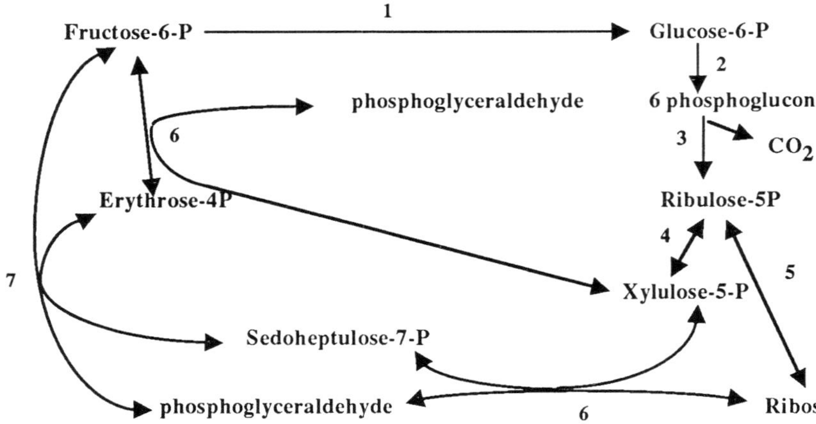

Figure 4.2. Pentose phosphate pathway. (1) Glucose phosphate isomerase, (2) glucose-6-phosphate dehydrogenase, (3) 6-phosphogluconate dehydrogenase, (4) epimerase, (5) isomerase, (6) transketolase, (7) transaldolase.

potato gene. No pleiotropic effects of the introduced gene were detected and it was concluded that the genetic modification caused a specific alteration in the activity of PFK(ATP). The introduced PFK(ATP) caused a small decrease in the amount of the hexose phosphates, glucose-1-phosphate (Glc-1-P), glucose-6-phosphate (Glc-6-P), and Fru-6-P, which precede the step catalyzed by PFK(ATP) in glycolysis. In addition, increased PFK(ATP) caused some substantial increases in the amounts of intermediates that occur after the enzyme in the pathway of glycolysis. To determine whether these alterations in enzyme activity and metabolites created changes in respiration, two types of tissue were examined: mature tubers which had been harvested for at least one week and aged discs of tuber. The former have a relatively low rate of respiration and the latter have a relatively high rate (ap Rees and Beevers, 1960). However, for both states of tissue, material from the control and PFK(ATP) transgenic plants had the same rate of respiration. Also, there was no evidence that the GM of PFK(ATP) activity had caused a redistribution of flux between the pentose phosphate pathway and glycolysis. Therefore, under two different environmental conditions despite the changes in amount of intermediates, there was no detectable change in the flux of carbon through glycolysis. The results strongly suggest that, at least for the conditions under which it was examined, the respiration of potato tubers is not significantly limited by PFK(ATP). Thus the hypothesis that this enzyme exerts appreciable control over respiration is untenable. CO2 production in anoxia was also examined and did not increase in the transgenic PFK(ATP) plants. This suggests that glycolysis *per se*, as opposed to respiration as a whole, is not controlled primarily by PFK(ATP).

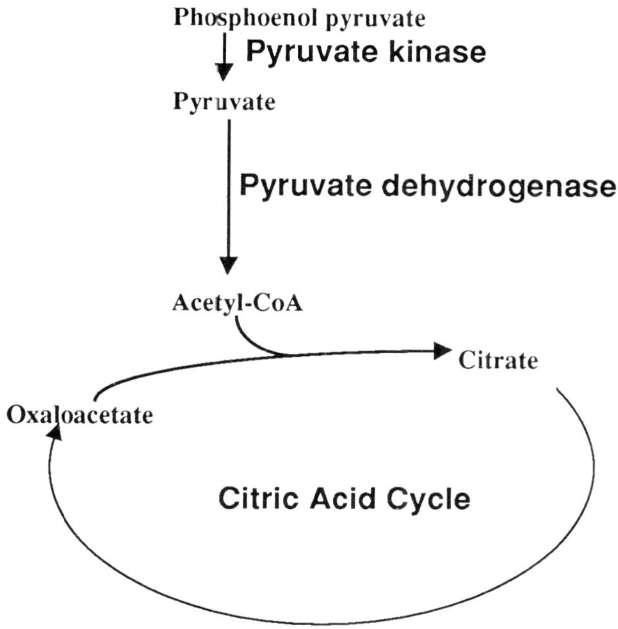

Figure 4.3. Entrance of pyruvate into the citric acid cycle.

A second enzyme that has been highlighted as important in the control of glycolysis is pyruvate kinase (Copeland and Turner, 1987). Pyruvate kinase catalyzes the conversion of phosphoenol pyruvate to pyruvate. The pyruvate provides a source of the acetyl CoA which is required for the normal operation of the citric acid cycle (Fig. 4.3). Transgenic plants have been produced with leaves which contained no measurable cytosolic pyruvate kinase (Gottlob-McHugh *et al.*, 1992). This suggests either that the conversion of phosphoenol pyruvate to pyruvate in the cytosol is not essential for leaf respiration or that there are alternative pathways. Dunford (1992) produced transgenic potato plants containing an antisense gene to pyruvate kinase and these also developed normal tubers.

The observations described above are in agreement with observations made in several other organisms. For example, eight different strains of yeast were produced by genetic manipulation to contain increased amounts of hexokinase, phosphoglucose isomerase, phosphoglucomutase, phosphofructokinase, pyruvate kinase or pyruvate decarboxylase (Schaaff *et al.*, 1989). There was no large change in ethanol production in any of the transgenic lines. The combination of the results discussed above lend support to the arguments of Kacser (1989) that control is shared amongst the steps in the pathway.

Sucrose Synthesis

The pathway of sucrose synthesis is shown in Fig. 4.4. However, during the development of a tuber, it is unclear how much of the sucrose translocated from the leaves is stored directly as sucrose and how much is synthesized from hexose phosphate. Experiments with radiolabeled compounds have indicated that sucrose is rapidly synthesized and broken down (Pollock and ap Rees, 1975; Davies, 1984; Davies and Oparka, 1985).

It is generally believed that sucrose is synthesized from UDP glucose (UDPG) and Fru-6-P by the action of sucrose phosphate synthase (SPS) (Huber *et al.*, 1992). SPS has been characterized from leaf tissues (Doehlert and Huber, 1983; Stitt *et al.*, 1988). The activity of the enzyme is affected by Glc-6-P and the activity has been inversely correlated with starch content (Huber, 1981). The activity of the enzyme is also thought to be altered by covalent modification. The substrate for SPS, UDPG, is formed by UDPG pyrophosphorylase (UGPase). The activity of the enzyme has been correlated with the accumulation of glucose during cold storage (Sowokinos, 1990).

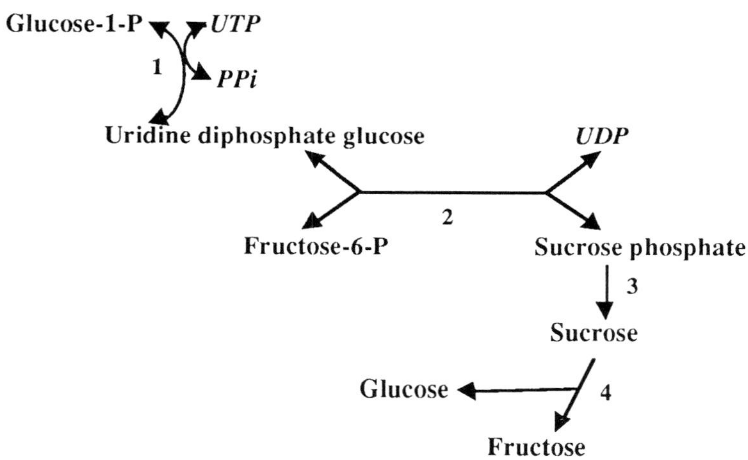

Figure 4.4. Sucrose and hexose synthesis. (1) UDP-glucose pyrophosphorylase, (2) sucrose phosphate synthase, (3) sucrose-6-phosphatase, (4) invertase.

The exact route between sucrose and hexose phosphate is not known. Sucrose synthase may operate in the degradative direction which would produce fructose and UDPG. Alternatively an invertase may cleave the sucrose to glucose and fructose. The free sugars produced in these reactions would then have to be phosphorylated with an appropriate kinase. During tuber growth there

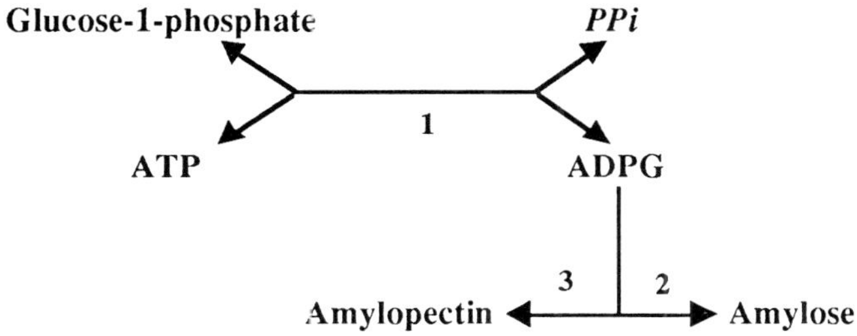

Figure 4.5. Starch biosynthesis. (1) ADPG pyrophosphorylase, (2) starch synthase, (3) starch synthase and branching enzyme.

is substantial activity of sucrose synthase and the activity declines during development (Morrell and ap Rees, 1986). In addition to sucrose synthase, potato tubers contain alkaline invertase, acid invertase, hexokinase and fructokinase (Davies and Oparka, 1985; Morrell and ap Rees, 1986). The view that invertases and hexokinase are involved in sucrose turnover is based on the observations that acid invertase activity is inversely correlated with sucrose content (Hatch and Glasziou, 1963; ap Rees, 1974; Davies *et al.*, 1989). Pressey (1967) reported the presence of an inhibitor of acid invertase in potato tubers. However, no evidence has been obtained that this protein is in the same intracellular compartment as acid invertase (Isla *et al.*, 1992). Clearly, negative evidence is not proof, but the ability to find the inhibitor in the cell wall and not the vacuole must cast doubt on its role *in vivo*.

Thus, there is at least circumstantial evidence that the activity of most of the enzymes involved in sucrose turnover may be regulated. However, it is quite unclear how control is partitioned between these enzymes. Recently, Zrenner *et al.* (1993) reported the production of transgenic plants containing an antisense gene to UGPase which showed a reduction in UGPase activity to 4% of the control plants. There was very little effect on the size of metabolic pools when the amount of extractable UGPase was greatly reduced. These experiments are discussed in more detail in Chapter 7 of this volume.

Starch Synthesis

The pathway of starch synthesis is shown in Fig. 4.5. The evidence for this pathway has been reviewed by ap Rees (1992). It has been argued (Preiss *et al.*, 1989; Morell *et al.*, 1987; Preiss, 1988), based on the properties of the enzyme, that ADP-glucose pyrophosphorylase (AGPase) is of overriding importance in regulating the synthesis of starch. However, Neuhaus and Stitt (1990) studied the regulation of starch synthesis in mutants of *Arabidopsis thaliana*. Under conditions of low and high light intensity, only 28 and 65%, respectively, of the flux control could be attributed to AGPase. Stark *et al.* (1992) used the *E. coli*

GlgC16 gene to alter the amount of AGPase in potato tubers. The *GlgC16* gene produces a mutant form of AGPase that is less responsive to allosteric effectors. Thus, the extractable activity of the enzyme is more likely to reflect the *in vivo* activity. As yet no detailed characterization of the transgenic plants has been reported. Recently, in similar experiments, Sweetlove and ap Rees (personal communication) have reported that incremental changes in the amount of enzyme do not necessarily cause a proportional increase in the amount of starch stored.

The synthesis of starch from ADPG involves both the formation of the α-1,4-glucan and its branching. This process involves more than one synthase and more than one branching enzyme and the precise pathway is not understood. The subject has been reviewed by Preiss (1988) and Bhattacharyya *et al.* (1993). Genetic evidence from the *r* mutant of peas indicates that an alteration in branching enzyme activity can alter the amount of starch synthesized (Bhattacharyya *et al.*, 1993).

Therefore it is not clear how the flux into starch is controled but genetically modified plants are being produced which will allow the question to be addressed directly.

Conclusion

The work discussed above suggests that with genetic modification of potato it is possible to alter the balance of metabolism such that metabolite pools change but steady state fluxes remain unaltered. The plants produced will help us understand how control of metabolism is shared amongst the steps of a pathway and how the potato tuber uses alternative pathways when one is inhibited.

Acknowledgements

I wish thank Dr. S. Coates, Professor T. ap Rees and Professor J. Sowokinos for their helpful comments on the manuscript.

References

Anderson, J.W. and Beardall, J. (1991) *Molecular Activities of Plant Cells.* Blackwell Scientific Publications, Oxford.

ap Rees, T. (1974) Pathways of carbohydrate breakdown in plants. *MTP International Reviews of Science. Plant Biochemistry* 11:89-127.

ap Rees, T. (1992) Synthesis of storage starch. In: Pollock, C.J. Farra, J.F. and Gordon, A.J. (eds), *Carbon Partitioning*. BIOS, Oxford, UK, pp. 115-131.

ap Rees, T. and Beevers, H. (1960) Pentose phosphate pathway as a major component of induced respiration of carrot and potato slices. *Plant Physiology* 35:839-847.

ap Rees, T., Fuller, W.A. and Wright, B.W. (1977) Measurements of glycolytic intermediates during the onset of thermogenesis in the spadix of *Arum* maculatum. *Biochimica et Biophysica Acta* 461:274-282.

Bhattacharyya, M., Martin, C. and Smith, A. (1993) The importance of starch biosynthesis in the wrinkled seed shape character of peas studied by Mendel. *Plant Molecular Biology* 22:525-531.

Bryce, J.H. and Hill, S.A. (1993) Energy production in plant cells. In: Lee, P.J. and Leegood, R.C. (eds), *Plant Biochemistry and Molecular Biology.* Wiley, Chichester, pp. 1-21.

Burrell, M.M., Mooney, P.J., Blundy, M., Carter, D., Wilson, F.M., Green, J., Blundy, K.S. and ap Rees, T. (1993) Genetic manipulation of 6-phosphofructokinase in potato tubers. *Planta* 194:95-101.

Copeland, L. and Turner, J.F. (1987) The regulation of glycolysis and the pentose phosphate pathway. In: Davies, D.D (ed.), *The Biochemistry of Plants.* Academic Press, NY, Vol. 11, pp. 107-128.

Davies, H.V. (1984) Sugar metabolism in stolon tips of potato during early tuberisation. *Zeitschrift fur Pflanzenphysiologie* 113:377-381.

Davies, H.V. and Oparka, K.J. (1985) Hexose metabolism in developing tubers of potato (*Solanum tuberosum* L.) cv Maris Piper. *Journal of Plant Physiology* 119:311-316.

Davies, H.V., Jefferies, R.A. and Scobie, L. (1989) Hexose accumulation in cold stored tubers of potato (*Solanum tuberosum* L.): the effects of water stress. *Journal of Plant Physiology* 134:471-475.

Dixon, W.L. and ap Rees, T. (1980) Identification of the regulatory steps in glycolysis in potato tubers. *Phytochemistry* 19:1297-1301.

Doehlert, D.C. and Huber, S.C. (1983) Regulation of spinach leaf sucrose phosphate synthase by glucose-6-phosphate, inorganic phosphate and pH. *Plant Physiology* 73:989-994.

Dunford, R. (1992) Pyruvate kinase and glycolytic control in potatoes. PhD thesis. Cambridge University, Cambridge, UK.

Faiz-ur-Rahman, A.T.M., Trewavas, A.J. and Davies, D.D. (1974) The Pasteur effect in carrot root tissue. *Planta* 118:195-210.

Gottlob-McHugh, S.G., Sangwan, R.S., Blakely, S.D., Vanlerberghe, G.C., Ko, K., Turpin, D.H., Plaxton, W.C., Miki, B.L. and Dennis, D.T. (1992) Normal growth of transgenic tobacco plants in the absence of cytosolic pyruvate kinase. *Plant Physiology* 100:820-825.

Hatch, M.D. and Glasziou, K.T. (1963) Sugar accumulation cycle in sugar cane. II Relationship of invertase activity to sugar content and growth rate in storage tissue of plants grown in controlled environments. *Plant Physiology* 38:344-348.

Hatzfeld, W.-D., Dancer, J. and Stitt, W.M. (1990) Fructose-2,6-bisphosphate, metabolites and coarse control of pyrophosphate: fructose-6-phosphate phosphotransferase during triose-phosphate cycling in heterotrophic cell-suspension cultures of *Chenopodium rubrum. Planta* 180:205-211.

Huber, S.C. (1981) Interspecific variation in activity and regulation of leaf sucrose phosphate synthase. *Zeitschrift fur Pflanzenphysiologie* 102:443-450.

Huber, S.C., Huber, J.L.A. and McMichael, R.W., Jr (1992) The regulation of sucrose synthesis in leaves. In: Pollock, C.J., Farrar, J.F. and Gordon, A.J. (eds), *Carbon Partitioning*. BIOS, Oxford, UK, pp. 1-26.

Isla, M.I., Leal, D.P., Vattuone, M.A. and Sampietro, A.R. (1992) Cellular localisation of the invertase, proteinaceous inhibitor and lectin from potato tubers. *Phytochemistry* 31:1115-1118.

Kacser, H. (1987) Control of metabolism. In: Davies, D.D. (ed.), *The Biochemistry of Plants*. Academic Press, London, Vol. 11, pp. 39-67.

Kobr, M.J. and Beevers, H. (1971) Gluconeogenesis in the castor bean endosperm. *Plant Physiology* 47:48-52.

Morrell, S. and ap Rees, T. (1986) Sugar metabolism in developing tubers of *Solanum tuberosum*. *Phytochemistry* 25:1579-1585.

Morell, M., Bloom, M., Larsen, R., Okita, T. and Preiss, J. (1987) Biochemistry and molecular biology of starch synthesis. In: Key, J.L., McIntosh, I. and Liss A.R. (eds), *UCLA Symposium on Molecular Biology of Starch Synthesis*. New Series, Vol. 62, pp. 227-242.

Neuhaus, H.E. and Stitt, M. (1990) Control analysis of photosynthate partitioning. *Planta* 182:445-454.

Pollock, C.J. and ap Rees, T. (1975) Effect of cold on glucose metabolism by callus and tubers of *Solanum tuberosum*. *Phytochemistry* 14:1903-1906

Preiss, J. (1988) Biosynthesis of starch and its regulation. In: Stumpf, P.K. and Conn, E. E. (eds), *The Biochemistry of Plants*. Academic Press, NY. Vol. 14, pp. 182-249.

Preiss, J., Cress,D., Hutny, J., Morell, M., Bloom, M., Okita, T. and Anderson, J. (1989) Regulation of starch synthesis. In: Whitaker, J.R. and Sonnet, P.E. (eds), *Biocatalysis in Agricultural Biotechnology*. Symposia Series 389. American Chemical Society, Washington, DC. pp. 84-92.

Pressey, R. (1967) Invertase inhibitor from potatoes. Purification, characterisation and reactivity with plant invertases. *Plant Physiology* 42:1780-1786.

Ruffner, H.P. and Hawker, J.S. (1977) Control of glycolysis in *Vitis vinifera*. *Phytochemistry* 16:1171-1175.

Schaaff, I., Heinisch, J. and Zimmerman, F.K. (1989) Overproduction of glycolytic enzymes in yeast. *Yeast* 5:285-290.

Stark, D.M, Timmerman, K.P, Barry, G.F., Preiss, J. and Kishore, G.M. (1992) Regulation of the amount of starch in plant tissues by ADP glucose pyrophosphorylase. *Science* 258:287-292.

Stitt, M., Wilke, I., Feil, R. and Heldt, H.W. (1988) Coarse control of sucrose phosphate synthase in leaves: alterations of kinetic properties in response to the rate of photosynthesis and the accumulation of sucrose. *Planta* 174:217-230.

Sowokinos, J. (1990) Stress induced alterations in carbohydrate metabolism. In: Vayda, M. and Park, W.D. (eds), *The Molecular and Cellular Biology of the Potato*. CAB International, Wallingford, UK, pp. 137-158.

Zrenner, R., Willmitzer, L. and Sonnewald, U. (1993) Analysis of the expression of potato uridine diphosphate glucose pyrophosphorylase and its inhibition by antisense RNA. *Planta* 190:247-252.

Chapter Five:

Transgenic Potatoes Changed in Carbohydrate Partitioning and Allocation

Lothar Willmitzer, Wolf Bernd Frommer, Jens Kossmann, Bernd Müller-Röber, Jorg Riesmeier, Uwe Sonnewald, Ulf-Ingo Flügge[†] and Hans Heldt[‡]

Institut für Genbiologische Forschung, Berlin, Germany. [†]Institut für Botanik, Würzburg, Germany. [‡]Institut für Botanik, Göttingen, Germany.

Growth and development of a plant is dependent upon the energy gained by fixing CO_2 into carbohydrates during photosynthesis. Primary sites for photosynthesis are leaf and, to a much lesser extent, stem tissues. Other organs such as roots, seeds or tubers do not contribute to carbon assimilation, but rather totally depend upon the photosynthetically active organs. Thus there is a net flow of carbohydrate from photosynthetically active tissues, representing the sources (defined as net exporters of fixed carbon) to photosynthetically inactive parts of the plant, representing the sinks (defined as net importers of fixed carbon).

Sink tissues can be divided into at least two different classes (Ho, 1988). The first class is comprised of metabolically highly active, mostly rapidly growing tissue, e.g. meristem tissue and immature leaves. These so-called utilization sinks use the imported carbohydrates mainly in respiratory processes and for the biosynthesis of cellular structures, such as the cell wall. The second class is comprised of storage tissue such as is found in tubers, seeds, fruits and, to a certain extent, in the stem. Most of the imported carbohydrates are deposited in the form of starch or sucrose to be used at a later developmental stage (e.g. during seedling development) as energy reservoirs. These sinks are called storage sinks.

Sink-source interactions represent an interesting subject to study for several reasons. First, due to the fact that sink-source interactions are coupled to the flow and distribution of the primary products of photosynthesis within the plant and thus control the major flow of energy throughout the plant, it is obvious that these interactions are central to plant development. In addition, these interactions continuously change during plant development. Each organ of a plant, at a certain developmental stage, represents a sink, whereas, at a different developmental stage it changes into a source tissue.

The Molecular and Cellular Biology of the Potato, Second Edition
Edited by W.R. Belknap, M.E. Vayda and W.D. Park

Sink-source relations have also been shown to be under environmental control. Thus, shading of some source leaves leads to a rapid establishment of new equilibria between exported and retained photoassimilates in the remaining photosynthetically active non-shaded source leaves. This is reflected by a new equilibrium between the amount of carbohydrate present in the form used for either transport (i.e. sucrose) or present in the form used for transitory storage (i.e. starch). Finally, the success of plant breeding during the last 60 years was not primarily due to a large increase in total amount of the biomass formed in various crop plants but rather to the preferential partitioning of this biomass into harvestable organs, such as seeds or tubers. In *Solanum demissum* tubers represent only 7% of the total weight of the plant, while in a modern *Solanum tuberosum* variety tubers represent more than 80% of the biomass (Ho, 1988). Similar data are well-established for other crops, such as soybean and wheat (Gifford *et al.*, 1984). A preferential partitioning of the biomass into harvestable organs has led to an increase in the harvest index and therefore to increased yield.

We are interested in identifying key parameters which might be controling the allocation and partitioning of photoassimilates in higher plants. To this end the following approach was used:

- identification of key steps which might be involved in controlling photoassimilate allocation and partitioning
- creation of transgenic plants which are modulated in this step by either (i) antisense RNA mediated down-regulation of the expression of the corresponding gene, or (ii) by ectopic expression of an (alien) gene
- analysis of these plants with respect to molecular, biochemical and physiological parameters.

Manipulation of the Triose-Phosphate Translocator Activity in Transgenic Plants

There has been considerable debate as to whether photosynthates are exported from the chloroplast into the cytosol as triose-phosphate or as hexoses. Furthermore it is an open question whether this export is rate limiting for photosynthesis. Biochemical data derived from spinach chloroplasts clearly indicate that triose-phosphate export is the major route for carbon channeled into sucrose for further transport into sink organs. Other transport activities could not account for sufficient translocation of metabolites across the chloroplast membrane during photosynthesis (Stitt, 1990).

In order to understand the function of the triose-phosphate translocator (TPT) protein, antisense experiments were performed in potato plants (Riesmeier *et al.*, 1993a). These experiments demonstrated the important role of this protein, since only a 30% reduction of the protein resulted in a 50% decrease of photosynthesis. This reduction of photosynthesis was accompanied by a shift of carbohydrate partitioning from sucrose to starch, and from triose-phosphate to 3-PGA.

These results indicate that the export of carbohydrates from the chloroplast into the cytosol might be rate limiting for photosynthesis. However, it should be stated that even in these plants no significant change in tuber yield was observed (Reismeier and Willmitzer, unpublished observation).

Future experiments using these plants should further elucidate the role of the TPT protein during photosynthesis and give further insights into carbon partitioning.

Manipulation of Sucrose Biosynthesis

The major controlling steps of sucrose biosynthesis in higher plants are the interconversion of fructose-l,6-bisphosphate and fructose-6-phosphate, the formation of sucrose-6-phosphate from UDP-glucose (UDPG) and fructose-6-phosphate, and the dephosphorylation of sucrose-6-phosphate yielding the final product, sucrose.

Three enzymes are involved in catalyzing the interconversion of fructose-l,6-bisphosphate and fructose-6-phosphate. The forward reaction is catalyzed by fructose-1,6-bisphosphatase (FBPase). The backward reaction, towards glycolysis, is controlled by phosphofructokinase (PFK). The third enzyme, a pyrophosphate-dependent phosphofructokinase (PFP), is able to catalyze the reaction in both directions. The reversible reaction catalyzed by PFP may be shifted towards sucrose synthesis by removal of the phosphate group donor inorganic pyrophosphate (PPi). In a recent study *Escherichia coli* inorganic pyrophosphatase (PPase) was introduced into the cytosol of transgenic tobacco and potato plants (Sonnewald, 1992). The expression of PPase in transgenic plants leads to a dramatic increase of soluble sugars in source leaves. Although an accumulation of starch is observed in older leaves, the ratio between soluble sugars and starch is increased up to 3-fold. This result suggests that expression of PPase leads to the expected alteration of carbon partitioning. The interconversion of fructose-l,6-bisphosphate and fructose-6-phosphate is shifted towards fructose-6-phosphate, i.e. in the direction of sucrose synthesis. In addition, the formation of UDPG catalyzed by the enzyme UGPase becomes irreversible due to removal of PPi. Based on an increase in free Pi in the cytosol, the export of triose-phosphates from the chloroplast is triggered. Growth and development of these plants are changed drastically. These plants are characterized by a stunted growth and loss of chlorophyll. Potato plants also showed an increase in tuber number, arguing for a reduction of sink capacity. As the PPase was expressed in all tissues, several explanations are possible for the observed changes. Expression in sink tissues may prevent the utilization of UDPG which is produced by the cleavage of sucrose by the enzyme sucrose synthase. In addition, PPase activity in companion cells may block phloem loading (Sonnewald, 1992).

Sucrose Loading into the Phloem

In photosynthetically active source tissues a surplus of fixed carbon is produced, which is transported to sink tissues, e.g. roots, seeds, tubers and sink leaves. Solutes are transported within the phloem from source to sink tissues. A key step is the transport of sucrose from mesophyll cells into the sieve element/companion cell complex (SECCC). The question whether phloem loading is exclusively symplastic, or if an apoplastic transport step is involved, has been the subject of numerous discussions (Lucas and Madore, 1988; Turgeon, 1989; Bush, 1993). The apoplastic loading model predicts an energy-dependent transport step of sucrose across the SECCC plasma membrane generating sucrose gradients with a 10-fold difference in concentration. In this model, transport is driven by a proton gradient generated by a plasmalemma H^+ATPase. In a variety of experiments a sucrose/proton symport activity was undoubtedly detected in plasma membrane vesicles and protoplasts of several plant species (Bush, 1993). There are two explanations for such a transport activity. First, the transport protein is part of a general sucrose retrieval system (Maynard and Lucas, 1982). Second, the protein is responsible for the active transport of sucrose into the SECCC. We have created transgenic plants which allowed a deeper insight into the *in vivo* processes related to the questions concerning phloem loading.

In a first approach, a yeast invertase was expressed in the apoplast of transgenic potato. Significant phenotypic changes were observed. Transformed potato plants showed stunted growth, suppressed root formation and the development of bleached and necrotic regions in older leaves. In addition, these leaves also accumulate high levels of carbohydrates and starch. These carbohydrates showed no diurnal turnover and the accumulation was accompanied by an inhibition of photosynthesis. This decrease in photosynthesis is not caused by a feedback regulation of sucrose synthesis, rather it is accompanied by an increase in osmotic pressure in the leaf. This suggests that increased levels of ABA, and thus stomatal closure, represents the underlying reason. If sucrose was loaded symplastically through plasmodesmata into the phloem, it would not enter the apoplastic space and would, thus, be inaccessible for the invertase. In this case no phenotypical or physiological changes would be expected in the transgenic plants. The fact that clear phenotypic changes were observed due to the apoplastic expression of the yeast invertase argues strongly for the transport of sucrose into the apoplast, and the subsequent import into the SECCC mediated by an energy-dependent transport step (von Schaewen *et al.*, 1990; Heineke *et al.*, 1992).

Reduction of the Expression of the Sucrose Transporter in Transgenic Potato Plants

A prerequisite for this approach is the isolation of the gene to be suppressed. Recently we cloned sucrose transporter cDNAs from spinach and potato by complementation of a yeast mutant (Riesmeier *et al.*, 1992, 1993b). The transporter proteins from both spinach (SoSUT) and potato (StSUT) showed characteristics consistent with those reported for the sucrose/proton symport in isolated plasma membrane vesicles. RNA *in situ* hybridization data show that the transporter gene is specially expressed in minor veins of mature leaves.

To investigate further the physiological function of the isolated cDNAs, potato plants were transformed with an antisense construct (Riesmeier *et al.*, 1994). The antisense inhibition leads to drastic phenotypic effects. The growth of the plants is strongly retarded, mature leaves are curled, bleached and accumulate anthocyanins at the rims, whereas sink leaves show no phenotypic changes. Measurement of the export of soluble sugars through the petiole provides direct evidence of reduction in flow of sucrose in the antisense plants. The reduced efflux strongly affects the supply of carbohydrates to sink organs, clearly visible as a reduced root system. Additionally, no tubers are formed in the case of plants showing the highest reduction of *StSut* mRNA. The plants were analyzed more intensively with respect to physiological changes. The soluble carbohydrate content in mature leaves increased up to 20-fold and the starch content up to 10-fold. The maximal photosynthesis rate decreased to 50% in strongly effected plants relative to wild-type plants. Fluorescence measurements show that the electron yield of photosystem II was inversely correlated with soluble sugars and starch in the leaves (Riesmeier *et al*, unpublished results).

The results observed are in good agreement with the postulated function of the sucrose transporter during phloem loading. The reduced sucrose transport capacity at the SECCC due to the antisense inhibition prevents to some extent phloem loading. This leads directly to a reduced carbohydrate supply for sink organs. The surplus of sucrose produced in source leaves cannot enter the phloem and thus accumulates in mature leaves. The leaf phenotype can be explained by osmotic stress due to the sugar accumulation. Taken together, the data presented here are consistent with the interpretation that in potato plants an apoplastic transport step, mediated by the isolated sucrose transporter, is necessary for sucrose loading of the phloem.

Approaches to Modulate Carbohydrate Metabolism in Sink Organs (Tubers)

In order to analyze the importance of ADP-glucose pyrophosphorylase (AGPase) for starch biosynthesis in tubers, we used the antisense RNA technique. To this end the cDNAs encoding the small subunit (AGPase B) and one of the large subunits (AGPase S; Müller-Röber *et al.*, 1990) were inserted in reverse orientation behind various promoters and transformed into potato. Initially, we fused the cDNAs to the constitutively expressed cauliflower mosaic virus (CaMV) 35S RNA promoter. When the large subunit cDNA was used, only a weak reduction in leaf-specific AGPase activity was detected. In tubers of these plants the inhibition was slightly better (relative to the AGPase level in untransformed plants); however, it did not reach the strong reduction observed when the small subunit cDNA was used in the antisense experiment. In some of the CaMV 35S-anti AGPase B transgenic plants the leaf AGPase decreased to 5-30% of the wild type activity. An even stronger reduction (down to ca. 2%) was seen in developing tubers (Müller-Röber *et al.*, 1992). The amount of residual AGPase B mRNA and protein correlated to the degree of inhibition in AGPase activity and was accompanied by a reduction in starch content in tubers down of 5-35%. Phenotypic changes occurred in the transgenic plants. When the plants were cultivated in the greenhouse, the most strongly inhibited plants showed a higher tendency to form axillary shoots and accumulated slightly more anthocyanins in the upper stem regions. Most interestingly, onset of flower formation occurred 2-4 weeks earlier in these plants. The overall structure of the flowers remained unchanged as judged by visual inspection. Both higher tendency of axillary shoot outgrowth and early flower formation may indicate a change of some signals (sugars?) important for these processes.

Inhibition of starch synthesis resulted in an additional type of developmental change. First, tuber number increased, reaching in extreme cases (e.g. when large pots were used to grow the plants) up to approximately 90 tubers per plant. Control plants typically produced around 10 tubers. Secondly, the stolon/tuber structure was altered. Instead of having a single tuber at the end of a stolon (which is normal for the wild-type) antisense plants frequently developed stolons with many (up to 10) but smaller tubers, indicating that AGPase activity might be an important determinator of tuber sink strength (for discussion see Müller-Röber *et al.*, 1992). The loss of starch in transgenic tubers was partially compensated for by an increase in the level of soluble sugars, mainly sucrose (up to approximately 30% of the tuber dry weight) and glucose (up to approximately 8%). The composition of the soluble carbohydrates was not always constant when different tubers from a plant or tubers from different plants (e.g. grown at various seasons) were analyzed. However, the general picture of a strong increase in the amount of soluble sugars was always observed. Due to the high level of sugars, AGPase-deficient tubers had relatively high water content. Taken together, despite an increase in tuber number and total tuber fresh weight, the total tuber dry weight was reduced by around 30-40% as compared with controls. Therefore, constitutive inhibition of starch synthesis resulted in a depression of tuber yield.

Approaches to Influence Starch Quality

Starch is synthesized by the concerted action of several isoforms of starch synthases which catalyze a polymerization reaction by transferring the glucosyl residue from ADP-glucose to α–1,4-glucans and branching enzymes which introduce α–1,6-glycosidic branch points into linear α–1,4-glucans. However, an involvement of additional hydrolytic, transglucanosilytic or phosphorolytic activities cannot be excluded (Preiss, 1991).

The starch synthases can be divided into two classes, the granule-bound (GBSS I and II) and the soluble starch synthases (SSS I and II). Within these classes, the existence of isoforms, the so-called primer-dependent and -independent starch synthases, have been reported for several plant species.

Branching enzymes also occur as two distinct isoforms (BE I and II) in most of the plants investigated so far. Branching enzymes act in a dual catalytic manner, i.e. they first hydrolyze an α–1,4-glucan and then transfer the released segment to an α–1 ,4-glucan at the C-6 position.

The importance of the starch synthase isoforms with respect to their function in starch biosynthesis can only be ascribed to the GBSS I protein, since it is the only form where plants are described which have a reduced content of this enzyme. It is the major protein isolated from starch granules. Absence of this enzyme in mutants (the so-called waxy mutants) or in transgenic plants which were transformed with a construct expressing the GBSS coding sequence in antisense orientation, results in the production of amylose-free starch (Visser *et al.*, 1991; Sonnewald *et al.*, unpublished).

Complete inhibition of branching enzyme activity in potato tubers should ideally result in the production of starch composed only of linear glucans. The antisense RNA technology usually results in the generation of mutants showing a range between zero and complete inhibition. This allows the selection of plants with an optimal degree of inhibition, where the amylose content is maximized while negative effects leading to yield penalties are minimized. Experiments were undertaken to suppress branching enzyme activity in transgenic potato plants through the expression of a cDNA for potato BE (Kossmann *et al.*, 1991) in antisense orientation. Surprisingly, an almost complete reduction of BE did not result in a modification of the amylose content (J. Kossmann, U. Sonnewald and L. Willmitzer, unpublished observations). Possible explanations are: (i) the degree of inhibition in the plants selected was not high enough, i.e. that a very minor amount of residual protein is sufficient to confer wild-type phenotype, or (ii) that another activity capable of branching linear glucans is present in potato tubers. If the latter is true, it raises the question of the *in vivo* function of potato BE. Sequence comparisons indicate that it is very closely related to BE I (Baba *et al.*, 1991) rather than BE II (Fisher *et al.*, 1993) from maize. No mutants have been reported from any plant species where an isoform corresponding to BE I from maize is lacking, therefore the function of the enzyme remains elusive. Recent biochemical data indicate that it might be responsible for the synthesis of infrequently branched polysaccharides, which then serve as a substrate for other isoforms of BE and the starch synthases for the synthesis of amylopectin (Guan and Preiss, 1993; Takeda *et al.*, 1993)

Conclusions

Due to its amenability to transformation and the large amount of physiological data, potato is one of the best suited plants for the study of carbohydrate partitioning and allocation. Techniques such as ectopic expression or antisense RNA-mediated inhibition are now routine in this plant. Thus the future will undoubtedly give a new insight into this important aspect of plant physiology at the molecular level.

References

Baba, T., Kimura, K., Mizuno, K., Etoh, H., Ishida, Y., Shida, O. and Arai, Y. (1991) Sequence conservation of the catalytic regions of amylolytic enzymes in make branching enzyme 1. *Biochemical and Biophysical Research Communications* 181:87-94.

Bush, D.R. (1993) Proton-coupled sugar and amino acid transporters in plants. *Annual Review of Plant Physiology and Plant Molecular Biology* 44:513-542.

Fisher, D.K., Boyer, C.D. and Hannah, L.C. (1993) Starch branching enzyme II from maize endosperm. *Plant Physiology* 102:1045-1046.

Gifford, R.M., Thorne, J.H., Hitz, W.D. and Giaquinta, R.T. (1984) Crop productivity and photoassimilate partitioning. *Science* 225:801-808.

Guan, H.P. and Preiss, J. (1993) Differentiation of the properties of the branching isozymes from maize (*Zea mays*). *Plant Physiology* 102:1269-1273.

Heineke, D., Sonnewald, U., Bussis, G., Gunter, K., Leidreiter, K., Wilke, I., Raschke, K., Willmitzer, L. and Heldt, H. (1992) Apoplastic expression of yeast-derived invertase in potato. *Plant Physiology* 100:301-308.

Ho, L.C. (1988) Metabolism and compartmentation of imported sugars in sink organs in relation to sink strength. *Annual Review of Plant Physiology and Plant Molecular Biology* 40:415.

Kossmann, J., Visser, R.G.F., Müller-Röber, B., Willmitzer, L. and Sonnewald, U. (1991) Cloning and expression analysis of a potato cDNA that encodes branching enzyme: evidence for co-expression of starch biosynthetic genes. *Molecular and General Genetics* 230:39-44.

Lucas, W.J. and Madore, M.A. (1988) Recent advances in sugar transport. In: *The Biochemistry of Plants.* Academic Press Inc., NY, Vol. 14, pp. 35-84.

Maynard, J.W. and Lucas, W.J. (1982) Sucrose and glucose uptake into *Beta vulgaris* leaf tissues. A case for general (apoplastic) retrieval systems. *Plant Physiology* 70:1436-1443.

Müller-Röber B., Kossmann, J., Hannah, L.C., Willmitzer, L. and Sonnewald, U. (1990) Only one of two different ADP-glucose pyrophosphorylase genes from potato responds strongly to elevated levels of sucrose. *Molecular and General Genetics* 224:136-146.

Müller-Röber, B., Sonnewald, U. and Willmitzer, L. (1992) Antisense inhibition of the ADP-glucose pyrophosphorylase in transgenic potatoes leads to sugar-storing tubers and influences tuber formation and expression of tuber storage proteins. *EMBO Journal* 11:1229-1238.

Preiss, J. (1991) Biology and molecular biology of starch synthesis and its regulation. In: Miflin B.J. and Miflin H.F. (eds), *Oxford Surveys of Plant Molecular and Cell Biology*. Oxford University Press, Oxford, Vol. 7. pp. 59-114.

Riesmeier, J.W., Willmitzer, L. and Frommer, W.B. (1992) Isolation and characterization of a sucrose carrier cDNA from spinach by functional expression in yeast. *EMBO Journal* 11:470-4713.

Riesmeier, J.W., Flugge, U.I., Schulz, B., Heineke, D., Heldt, H.W., Willmitzer, L. and Frommer, W.B. (1993a) Antisense repression of the chloroplast triose phosphate translocator affects carbon partitioning in transgenic potato plants. *Proceedings of the National Academy of Sciences USA* 90:6160-6164.

Riesmeier, J.W., Hirner, B. and Frommer, W.B. (1993b) Potato sucrose transporter expression in minor veins indicates a role in phloem loading. *Plant Cell* 5:1591-1598.

Riesmeier, J.W., Willmitzer, L. and Frommer W.B. (1994) Evidence for an essential role of the sucrose transporter in phloem loading and assimilate partitioning. *EMBO Journal* 13:1-7.

Sonnewald, U. (1992) Expression of *E. coli* inorganic pyrophosphatase in transgenic plants alters photoassimilate partitioning. *The Plant Journal* 2:571-581.

Stitt, M. (1990) The flux of carbon between the chloroplast and the cytosol. In: Dennis, D.T. and Turpin, D.H. (eds), *Plant Physiology, Biochemistry and Molecular Biology*. Longman Scientific and Technical, London, UK, pp. 309-326.

Takeda, Y., Guan, H.P. and Preiss, J. (1993) Branching of amylose by the branching isoenzymes of maize endosperm. *Carbohydrate Research* 240:253-263.

Turgeon, R. (1989) The sink-source transition in leaves. *Annual Review of Plant Physiology and Plant Molecular Biology* 40:119-138.

Visser, R.G.F., Somhorst, I., Kuipers, G.J., Ruys, N.J., Feenstra, W.J. and Jacobsen, E. (1991) Inhibition of the expression of the gene for granule-bound starch synthase in potato by antisense constructs. *Molecular and General Genetics* 225:289-296.

von Schaewen, A., Stitt, M., Schmidt, R., Sonnewald, U. and Willmitzer, L. (1990) Expression of a yeast-derived invertase in the cell wall of tobacco and *Arabidopsis* plants leads to accumulation of carbohydrate and inhibition of photosynthesis and strongly influences growth and phenotype of transgenic tobacco plants. *EMBO Journal* 9:3033-3044.

Chapter Six:

Control of Sugar Balance in Potato Tubers

H.V. Davies and R. Viola
Department of Cellular and Environmental Physiology, Scottish Crop Research Institute, Invergowrie, UK.

The major soluble carbohydrates of the potato tuber are the hexoses glucose and fructose and the non-reducing disaccharide sucrose. The concentrations and relative proportions of the individual sugars are variable both between tubers on the same plant and between plants growing in close proximity. Nevertheless, the various phases of tuber growth and development are generally characterized by fairly distinct changes in sugar profiles. For example, the rapid phase of tuber growth is characterized by a decline in the concentration of all soluble sugars, more as a result of increased tuber growth than the apparent rate of sucrose influx (Burton, 1989 and references therein). Significant modifications can ensue in the post-harvest period, particularly in response to low temperatures but also following prolonged storage, resulting in an irreversible accumulation of hexose. The extent of such modifications is influenced by environmental and cultural variables experienced during tuber growth e.g. nitrogen and water regime, and by the genetic background of the clones or cultivars in question i.e. there is a typical genotype x environment interaction. Arguably, the genetic component is most important: breeding programs such as those at the USDA and the Scottish Crop Research Institute have given rise to clones which do not accumulate significant quantities of hexoses at low temperature, to clones with extremely high ratios of glucose:fructose in mature tubers and to those with high sucrose:hexose or low sucrose:hexose ratios (Mackay, personal communication). Some might consider this a biochemist's dream, others a nightmare. Whichever attitude prevails, the road to potato improvement offers many challenges and begs a multidisciplinary approach with more than a modicum of lateral thinking.

This chapter describes some aspects of metabolism which contribute to control of sugar balance in tubers at several stages in its life cycle.

The Molecular and Cellular Biology of the Potato, Second Edition
Edited by W.R. Belknap, M.E. Vayda and W.D. Park

Tuberization

Burton and Wilson (1970) reported that the reducing sugar (RS) and sucrose contents of stolon tips of cultivar Record were 1.3 and 0.09% of fresh weight (FW), respectively, with an RS:sucrose ratio of 48. In very small tubers the sucrose content increased to *ca.* 0.46% of FW and the average RS:sucrose ratio was 3.3. Around 5 days after tuber initiation RS and sucrose contents were *ca.* 0.56 and 0.95% of FW and the RS:sucrose ratio was 0.7. Davies (1984) also reported substantial decreases in the RS:sucrose ratio during tuberization, which apparently arose as a consequence of a rapid decline in RS rather than from an increase in sucrose. More recently, Ross *et al.* (1994) separated events occurring in the subapical, tuberizing stolon region (zone 1) compared with the adjacent non-tuberizing stolon tissue (zone 2). They showed that RS concentrations declined in both zones as tubers formed but confirmed an increase in sucrose content (Fig. 6.1). The reasons for sucrose accumulation around tuberization are not clear. However, it has been argued that tuberization is preceded by a cessation of stolon elongation growth and, presumably, by a reduced demand for energy at a time when, under longer daylength, more photoassimilate is available for transport. Under such circumstances sucrose may accumulate, particularly when this developmental switch is accompanied by a substantial decrease in invertase activity (Ross *et al.*, 1994: Fig. 6.1E and F). Alternatively, if there is significant symplastic unloading via plasmodesmata (as indicated by the work of Oparka, 1986), then modifications to plasmodesmatal frequency or functioning may be important. Whatever the mechanism, it may be argued that the elevated sucrose concentrations may contribute to the onset of rapid starch biosynthesis due to the effect of sucrose on the expression of genes encoding enzymes such as sucrose synthase and ADP-glucose pyrophosphorylase (AGPase) (Salanobout and Belliard, 1989: Müller-Röber *et al.*, 1990).

Tuber formation is characterized not only by modifications to the RS:sucrose balance but also by substantial changes in the glucose:fructose ratio. The glucose:fructose ratio is around 2:1 in the non-tuberizing subapical stolon region, increasing to *ca.* 10:1 during the early stages of tuberization (Ross *et al.*, 1994). Ratios as high as 100:1 have been recorded (Davies, unpublished data). Davies and Oparka (1985) identified fructokinase (fructose+NTP -----> fructose-6-P + NDP) activity in young tubers and explained the high glucose:fructose ratio by the high activity of fructose-specific compared with glucose-specific hexose kinase. Ross *et al.* (1994) subsequently demonstrated a rapid and substantial increase in fructokinase (FK) activity in tuberizing stolon tips (see Fig. 6.1D), correlating with an increased glucose:fructose ratio (Fig. 6.1A). Increased fructokinase activity coincides with elevated levels of sucrose synthase ('Susy') [sucrose + UDP <-------> UDPG +fructose] as shown in Fig. 6.1C. This is not entirely surprising since FK may be considered to be its 'sister' enzyme such that an increase in the proportion of incoming sucrose cleaved by Susy is linked with a high capacity to phosphorylate one end product. This minimizes the potential for end product inhibition of Susy. UDP-glucose (UDPG)

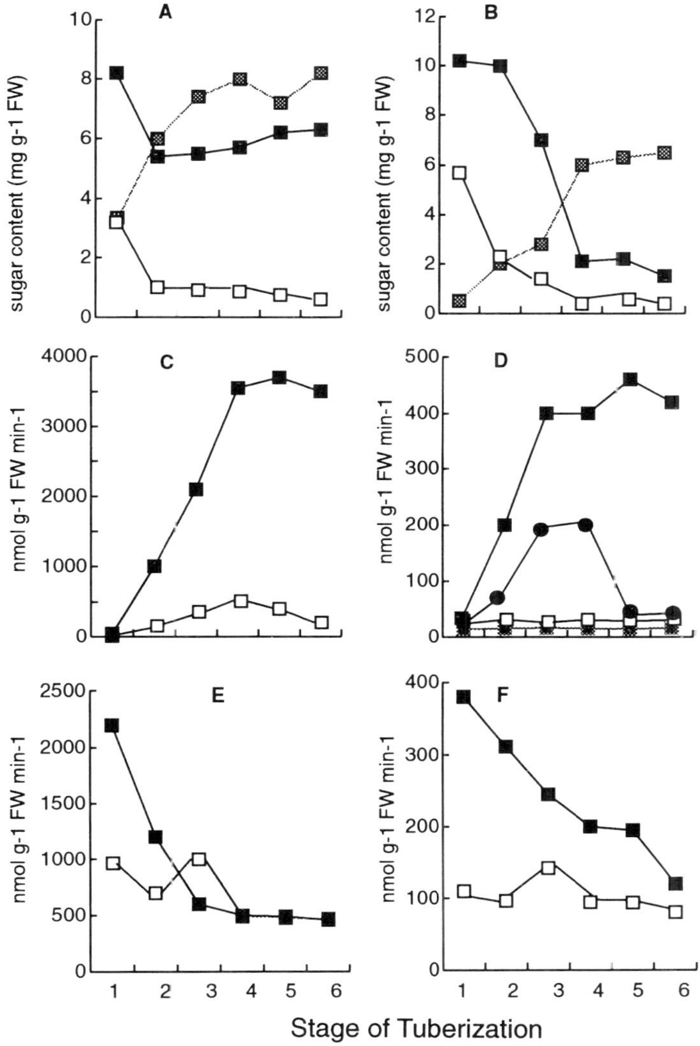

Figure 6.1. Sugar content and enzyme activities in the tuberizing subapical stolon region (zone 1) and the adjacent non-tuberizing region (zone 2). Stage 1: non-tuberized stolon; stage 2: tuber swelling 1-4 mm greater than stolon diameter, stages 3-6: tuber swelling 4-8, 8-12, 12-16 and 16-20 mm greater than stolon diameter, respectively. (A) sucrose (▩), glucose (■), and fructose (□) levels in zone 1. (B) sucrose, glucose and fructose in zone 2. (C) Sucrose synthase activity in zone 1 (■) and zone 2 (□). (D) Fructokinase activity in zones 1 (■) and 2 (●), glucokinase activity in zones 1 (□) and 2 (▩). (E) Acid invertase in zones 1 (■) and 2 (□). F: Alkaline invertase in zones 1 (■) and 2 (□).

pyrophosphorylase activity also increases during tuberization (Hawker *et al.* 1979) and appears to be present in considerable excess. Down-regulating its activity to only *ca.* 5% of that found in wild-type tubers has no significant effect on net accumulation of starch (Zrenner *et al.*, 1993). The combined activities of FK and UDPG pyrophosphorylase ensure net flux of UDPG and fructose in the direction of starch biosynthesis. This is an important point since Susy catalyzes a reversible reaction *in vivo*.

Gardner *et al.* (1992) purified and characterized the tuber FKs. Three isoforms were identified in young developing tubers by 2-D immunoblotting and analytical scale ion exchange chromatography. These were nominated FKs I, II and III. Renz *et al.* (1993) confirmed the presence of three tuber FKs, two forms with high activity in tubers and a third with highest activity in leaves. Gardner *et al.* (1992) showed that the molecular mass of each of the co-purifying native tuber proteins was *ca.* 70 kDa. The proteins were comprised of two subunits of 36 kDa with no evidence for a monomeric, 72 kDa polypeptide, as occurs in pea seed (D.D. Randall, University of Missouri, personal communication). A V8 digest of the gel purified tuber protein (a mixture of FKs I, II and III) yielded sequencable 10 kDa and 5 kDa internal fragments. Certain cycles of the Edman degradation yielded three amino acids again indicating the presence of three indistinct proteins. The amino acid sequences obtained showed no similarity to published sequences for hexokinases from yeast and mammals nor to mammalian glucokinase. However, significant homologies existed with the FK from *Vibrio alginolyticus* and *Klebsiella pneumoniae* (Fig. 6.2). Fructokinases (6-phosphotransferases) have been reported in bacteria and plants but not in yeast and mammalian systems. Plant FKs may have evolved from bacterial proteins rather than from yeast or mammalian hexose kinases.

Deduced amino acid sequence comparison of the potato FK gene with the *scrK* genes of *Vibrio alginolyticus* show 39% identity and 63% similarity using the GAP alignment package. Expression studies have revealed a 1.2 kb transcript in swelling stolons which increases during tuber development (Smith, Taylor and Davies, unpublished data). Southern analysis indicates low copy number. The cDNA contains an open reading frame of 957 bp, an ATG initiation codon at nucleotide 11 and a TAG termination codon at 968. A polyadenylation signal is present at 1119. The gene encodes a polypeptide of 323 amino acids with a calculated molecular weight of 34,392 daltons.

FKs I and II are inhibited by fructose concentrations in excess of 1 mM (*ca.* 50% inhibition at 2 mM) (Gardner *et al.*, 1992): FK III is not inhibited by concentrations as high as 16 mM *in vitro*. Substrate inhibition of the various FKs obviously depends on the subcellular localization of fructose. Morrell and ap Rees (1986) reported fructose levels between 1 and 5 mmol g^{-1} FW in developing tubers. Davies and Oparka (1985) recorded values between 0.12 and 0.29 mmol g^{-1} FW. Assuming an equal distribution of fructose between cytosol and vacuole this provides concentrations in the range *ca.* 0.1-5 mm. If all the fructose is contained within the cytosol the range could be as high as 1-50 mM. This would be high enough to cause inhibition of FKs I and II and to

```
Potato Peptide (Mr 5,000)    F M F Y R N P S A D E L L T P D E L N T F
                             |   |     ● | | | | ●   ● ●   ● ● ●     |
Vibrio alginolyticus         F T F M V K P S A D Q F M S V E D M G N F

Potato Peptide (Mr 10,000)   L L T L S G I D I G - E A G N N L D N D G
                                   | ● ● ● ● ● ●   | | | ●   | ●   | ●
Vibrio alginolyticus         S A H L K A M E V A K E A G A L L S Y D P

Potato Peptide (Mr 5,000)    F M F Y R N P S A D E L L T P D E L N T F S
                             |   |       | | | |   |   | ● ● |       ●
Klebsiella pneumonia         F T F M V R P S A D L F L Q P E D L P P F A
```

Figure 6.2. Partial amino acid sequence of potato tuber fructokinase protein (V8 digest of a mixture of FKs I, II and III) and fructokinases from *Vibrio alginolyticus* and *Klebsiella pneumoniae.* | denotes identical amino acids. ● denotes conservative substitutions.

determine the reversibility of the reaction catalyzed by Susy *in vivo*. This scenario has been put forward by Geigenberger and Stitt (1993). In support of their hypothesis is the fact that the maximum catalytic activity of FK is substantially higher than the rate of starch biosynthesis in tubers. Until the concentrations of individual sugars in the vacuole and cytosol are established the potential role of substrate inhibition cannot be determined with any certainty. Substantial inhibition of enzyme activity by physiological concentrations of fructose has been reported for FKs from pea seed (Copeland *et al.* 1978), spinach leaves (Baldus *et al.* 1981) and maize kernels (Doehlert, 1989). Others, for example from avocado, are, like potato FK III, unaffected by high fructose concentrations (Copeland and Tanner, 1988).

It could be argued that what fructose there is in developing tubers resides in the vacuole as a product of acid invertase activity. This would indicate that the phosphorylative capacity of FK is sufficient to prevent any fructose accumulation in the cytosol and that the activity of Susy is irreversible *in vivo*. Alternatively, fructose may reside in the cytosol as a result of the combined activities of alkaline invertase (when present) and Susy.

Potato FKs have a high specificity for fructose, with K_ms ranging between 0.041 and 0.128 mM. There is, effectively, no activity with glucose or other hexoses. The preferred nucleoside triphosphate (NTP), as calculated from V_{max}/K_m, is GTP for FK I, closely followed by ATP. ATP is the preferred NTP for FKs II and III. The fact that UTP can support FK activity in crude potato extracts (Xu *et al.*, 1989) led to the hypothesis that sucrose degradation via Susy is integrated with fructose phosphorylation utilizing UTP. The UTP is generated from UDPG via PPi-driven UDPG pyrophosphorylase. The relative concentrations of ATP and UTP in developing tubers are not known, but it is postulated that the UTP content is substantially lower than that of ATP (Renz and Stitt, 1993). This, together with the lower affinity of the tuber FK for UTP compared with ATP, makes it unlikely that FK activity is responsible for the synthesis of substantial quantities of UDP.

All three FKs are inhibited by ADP competitively to ATP (Renz and Stitt, 1993). However, the extent of such inhibition *in vivo* (when the ATP:ADP ratio is relatively high) is unknown. Fructose-6-phosphate inhibits FK activity non-competitively, but again the concentration required to inhibit activity extensively *in vitro* casts doubt over the importance of fructose-6-phosphate as the sole, major regulatory metabolite of FK *in vivo* (Gardner *et al.* 1992; Renz and Stitt, 1993). Regulation of activity *in vivo* may result from a combination of factors.

Glucose phosphorylation in tubers is mediated via the action of hexose kinases with a high affinity for glucose and mannose but a very low affinity for fructose (Gardner, 1992; Renz and Stitt, 1993). ATP is the most effective phosphoryl donor and ADP can inhibit activity via competition with ATP. Again this may not be significant under aerobic conditions, however, inhibition by glucose-6-P *in vivo* may be.

From data published to date, tubers appear to lack a true hexokinase, i.e. a protein equally effective at phosphorylating both of the major tuber hexoses. Furthermore, the kinetic properties of individual FKs and glucokinases can be significantly different. The relative contributions of each isoform to hexose phosphorylation may vary depending on the prevailing metabolic status of the tuber.

Developing Tubers

Carbohydrate metabolism during tuber bulking is clearly dominated by starch deposition and, in general, is accompanied by a decline in both hexose and sucrose concentrations (Richardson *et al.*, 1989). This can be explained in terms of a relatively constant rate of sucrose influx accompanied by an increase in tuber size as the plants and tubers mature. Susy continues to be the dominant sucrolytic enzyme during this phase of development - acid invertase activity is *ca.* 200-fold lower. Some controversy still surrounds the neutral or alkaline invertase, Merlo *et al.* (1993) reporting almost equal activities of acid and alkaline invertases in both developing and cold stored tubers, Ross *et al.* (1994) describing a rapid and irreversible loss in alkaline invertase activity and alkaline invertase protein during tuberization. Thus, in terms of sucrose hydrolysis, the sugar balance in developing tubers is set principally by sucrose synthase and again by its 'sister' enzyme FK. As the growing season progresses there is a substantial decline in both Susy and FK, until Susy becomes almost

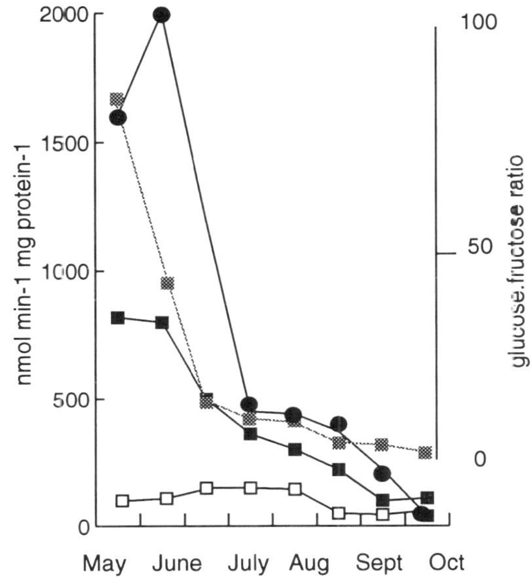

Figure 6.3. Developmental changes in sucrose synthase (●), fructokinase (■) and glucokinase (□) activities in glasshouse grown potato tubers. The glucose:fructose ratio is also shown (▩).

undetectable and glucose phosphorylating potential approximates fructose phosphorylating potential (Fig. 6.3). The glucose:fructose ratio is reduced from *ca.* 80:1 to 2:1 as these changes occur.

In developing tubers, the control of sucrose content is an important determinant of sink strength. However, even in developing tubers which are actively converting sucrose into starch, substantial labeling of sucrose is observed when labeled glucose is supplied (Oparka *et al.* 1990). This implies the operation of a 'cycling' mechanism of sucrose breakdown and resynthesis. Developing tubers contain substantial sucrose phosphate synthase (SPS) activity (Merlo *et al.*, 1993). Using ^{13}C NMR the pattern of isotopic distribution between the two hexosyl moieties of sucrose when specifically labeled glucose is supplied to developing tuber discs seems to indicate that the main pathway of sucrose biosynthesis is, indeed, via SPS (Viola *et al.*, 1991). However, Geigenberger and Stitt (1993) observed asymmetrical labeling of sucrose (i.e. considerably more in the glucosyl moiety) when labeled glucose and unlabeled fructose were supplied together. This would indicate that the reaction catalyzed by sucrose synthase is readily reversible *in vivo* and, for this reason, it is unlikely to be a major determinant of sink strength. The data presented by Zrenner and Sonnewald (1993) indicate that down-regulating Susy inhibits starch

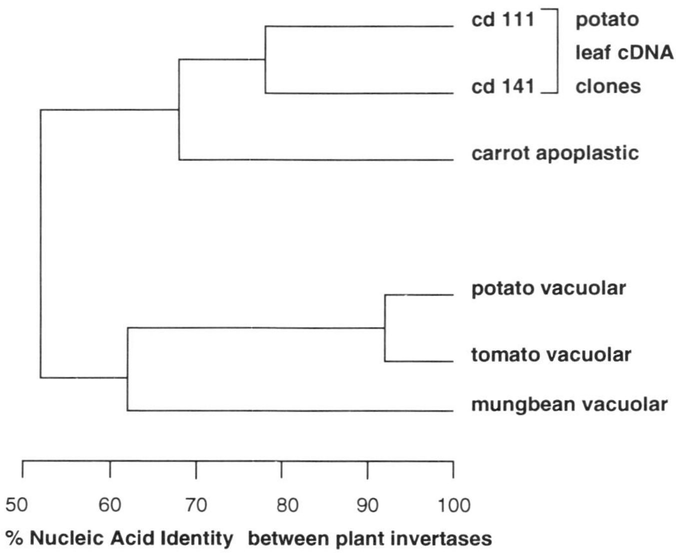

Figure 6.4. Percent nucleic acid identities between genes encoding apoplastic and vacuolar invertases.

synthesis but does not result in sucrose accumulation. This is an intriguing result and one which does imply an important role for Susy in regulating sucrose import and sink strength *per se*. The question of which pathway of sucrose synthesis prevails in developing tubers (via SPS or Susy?) is still unresolved but has obvious relevance. It has to be pointed out that increased carbon allocation to sucrose occurs whenever starch biosynthesis is down-regulated in tubers e.g. in response to fluoride treatment (Viola and Davies 1991), expression of antisense ADP glucose pyrophosphorylase genes (Müller-Röber *et al.*, 1992), or following tuber removal from the mother plant (Oparka *et al.*, 1990). Fluoride treatment and tuber excision both produce significant increases in hexose-phosphate content (Viola and Davies, unpublished data). This is a result of reduced flux into starch. Under such circumstances flux through SPS should increase dramatically as a result of increased concentrations of substrate (particularly fructose-6-phosphate) and allosteric activator (glucose-6-phosphate). The latter also reduces the K_m for fructose-6-phosphate. SPS activity in developing potato tubers increases over an order of magnitude when assayed in the presence of 20 mM glucose-6-phosphate and 4 mM fructose-6-phosphate compared with 10 mM glucose-6-phosphate and 2 mM fructose-6-phosphate (Merlo *et al.*, 1993).

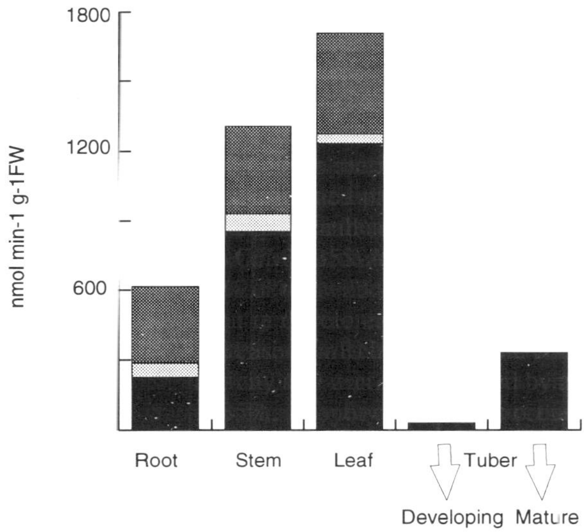

Figure 6.5. Recovery of acid invertase activity from root, leaf, stem and tubers (developing and mature) following sequential extractions of tissue with 50 mM acetate buffer, pH 5 (■) and acetate buffer containing 1 M NaCl (▩). (▩)- activity remaining bound to tissue pellet.

Mature Tubers

The excision of young developing tubers from the mother plant causes a substantial and relatively rapid increase (within *ca.* 2 days) in both hexose concentrations and acid invertase activity together with a massive decrease in Susy (Ross and Davies, 1992). These changes occur naturally and progressively as plants senesce and as the tubers undergo the transition from a sink for carbon to a source of carbon during storage and subsequent germination or sprouting. Sugar metabolism in stored tubers is well researched, particularly in relation to low temperature sweetening (see review by Davies and Viola, 1992). Starch breakdown, sucrose turnover and respiratory metabolism have been the principal factors addressed, but the relative contributions of each are unknown. However, results presented by Zrenner and Sonnewald (1993) point to an important role for soluble acid invertase, down-regulating expression of the gene restricting hexose production at the expense of sucrose accumulation.

Five invertase genes have been cloned so far from potato. At the SCRI four genes encoding apoplastic invertases have been isolated, (Hedley *et al.*, 1993; Machray *et al.*, 1993) two genomic and two cDNA clones (the latter from leaves), and, in addition a gene encoding a vacuolar invertase has been described by described by Zrenner and Sonnewald (1993). Fig. 6.4 shows the % identity at the nucleic acid level between the two potato leaf cDNA clones (cd111 and

Table 6.1 Partitioning of ^{14}C-labeled tuber storage parenchyma following the supply of [U-^{14}C]glucose (74 KBq ml^{-1}) to discs excised from mature tubers of cvs Record and Brodick stored for 2 months at 3°C or 15°C.

	^{14}C per fraction as % of total metabolized			
	Potato cultivar and incubation temperature			
Fraction	Record 15°C	Brodick 15°C	Record 3°C	Brodick 3°C
CO_2	3.7	4.9	2.6	2.3
Basic compounds	8.2	10.7	3.7	3.9
Organic acids	8.3	7.5	4.1	5.4
Non-starch pellet	7.6	9.2	5.2	4.7
Total	27.8	32.3	15.6	16.3
Phosphate-esters	3.2	1.9	3.3	2.1
Starch	24.5	22.8	38.7	54.8
Sucrose	30.9	27.6	30.8	17.6
Hexose	13.6	15.4	11.6	9.2
Total	72.2	67.7	84.4	83.7
^{14}C recovered in fractions as % of ^{14}C incorporated	86.3	80.9	93.4	88.4
Ratio of ^{14}C recovered in starch and sucrose	0.8	0.8	1.3	3.1

Incubation with radiolabel was carried out at the storage temperature. Values represent percentage of total label metabolized (^{14}C incorporated by the tissue minus that still present as the original label supplied). When calculation partitioning to hexoses, the proportion of label recovered in fructose was doubled. Data represent average values of four replicates each consisting of 20 discs randomly selected from four tubers. Standard errors are omitted for clarity, but did not exceed 15% of the values reported.

cd141) and the carrot apoplastic equivalent and between the gene encoding the potato vacuolar protein and equivalents from tomato and mungbean. RT-PCR has revealed that cd111 is expressed in sink leaves (net importers of carbon) whereas cd141 is expressed in source and sink leaves, stems and roots and tuber (although at very low levels in the latter). The gene encoding the vacuolar invertase is expressed in most, if not all, organs so far studied. However, physiological experiments have failed to demonstrate, unequivocally, the presence of a native apoplastic invertase in tubers or its involvement, if any, in regulating phloem unloading and tuber sugar balance. Fig. 6.5 shows the recovery of acid invertase activity following sequentially extractions of root, stem, leaf and tuber tissue in buffer and high salt. With the exception of developing tubers, high salt extracts a small amount of activity which is presumably ionically bound to the cell wall. Again with the exception of tubers, a considerable amount of activity remains bound to the tissue pellet and no treatments have yet been found which solubilize the protein (Davies, unpublished data). The presence of an apoplastic invertase in leaves raises the possibility that in wild-type plants a proportion of sucrose loaded into the phloem via the apoplast is degraded. Over-expressing high levels of yeast invertase in the leaf apoplast has dramatic effects on phloem loading and leaf metabolism in the transgenics (Heineke *et al.*, 1992), but the role and extent of functioning of the native apoplastic enzyme must be ascertained.

While acid invertase is the major if not sole sucrolytic enzyme in stored tubers (the natural decline in the extractable activity of Susy can be rapid and complete or partial and prolonged) the glucose:fructose ratio is rarely 1:1. Ratios of 2:1 or 3:1 are common. Ratios as high as 10:1 are not unknown. The ratio may be genotype-dependent in some cases (Mackay, pers. comm.). One could again invoke FK activity to explain such a ratio, but net starch breakdown, if it occurs partially or completely via amylases, provides a logical and additional explanation. The pathway of starch breakdown in tubers is still very much open to question. Equally important in restricting the availability of carbon for sucrose, and ultimately hexose, accumulation is the starch synthesizing potential of the stored tuber. Considerable incorporation of label from [^{14}C]glucose into starch can occur in discs excised from stored tubers. Since a net flux of carbon from starch to soluble sugars prevails in stored potatoes, this indicates that starch turnover can be relatively substantial. In pulse experiments the partitioning of incorporated radiolabel between starch and sucrose appears to be temperature- and genotype-dependent (Table 6.1). At low ('sweetening') temperatures, more label is partitioned to sucrose than to starch. The importance of starch turnover in determining the tuber's sugar balance in storage is currently being assessed with a range of genotypes exhibiting different rates of hexose accumulation at low temperature.

Conclusions

The sugar balance of the potato tuber is governed by many inter-related metabolic pathways and is subject to significant genetic and environmental control. Tuber initiation, growth and development and post-harvest physiology are characterized by significant modifications in hexose and sucrose concentrations and in the ratios of hexoses:sucrose and glucose:fructose. In part these can be explained by switches in the pathway of sucrolysis from one dominated by invertase in elongating stolons, to one dominated by sucrose synthase during the rapid growth phase when starch reserves are being laid down. Acid invertase is the dominant and sometimes the sole sucrolytic enzyme in the post-harvest period, treatments which interrupt the supply of assimilate to tubers also modifying the balance between these enzymes. Fructokinase appears to play a key role in regulating the fructose content of developing tubers and hence the glucose:fructose ratio, but in the post-harvest phase starch breakdown is likely to contribute towards maintaining the glucose:fructose ratio significantly in excess of unity. Ongoing work is highlighting the importance of starch turnover in regulating the availability of carbon for sucrose biosynthesis and hexose formation in mature tubers.

Acknowledgements

The authors would like to thank the Scottish Office Agriculture and Fisheries Department, ECSA Research Ltd and the European Commission for financial support. From SCRI the research contributions of Heather Ross, Gordon Machray, Peter Hedley, Robbie Waugh, Karl Oparka, Lindsay Burch, Andy Gardner, Mark Taylor and Susan Jarvis are also acknowledged.

References

Baldus, B., Kelly, G.J. and Latzko, E. (1981) Hexokinases of spinach leaves. *Phytochemistry* 20:1811-1814.

Burton, W.G. (1989) *The Potato*. 3rd edition. Longman, Scientific and Technical, UK, pp. 286-336.

Burton, W.G. and Wilson, A.R. (1970) The apparent effect of the latitude of the place of cultivation upon the sugar content of potatoes grown in Great Britain. *Potato Research* 13:269-283.

Copeland, L., Harrison, D.D. and Turner, J.F. (1978) Fructokinases (fraction III) of pea seeds. *Plant Physiology* 62:291-294

Copeland, L. and Tanner, G.J. (1988) Hexose kinases of avocado. *Physiologia Plantarum* 74:531-536.

Davies, H.V. (1984) Sugar metabolism in stolon tips of potato during early tuberisation. *Zeitschrift fur Pflanzenphysiologie* 113:373-376.

Davies, H.V. and Oparka, K.J. (1985) Hexose metabolism in developing tubers of potato (*Solanum tuberosum* L.) cv Maris Piper. *Journal of Plant Physiology* 119:311-316.

Davies, H.V. and Viola, R. (1992) Regulation of sugar accumulation in stored potato tubers. *Postharvest News and Information* 3:97N-100N.

Doehlert, D.C. (1989) Separation and characterization of four hexose kinases from developing maize kernels. *Plant Physiology* 89:1042-1048.

Gardner, A. (1992) Purification and characterisation of hexose kinases from potato tubers (*Solanum tuberosum* L.) and developing faba bean (*Vicia faba* L.) cotyledons. PhD thesis, University of Dundee, pp. 42-85.

Gardner, A., Davies, H.V. and Burch, L.R. (1992) Purification and properties of fructokinase from developing tubers of potato (*Solanum tuberosum* L.). *Plant Physiology* 100:178-183.

Geigenberger, P. and Stitt, M. (1993) Sucrose synthase catalyses a readily reversible reaction *in vivo* in developing potato tubers and other plant tissues. *Planta* 189:329-333.

Hawker, J.S., Marschner, H. and Krauss, A. (1979) Starch synthesis in developing potato tubers. *Physiologia Plantarum* 46:25-30.

Hedley, P., Machray, G.C., Davies, H.V., Burch, L. and Waugh, R. (1993) cDNA cloning and expression of a potato (*Solanum tuberosum*) invertase. *Plant Molecular Biology* 22:917-922.

Heineke, D., Sonnewald, U., Bussis, D., Gunter, G., Leidreiter, K., Wilke, I., Raschke, K., Willmitzer, L. and Heldt, H.W. (1992) Apoplastic expression of yeast-derived invertase in potato. Effects on photosynthesis, leaf solute composition, water relations and tuber composition. *Plant Physiology* 100:301-308.

Machray, G.C., Hedley, P., Burch, L., Meyer, R., Davies, H.V. and Waugh, R. (1993). Characterisation of potato invertase gene families. In:*Proceedings of the 3rd International Symposium on the Molecular Biology of the Potato.* Santa Cruz, USA, 25-30 July, p.19 (Abstract).

Merlo, L., Geigenberger, P., Hajirezaei, M. and Stitt, M. (1993) Changes in carbohydrates, metabolites and enzyme activities in potato tubers during development, and within a single tuber along a stolon apex gradient. *Journal of Plant Physiology* 142:392-402.

Morrell, S. and ap Rees, T. (1986) Sugar metabolism in developing tubers of *Solanum tuberosum. Phytochemistry* 25:1579-1595.

Müller-Röber, B., Kossman, J., Hannah, C.L., Willmitzer, L. and Sonnewald, U. (1990) One of two different ADP-glucose pyrophosphorylase genes from potato responds strongly to elevated levels of sucrose. *Molecular and General Genetics* 224:136-146.

Müller-Röber, B., Sonnewald, U. and Willmitzer, L. (1992) Inhibition of the ADP glucose pyrophosphorylase in transgenic potatoes leads to sugar-storing tubers and influences tuber formation and expression of tuber storage protein genes. *EMBO Journal* 11:1229-1239.

Oparka, K.J. (1986) Phloem unloading in the potato tuber. Pathways and sites of ATPase. *Protoplasma* 131:201-210.

Oparka, K.J., Davies, H.V., Wright, K.M., Viola, R. and Prior, D.A.M. (1990) Effect of sink removal on sugar uptake and starch synthesis by potato-tuber storage parenchyma. *Planta* 182:113-117.

Renz, A. and Stitt, M. (1993) Substrate specificity and product inhibition of different forms of fructokinases and hexokinases in developing potato tubers. *Planta* 190:166-175.

Renz, A., Merlo, L. and Stitt, M. (1993) Partial purification of three fructokinases and three hexokinases from potato tubers, which show differing organ and developmental specificity. *Planta* 190:156-165.

Richardson, D.L., Davies, H.V. and Ross, H.A. (1989) An investigation into the factors influencing sugar levels in UK grown potatoes (cv Record). *Potato Research* 33:235-239.

Ross, H.A. and Davies, H.V. (1992) Effects of sink removal and sucrose flux on sucrose-degrading enzymes. *Plant Physiology* 98:287-293.

Ross, H.A., Davies, H.V., Burch, L.R., Viola, R. and Macrae, D. (1994) Developmental changes in carbohydrate content and sucrose degrading enzymes in tuberising stolons of potato (*Solanum tuberosum* L.). *Physiologia Plantarum* 90:748-756.

Salanobout, M. and Belliard, G. (1989) The steady state level of potato sucrose synthase mRNA is dependent on wounding, anaerobiosis and sucrose concentration. *Gene* 84:181-185.

Viola, R. and Davies, H.V. (1991) Fluoride-induced inhibition of starch biosynthesis in developing potato (*Solanum tuberosum* L.) tubers is associated with pyrophosphate accumulation. *Plant Physiology* 97:638-643.

Viola, R., Davies, H.V. and Chudeck, A.R. (1991) Pathways of starch and sucrose biosynthesis in developing tubers of potato (*Solanum tuberosum* L.) and seeds of faba bean (*Vicia faba* L.). *Planta* 183:203-208.

Xu D-P, Sung, S-S.J, Lobata, T., Kormanik, P.P. and Black, C.C. (1989) Characterization of sucrolysis via the uridine diphosphate and pyrophosphate-dependent sucrose synthase pathway. *Plant Physiology* 80:635-642.

Zrenner, R. and Sonnewald, U. (1993) Molecular approaches to influence sucrose metabolism in transgenic plants. In:*Proceedings of the 3rd International Symposium on the Molecular Biology of the Potato.* Santa Cruz, USA, 25-30 July, p.13 (Abstract).

Zrenner, R., Willmitzer, L. and Sonnewald, U. (1993) Analysis of the expression of potato uridinediphosphate-glucose pyrophosphorylase and its inhibition by antisense RNA. *Planta* 190:247-252.

Chapter Seven:

Post-Harvest Regulation of Sucrose Accumulation in Transgenic Potatoes: Role and Properties of Potato Tuber UDP-Glucose Pyrophosphorylase

Joe Sowokinos
Department of Horticultural Science, University of Minnesota, St Paul, MN 55108, USA.

A widespread response of plants exposed to low temperatures is the accumulation of free sugars. The accumulation of fructans, sucrose, glucose, fructose or other free saccharides is species dependent (Guy *et al.*, 1992). Starch in potato tubers (*Solanum tuberosum* L.) is frequently converted to undesirably high concentrations of reducing sugars (i.e. glucose and fructose) as a result of stresses experienced during growth and/or storage (Sowokinos, 1990a and references therein). The adverse influence of these reducing sugars on the processing quality of potatoes used for french fries and chips (potato crisps) is well known.

The mechanism(s) responsible for the stress-induced sweetening of potato tissue at the molecular level is still unexplained. This is not surprising since the production of free sugars in potato cells is certainly not controlled by a single factor, but involves the interaction of several pathways of carbohydrate metabolism, i.e. starch synthesis, glycolysis, mitochondrial respiration and gluconeogenesis. In addition, various levels of cellular control including (i) hormones, (ii) membrane structure and function, (iii) compartmentalization and concentration of key ions, substrates, and enzymes and (iv) enzyme activity and/or synthesis, operate in a synergistic manner to regulate sweetening (Sowokinos, 1990b). Therefore the use of traditional breeding and/or molecular engineering techniques to increase cold-insensitivity in existing or new potato clones has proven to be a formidable task. There is an urgent need to better understand which specific genes and gene products function to regulate the overall process of sweetening so that sensible strategies for crop improvement can be developed.

Of primary concern in this chapter will be the properties and role of UTP-glucose-1-phosphate uridylyltransferase (UGPase, EC 2.7.7.9) in catalyzing the first reaction unique to the sucroneogenic pathway via the formation of UDP-glucose (UDPG). The importance of UDPG pyrophosphorylase (UGPase) to

The Molecular and Cellular Biology of the Potato, Second Edition
Edited by W.R. Belknap, M.E. Vayda and W.D. Park

potato growth and development will be briefly reviewed. Data will be presented that show a coordinated expression of UGPase RNA with the levels of RNA of several other enzymes involved in the conversion of sucrose to starch in developing potato tubers. Initial results from transgenic potato tubers expressing sense and antisense constructs of UGPase (cv. Desirée) will be discussed in relation to post-harvest sucrose content from 4° and 12°C storage. Estimates of metabolic control coefficients (C values) relating UGPase activity to sucrose accumulation, from initial short-term storage experiments (6 weeks), suggested that UGPase can partially control sucroneogenesis in stored potato tubers.

Purification, Molecular Mass and Localization of Potato UGPase

Purification and characterization of UGPase was undertaken as a prelude to cloning and genetic manipulation of its activity in potato tubers. UGPase has been purified from several mammalian tissues, bacteria, yeast, slime mold, maize, and mung bean (Turnquist and Hansen, 1973 and refs. therein) as well as from other plant tissues (Hondo *et al.*, 1983; Vella and Copeland, 1990; Gustafson and Gardner, 1972) including the potato (Nakano *et al.*, 1989; Sowokinos *et al.*, 1993). The fact that this enzyme is ubiquitous in nature is not surprising since a major product of the reaction, namely UDPG, has a central role as a glucosyl donor in numerous cellular reactions (Ginsburg, 1964). Potato tuber UGPase (cv. Norchip) was purified to near homogeneity to a specific activity of 1099 international units/mg protein, using salt precipitation along with hydrophobic, Bio-Gel, anionic and cationic chromatography steps performed on samples extracted from potato acetone powder (Sowokinos *et al.*, 1993).

The molecular mass of the potato UGPase was determined to be 53 kDa by denaturing SDS-PAGE (Fig. 7.1) and gel-filtration chromatography. Polyclonal antisera produced against the purified enzyme revealed the presence of only a single band on the immunoblot against total tuber protein separated by SDS-PAGE (Fig. 7.2, lane 4). This band aligned with the 53 kDa subunit of the purified enzyme (lane 3). Previously, Nakano *et al.* (1989) indicated that potato UGPase probably exists as a monomeric polypeptide with an approximate molecular mass of 50 kDa. Based on the coding sequence of a cDNA for UGPase isolated from a cDNA library of immature potato tissue, Katsube *et al.* (1990) predicted a subunit mass of 51.8 kDa. The potato tuber monomeric subunit mass agrees with the reported values of 53 and 55 kDa for the enzyme from *Typha latifolia* pollen (Hondo *et al.*, 1983) and slime mold (Franke and Sussman, 1971), respectively. Although a monomeric nature for native potato UGPase is certainly possible, a multimeric structure would be consistent with the forms reported for UGPase from slime mold (Franke and Sussman, 1971), animal tissues (Turnquist and Hansen, 1973) and bacteria (Nakae, 1971). The enzyme from slime mold, which shows extensive amino acid homology to UGPase from potato (Katsube *et al.*, 1990), is a multimer. Both monomeric and dimeric

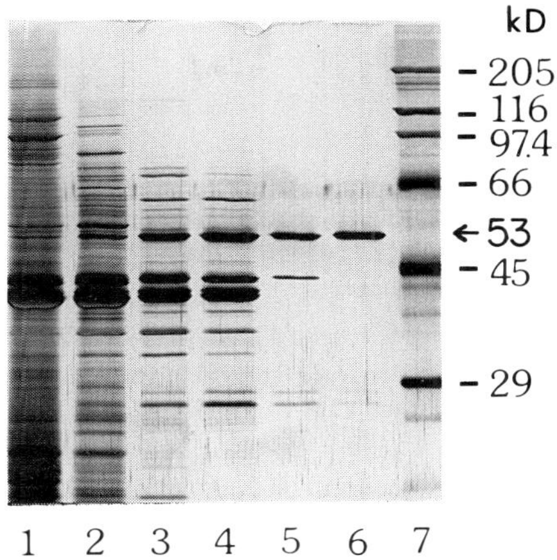

Figure 7.1. SDS-PAGE of purified potato tuber UGPase stained with silver. Lanes contain the following fractions: 1, acetone powder (25 μg); 2, ammonium sulfate precipitate (15 μg); 3, phenyl-Sepharose (10 μg); 4, Bio-Gel (10 μg); 5, DEAE-Sephacel (1 μg); 6, CM-Sepharose (1 μg); and 7, molecular mass standards (8 μg each). (Sowokinos *et al.*, 1993.)

forms of the enzyme, which are capable of interconversion, have been isolated from *Salmonella typhimurium* (Nakae, 1971). These different forms of the enzyme also differ in their catalytic properties.

Immunoblots of protein from various potato tissues revealed the presence of UGPase in tubers, stolons, leaves, stems, callus and suspension-cell cultures (Sowokinos *et al.*, 1993). Lower amounts of protein were present in leaf followed by only a trace in root tissue. Samples selected from actively growing potato plants were also assayed for UGPase activity. Table 7.1 indicates that the enzyme level, based on specific activity, was similar in tubers, stolons and stems, but was markedly lower in leaf and root tissue. Recovery of enzyme units following addition of purified UGPase to root tissue prior to homogenization indicated unequivocally that the enzyme was not destroyed during the extraction procedure. Northern blot analysis of various potato tissues during development and storage revealed the presence of UGPase transcript in all samples (Fig. 7.3) (Spychalla *et al.*, 1994). The root tissue, which showed the least amount of UGPase protein and activity, also contained the least amount of UGPase transcript. This was substantiated by the results of Zrenner *et al.* (1993) which showed a low level of UGPase RNA in roots compared with other tissues of the potato.

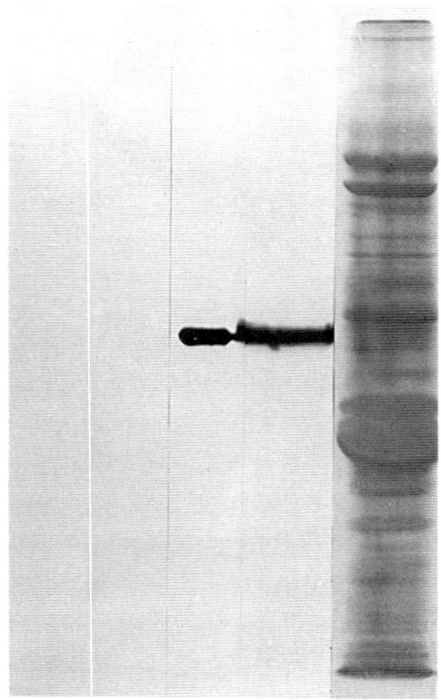

Figure 7.2. Specificity of anti-UGPase serum as demonstrated by reaction with a Western blot developed from SDS-PAGE of the final CM-Sepharose and total tuber protein fractions. Lanes 1 and 3 contain 0.2 μg of purified UGPase. Lanes 2 and 4 contain 100 μg of total tuber protein. Lanes 1 and 2 were probed with preimmune serum (1:1000) and lanes 3 and 4 were probed with whole rabbit anti-UGPase (1:1000). Lane 5 contains 100 μg of total tuber protein stained with Coomassie blue R-250. (Sowokinos *et al.*, 1993.)

Table 7.1. UGPase activity in various tissues selected from immature potatoes (cv. Norchip).

Tissue	Units[a]/g	Protein/g (mg)	Specific activity (units mg^{-1} protein)
Tubers	34.77 ± 1.29[b]	9.00 ± 0.25	3.86 ± 0.14
Roots	5.84 ± 0.09	4.92 ± 0.27	1.18 ± 0.02
Leaves	6.48 ± 0.08	31.64 ± 0.44	0.21 ± 0.01
Stolons	16.20 ± 0.51	4.94 ± 0.11	3.28 ± 0.10
Stems	11.54 ± 0.39	2.87 ± 0.12	4.04 ± 0.14

[a]Unit, the amount of enzyme necessary to catalyze the formation of 1 μmol of Glc-1-P per min.
[b]Mean ± SE for five experiments (Sowokinos *et al.*, 1993).

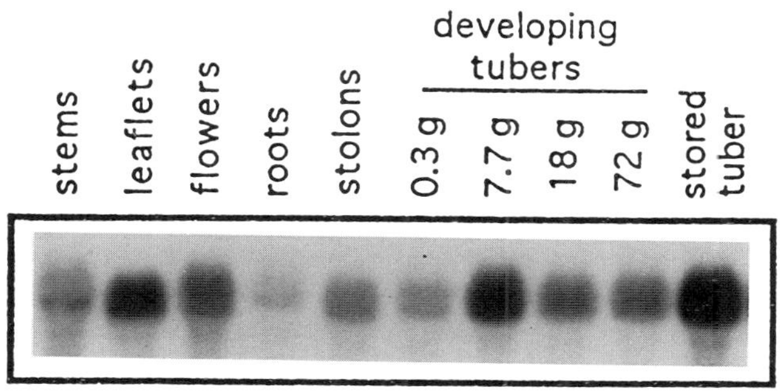

Figure 7.3. Distribution of UGPase RNA in total RNA of various potato tissues. Each lane contained 25 μg of total RNA extracted from glasshouse grown potato plants. (Spychalla *et al.*, 1994.)

Role of UGPase in Potato Growth

An actively growing potato tuber utilizes imported sucrose from leaves to supply the carbon and energy for cellular growth and starch biosynthesis. The potato storage cell receives sucrose from the phloem without prior hydrolysis by either acid or alkaline invertases during translocation (Hardy and Norton, 1968; Oparka and Prior, 1987). If the two hexose moieties of sucrose (i.e. fructose and glucose) are to be routed to starch formation, they must first enter the hexose-phosphate pool. Sung *et al.* (1989) explain that this conversion is accomplished sequentially by two enzymes, the first being sucrose synthase which catalyzes the formation of the high energy glucose-containing nucleotide UDPG (reaction 1, below). UDPG can be derivatized to other nucleoside diphospho-sugars that are used in the synthesis of pectic substances, hemicelluloses, glycolipids and other assorted glycosylated molecules (Goodwin and Mercer, 1983). For UDPG to be converted to hexose-phosphate, one of

$$\text{sucrose} + \text{UDP} \longleftrightarrow \text{fructose} + \text{UDPG} \quad (1)$$

$$\text{UDPG} + \text{PPi} \longleftrightarrow \text{Glc-1-P} + \text{UTP} \quad (2)$$

Key reactions entering sucrose into metabolism

the α–ß phosphodiester bonds must be cleaved. A cytosolic enzyme capable of catalyzing this conversion is UGPase (reaction 2, above). The major portion of the UDPG formed by sucrose synthase will be converted into Glc-1-P via this reaction (Morrell and ap Rees, 1986). Pyrophosphate (PPi) is utilized in this reaction to form a molecule of UTP and Glc-1-P. Potatoes contain sufficient PPi

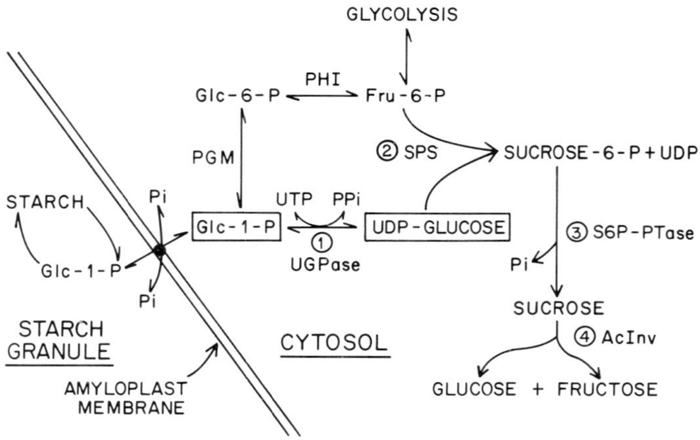

Figure 7.4. Enzyme reactions directly involved in free sugar formation in potato cells: 1, UDP-glucose pyrophosphorylase (UGPase); 2, sucrose 6-phosphate synthase (SPS); 3, sucrose 6-phosphate phosphatase; and 4, acid invertase (AcInv). Other enzymes represented include phosphoglucomutase (PGM) and phosphohexoseisomerase (PHI).

to allow UGPase to convert UDPG to Glc-1-P during rapid tuber growth (Mohanty and ap Rees, 1992). The Glc-1-P formed may then cross the amyloplast membrane directly en route to support starch synthesis (ap Rees and Morrell, 1990; Viola *et al.*, 1991). Glc-1-P can also enter respiratory pathways following conversion by phosphoglucomutase and phosphohexoseisomerase to Fru-6-P (Mares *et al.*, 1985).

Using antisense technology, Zrenner *et al.* (1993) indicated that carbohydrate metabolism was not substantially affected when potato plants had a 96% reduction in UGPase activity compared with wild-type plants. They concluded that only 4% of activity was sufficient for the enzyme to function normally during plant growth and development. A ratio change of UDPG:Glc-1-P from 14 to 27 in U-49 (96% inhibited plant) compared to control tubers, suggested a slight reduction in the flow of carbohydrates into starch. This slight reduction considered over the several weeks that tubers develop, had no long-term effect on sugar metabolism in mature tubers. The reason for excess levels of this enzyme during potato growth is not clear at this time.

Importance of UGPase in the Sucrose Formation Pathway

The enzyme UGPase occupies an important position in carbohydrate metabolism by catalyzing the reversible conversion of the metabolites Glc-1-P and UDPG (Fig. 7.4, reaction 1). Co-substrates in the synthesis and pyrophosphorolysis of

UDPG are UTP and inorganic pyrophosphate (PPi), respectively. During sweetening, the products of the UGPase reaction, namely UDPG and PPi, may have regulatory roles in directing carbon flux into the pathways of glycolysis, starch synthesis and/or hexose formation (Sowokinos, 1990a; Jelitto *et al.*, 1992).

UGPase is the only reported enzyme that is capable of forming the cytosolic metabolite UDPG. The enzyme SPS (Fig. 7.4, reaction 2), which is instrumental in regulating sucrose formation in leaf (Huber and Huber, 1990 and references therein) and non-photosynthetic tissues (Sowokinos, 1990a), requires high concentrations of UDPG (K_m = 2.5 mM) to achieve 50% of its maximum catalytic activity (Murata, 1972). In plants, the *in vivo* concentration of UDPG is 10- to 20-fold lower in relation to its K_m (1-7 mM) for SPS (Morrel and ap Rees, 1986; Sicher, 1986). In the cold, when hexose-phosphates are elevated in potato tubers (Barker, 1968; Isherwood, 1976) a 'coarse metabolic control' between clones via UGPase could supply different levels of UDPG and directly affect the rate of the SPS reaction and thus sucrose synthesis. Bruneau *et al.* (1991) observed that diurnal fluctuation of SPS in corn leaves is due to modification of its activity due to metabolites (i.e. substrates and allosteric effectors) rather than by modulation of the number of enzyme molecules (i.e. synthesis). Sicher (1986) also indicated that the availability of UDPG may be limiting the SPS reaction in leaf tissues. A similar type of regulation may be occurring during sucrose synthesis in non-photosynthetic potato tissue. In genetically diverse potato clones the amount of tuber UGPase activity in tubers stored for nine months at 3°C was found to be correlated (r = 0.920) with the glucose content (product of invertase action on sucrose) (Sowokinos, 1990a).

PPi, the other product of UGPase when the tuber is in the sucroneogenic mode, participates as a substrate in the glycolytic reaction catalyzed by PPi-dependent phosphofructokinase (PPi-PFK). This enzyme is thought to be the primary route by which hexose-phosphates enter glycolysis in cold stored tubers (Claassen *et al.*, 1991). PPi also participates in the tonoplast PPi-dependent proton pump reaction. Comparison of mass action ratios and equilibrium constants for UGPase, PPi-PFK and the tonoplast enzyme indicated that each catalyzes near-equilibrium reactions (Weiner *et al.*, 1987; Roberts, 1990). This suggested that the intracellular pool of PPi may be stabilized by a continuum of near-equilibrium reactions. Dancer and ap Rees (1989) found in hypoxic pea root tips, where the biosynthesis of PPi was inhibited, that PPi levels were unchanged relative to aerobic tissues.

Cellular Regulation of UGPase

Most enzymes are subject to regulation at the level of synthesis (i.e. coarse metabolite control) and/or at the catalytic level (i.e. fine metabolic control). Both mechanisms have been shown to influence UGPase activity, but controversy exists as to which is more prominent among tissues (Turnquist and Hansen,

1973). Although variations in UGPase enzyme level in higher plants have been associated with a change in capacity for starch and sucrose synthesis (Turner, 1969; Tovey and Roberts, 1970; Tsai *et al.*, 1970), substrate and catalytic effector molecules usually determine reaction rates.

$$\textbf{Glc-1-P + UTP} \longleftrightarrow \textbf{UDPG + PPi}$$

If the concentration of UGPase is not limiting, the reaction will proceed as fast as co-substrates are made available (Wright, 1970). This is true ideally, but product inhibition via UDPG occurs decreasing the reaction rate by competing with the substrate UTP (Gustafson and Wright, 1971). Nakano *et al.* (1989) conducted product inhibition studies to determine the sequence of the binding of substrates and release of products from the potato enzyme during catalysis (Table 7.2). The simplest mechanism suggested was sequential in nature, where the nucleotide substrate was the first to bind in either direction followed by either Glc-1-P or PPi. The first products released are Glc-1-P or MgPPi, while the nucleotide products MgUTP or UDPG are the last to leave the enzyme. This reaction mechanism was similar to that suggested from other sources (Tsuboi *et al.*, 1969; Gustafson and Gardner, 1972: Hondo *et al.*, 1983).

During differentiation of slime mold (*Dictyostelium discoideum*), UGPase increases and elevated levels of UDPG serve as a precursor of cell wall material (Wright, 1968). It was determined, however, that the increase in UGPase concentration *in vivo* could not account for the enhanced rate of synthesis of UDPG. Calculating the steady state level of substrates and products, as well as related kinetic constants for UGPase, it was determined that the major controling factor in this system was the availability of Glc-1-P. The metabolic importance of UGPase is demonstrated in mutant strains of *Dictyostelium* which have reduced activities of this enzyme. Morphogenesis in these strains is arrested where stored glycogen is to be mobilized into UDPG equivalents for use in cellulose formation (Diamond *et al.*, 1976).

Biochemical properties of certain potato clones which demonstrate a certain degree of cold-insensitivity to sweetening have been examined in storage. Generally, the UGPase activity of lower sugar clones decreases at some point during the storage period. Fig. 7.5 shows a decrease in UGPase activity in ND860-2 and Norchip, after 2 months in storage, compared with no decrease in activity in the high sugar clone, Red Pontiac. Change in UGPase activity observed in storage was not due to the presence or absence of a catalytic effector or to a change in the stability of the enzyme. From year to year, the onset of the decrease in UGPase activity does not necessarily coincide with changes in sugar levels among genetically diverse potato clones during storage. This indicated that other factors, besides the absolute level of UGPase, are important in controlling a clone's ability to form UDPG, sucrose and eventually reducing sugars.

Table 7.2. Types of product inhibition in the potato tuber UDP-glucose pyrophosphorylase reaction under different reaction conditions[1].

Substrate	Inhibitor	Substrate concentration fixed at K_m	Substrate concentration fixed at Saturation
UDP-glucose	Glucose-1-P	Noncompetitive	Uncompetitive
UDP-glucose	UTP	Competitive	Competitive
PPi	Glucose-1-P	Noncompetitive	Noncompetitive
PPi	UTP	Noncompetitive	(Not determined)

1 From Nakano *et al.*, 1989 with permission.

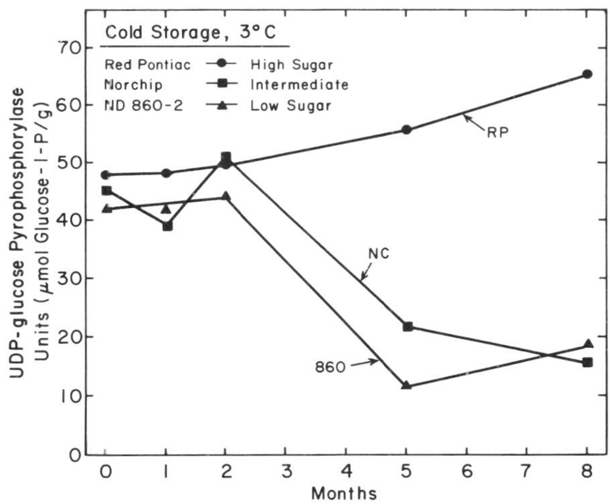

Figure 7.5. Change in UGPase activity in three potato clones stored for 8 months at 3°C.

Kinetics and Effectors of UGPase in the Pyrophosphorolysis Direction

$$\text{UDPG} + \text{PPi} \longleftrightarrow \text{Glc-1-P} + \text{UTP}$$

UGPase purified from two different potato clones yielded similar kinetic properties when the reaction was measured in the direction of Glc-1-P formation (Nakano *et al.*, 1989; Sowokinos *et al.*, 1993). Individually, enzyme, UDPG, PPi and Mg^{2+} were absolutely necessary for the reaction to proceed. ADP-glucose (ADPG) did not serve as a substrate and the purified enzyme was not activated by 3-P-glycerate. Mg•PPi was shown to be the actual substrate and maximum reaction velocities (V_{max}) were obtained when Mg^{2+} was added at approximately the same concentration as PPi (Nakano *et al.*, 1989). V_{max} occurred in the presence of 5 mM Mg^{2+}, whereas 50% of the V_{max} was noted with 0.8 mM Mg^{2+} (Sowokinos *et al.*, 1993). The pH curve for UGPase was fairly broad between pH 7.5 and 9.0; an optimum occurred near pH 8.0 and 8.5. Inorganic phosphate (Pi) inhibited Glc-1-P formation 50% at a concentration of 6 mM. Lesser inhibitions were observed with Fru-2,6-bisP (40%), Fru-6-P (27%) and 3-P-glycerate (20%) at relatively high concentrations of 2 µM, 4 mM, and 0.4 mM respectively (Sowokinos *et al.*, 1993). A further increase in the concentration of these metabolites and effectors did not enhance their percent inhibition. The K_m values for the pyrophosphorylase co-substrates UDPG and PPi were determined from the results of initial reaction velocity studies. Double reciprocal plots of the enzyme activity against the substrate concentrations were all linear (Nakano *et al.*, 1989; Sowokinos *et al.*, 1993). Calculated K_m values were low ranging from 0.11 to 0.14 mM. A high V_{max} between 800 and 900 IU/mg protein was also noted. Given the high levels of UGPase in developing potatoes along with its low K_m values, high V_{max}, and the K_{eq} of the UGPase reaction in plants, which lies in the direction of Glc-1-P synthesis (Turnquist and Hansen, 1973), the conversion of UDPG to support starch synthesis and glycolytic reactions must occur rapidly in developing potato tubers (Sowokinos *et al.*, 1993).

Kinetics and Effectors of UGPase in the UDPG Synthesis Direction

$$\textbf{UTP + Glc-1-P} \longleftrightarrow \textbf{PPi + UDP-Glc}$$

Enzyme, UTP, Glc-1-P and Mg^{2+} were absolutely necessary for the synthesis of UDPG to occur. Optimum reaction rate was obtained at pH 8.5. The nucleotides ATP, CTP and GTP could not substitute for UTP in the reaction. The Mg•UTP complex was shown to be actual substrate and unchelated UTP strongly inhibited the reaction (Nakano *et al.*, 1989). Inhibition with free UTP was similar to that reported for the enzyme from erythrocytes which was also strongly inhibited by the product UDPG (K_i = 0.015 mM) (Tsuboi *et al.*, 1969). These investigators demonstrated a K_i of 0.160 mM (UDPG) for the mung bean UGPase. They suggested the 11-fold higher value for the plant enzyme indicated a likely difference in biological significance compared with the mammalian enzyme. The inhibition constant for the product UDPG with purified potato UGPase is presently being determined. V_{max} occurred in the presence of 10 mM Mg^{2+}, whereas 50% of the reaction was observed with 5.4 mM Mg^{2+}.

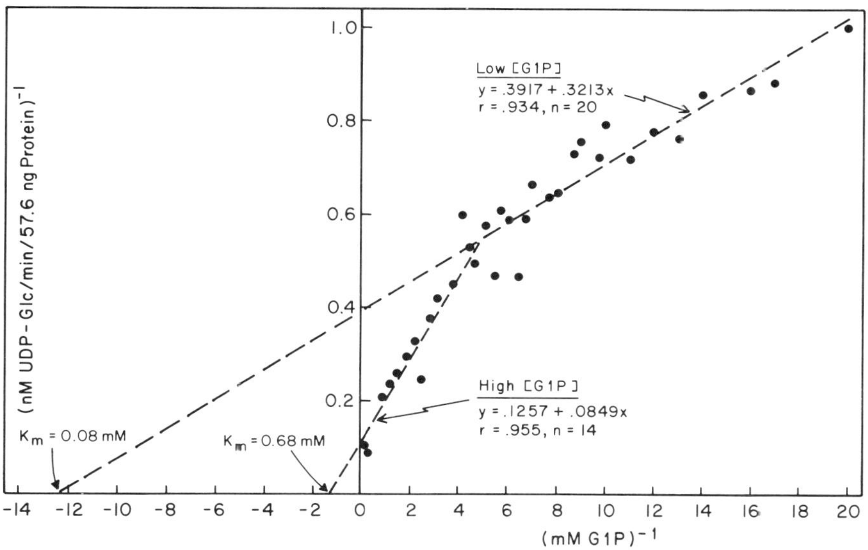

Figure 7.6. A Lineweaver-Burk plot of purified UGPase reaction in the direction of UDPG (UDP-Glc) synthesis. The reciprocal of each reaction velocity was plotted against the reciprocal of Glc-1-P concentration (range 0.05-5.0 mM). (Sowokinos *et al.*, 1993.)

High concentrations of Fru-2,6-bisP or 3-P-glycerate did not activate or inhibit synthesis of UDPG. Inorganic phosphate maximally inhibited the synthesis reaction only 30% at a high concentration of 12 mM (Sowokinos *et al.*, 1993). These compounds, therefore, were selective by partially inhibiting the reaction only in the pyrophosphorolysis direction.

Although most of the kinetic properties described thus far have been similar for UGPase purified from two potato cultivars using different chromatographic procedures (Nakano *et al.*, 1989; Sowokinos *et al.*, 1993), they differed markedly in respect to the enzymes affinity for the co-substrates Glc-1-P and UTP. Nakano and coworkers (1989) reported linear reaction kinetics with both substrates with their 50 kDa monomeric enzyme. Fig. 7.6 shows a Lineweaver-Burk plot of purified UGPase reaction in the direction of UDPG synthesis obtained from cv. Norchip potatoes. The concentration of Glc-1-P was varied from 0.5 to 5.0 mM. A biphasic substrate saturation curve was obtained with Glc-1-P as a substrate that could be explained by two separate regression equations (Sowokinos *et al.*, 1993). At low or physiological concentrations of Glc-1-P (0.05-0.20 mM, represented by reciprocals ranging from 5 to 20 in Fig. 7.6) one reaction slope was calculated yielding a K_m value of 0.08 mM. As the Glc-1-P concentration increased above 0.20 mM, the K_m of UGPase for Glc-1-P was increased 8.5-fold from 0.08 to 0.68. The V_{max} also increased in the presence of excess Glc-1-P. A similar pattern of biphasic kinetics was also noted when the

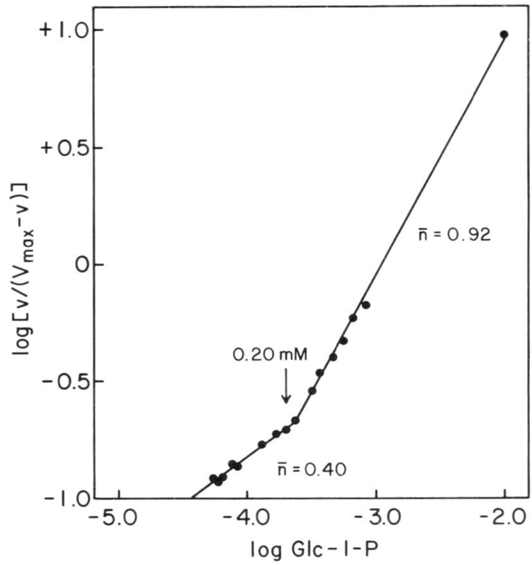

Figure 7.7. A Hill equation replot of the data in Fig. 7.6. (Sowokinos *et al.*, 1993.)

co-substrate concentration of UTP was varied. These reaction kinetics were similar in nature to those shown by Nakae and Nikaido (1971) for UGPase from *S. typhimurium*. They observed multiple forms of UGPase that demonstrated real differences in their catalytic properties. An enzyme form IIIb (a dimer, mol mass 80 kDa) demonstrated linear kinetics in the direction of pyrophosphorolysis and biphasic substrate saturation curves in the direction of UDPG synthesis. They explained the increased K_m values observed as a 'negative cooperative effect' as described by Conway and Koshland (1968), where the binding of one substrate molecule makes the binding of another molecule of the same substrate at a second binding site more difficult. Hill coefficients ($\bar{n}$ = estimated no. of binding sites) less than 1.0 (Fig. 7.7) supported a negative cooperative effect with the potato enzyme with respect to the binding of Glc-1-P. Corresponding $\bar{n}$ values increased from 0.40 to 0.92 as the concentration of Glc-1-P increased above 0.20 mM. Non-denaturing PAGE of purified potato UGPase (Fig. 7.8, lane 4) and molecular weight standards, suggested that the enzyme might exist as a dimer *in vitro*. Although additional non-denaturing gels need to be run with various polyacrylamide concentrations to prove that a dimeric form of potato UGPase can indeed exist, resultant biphasic kinetic data at least suggest the existence of a dimer.

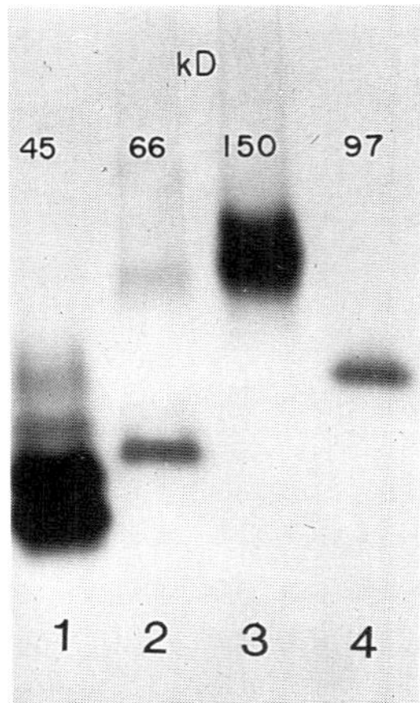

Figure 7.8. A non-denaturing poly acrylamide gel run with a 1.0 mm thick slab gel containing 7.5% acrylamide. Lanes 1, 2 and 3 contained 3 µg of the protein standards ovalbumin, bovine serum albumin and alcohol dehydrogenase, respectively. Lane 4 contained 0.075 µg of purified potato UGPase. Proteins were identified using silver stain.

Sucrose Induction of UGPase RNA Synthesis

Sucrose provided exogenously to plants has been shown to affect fundamental processes like photosynthesis, growth, cellular differentiation, flowering, and tuberization (Lecharny and Nguyen-Quoc, 1991). Potato genes responsive to sucrose have included the class 1 patatin gene (Jefferson *et al.*, 1990; Rocha-Sosa *et al.*, 1989; Wenzler *et al.*, 1989), one of the two ADPG pyrophosphorylase genes (i.e. AGPase S) (Müller *et al.*, 1990), proteinase inhibitor II (Johnson and Ryan, 1990), starch branching enzyme (Kossmann *et al.*, 1991) and sucrose synthase (Salanoubat and Belliard, 1989). It has been suggested that sucrose may coordinate the deposition of patatin and starch during tuber formation (Rocha-Sosa *et al.*, 1989). Kossmann *et al.* (1991), in studying the expression pattern of the gene coding for potato branching enzyme, raised the possibility that enzymes involved in the starch biosynthetic pathway are coordinately regulated at the transcriptional level. Fig. 7.9 shows that UGPase transcript could also be induced by sucrose treatments of excised potato leaves (Spychalla *et al.*, 1994). Twenty-four hour incubations in darkness with 75 mM and 150 mM sucrose had little effect on mRNA levels while UGPase message was noticeably increased when 225 mM and 300 mM treatments were used. The 24h treatments resulted in no observable change in UGPase activity.

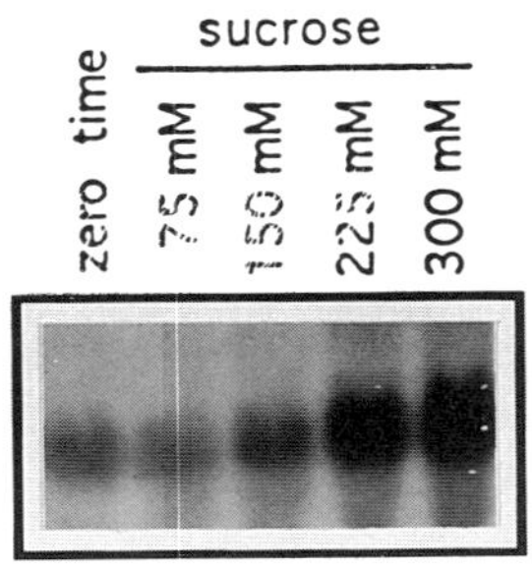

Figure 7.9. Sucrose induction of UGPase in axenic potato leaves. Each lane contained 25 μg of total RNA extracted from leaves incubated with various concentrations of sucrose plus 10 mM glutamine for 24 h in the dark. (Spychalla *et al.*, 1994.)

Cloning and Sequencing of Potato UGPase

A potato tuber cDNA expression library (cv. Desirée) in λgt11 of Holdsworth *et al.* (1992) was probed with polyclonal anti-UGPase sera of Sowokinos *et al.* (1993). The longest isolated clone (UP2) was subcloned into pUC119 and sequenced in both directions according to the methods of Murphy and Ward (1990). The cDNA contained 1701 bp and an open reading frame (ORF) of 1431 bp (Spychalla *et al.*, 1994). The length of the ORF and the position of the polyadenylation signal corresponded exactly to that of the potato tuber UGPase cDNA clone U4a isolated by Katsube *et al.* (1990). However, the coding region of UP2 differed from U4a at 26 bases.

These base changes resulted in five differences in the deduced amino acid sequence, of which two could lead to a protein with a different charge form. U4a codes for Thr, Tyr, Lys, Lys and Val while UP2 codes for Ala, Asp, Asn, Glu and Ile at amino acid positions 5, 30, 82, 445 and 450, respectively. The 3' non-translated region of UPa also differed appreciably from that of U4a.

Structure and Organization of the UGPase Gene

Reverse genetics and PCR-based cloning techniques were used to isolate both cDNA and genomic clones of UGPase. Katsube *et al.* (1990) designed synthetic oligonucleotides based on published sequences of peptide fragments (Nakano *et al.*, 1989), and screened a cDNA library of potato tuber mRNAs. They identified a cDNA containing a complete coding region of 1431 nucleotides. 5' to this sequence was a 129 nucleotide leader sequence and 3' of the sequence was a 174 nucleotide untranslated region containing a polyadenylation signal. PCR-cloning

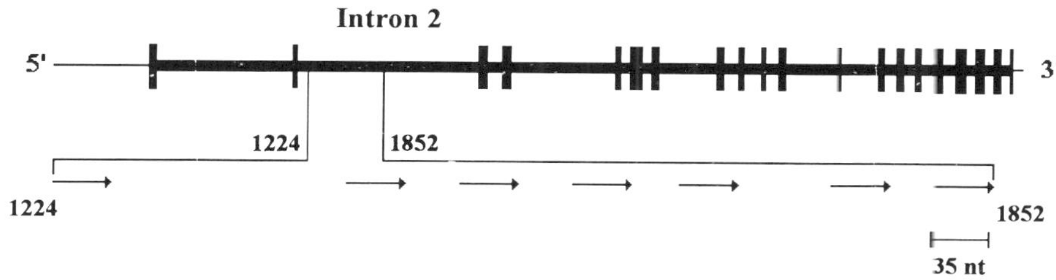

Figure 7.10. The structure of the UGPase gene of potato. The vertical bars represent the exons and the thin horizontal bars are the introns. The arrows below the second intron represent the location of the seven direct repeats.

of a UGPase cDNA from cv. Desirée by Spychalla *et al.* (1994) produced a sequence that was similar but not identical to the sequence described by Katsube *et al.* (1990). Although the 5' end of the two sequences were identical, nucleotide differences resulted in five amino acid differences.

Synthetic oligonucleotides based on the UGPase peptide fragments described by Nakano *et al.* (1989) were synthesized as primers for PCR amplification of potato genomic DNA (Borovkov *et al.*, 1993). A 2.4 kb fragment was sequenced and found to contain sequences that when translated matched the published peptide sequences. Two overlapping clones were selected from a genomic library of cv. Lemhi DNA using the 2.4 kb fragment as a probe.

Sequencing of these clones revealed that the UGPase gene is highly fragmented, containing 20 exons and 19 introns. The exons range in size from 24 to 124 nucleotides, and the introns range in size from 77 nucleotides to 1589 nucleotides. Intron 2, the largest intron, contains seven direct repeats, six of them arranged in tandem (Fig. 7.10). The 870 nucleotides 5'-non-coding region of the gene contains three possible TATA boxes. 5' mapping of the UGPase mRNA suggests that two distinct mRNA species may be transcribed. Sequence analysis suggested that transcription of these mRNAs would begin 30-40 nucleotides downstream of the putative TATA boxes. Finally, a comparison of the 5' ends of the cDNA and the genomic clones revealed the first 87 nucleotides 5' of the ATG-translation start site were identical, but beyond that site the two clones diverged.

Southern hybridization analysis suggested that UGPase is represented as a single or low copy number gene in the potato genome. Heterologous hybridization of a potato UGPase probe with conditions requiring 84% homology (65°C, 1 x SSC) was to single *Eco*RI and *Hin*dIII fragments of common bean, sunflower, wheat and tobacco DNA. These results indicate that UGPase is represented as a low copy number sequence in the genomes of many plant species.

The genomic UGPase of potato was compared with other known UGPase sequences. As expected, the nucleotide sequence of the coding regions of the gene was nearly identical to the cDNA sequence reported by Katsube *et al.* (1989). The two sequences differed by three nucleotides and only a single amino acid. The amino acid sequence was next compared with UGPase sequences from slime mold (Ragheb and Dottin, 1987), human (Peng and Chang, 1993; Genbank accession no. L14430) and bovine (Konishi *et al.*, 1993; Genbank accession no. L14019) UGPase. The slime mold sequence was the most divergent of the four. The potato, human and bovine sequences were 43.6% identical when gaps were not considered. Regions of greater identity were noted, specifically stretches of 41 and 159 amino acids which were 75.6% and 70.4% identical, respectively. Presumably these conserved domains reflect regions of structural importance.

Gene Construction and Potato Transformation

The antisense UGPase gene consists of the 5'-most 0.55 kb *Eco*RI fragment of the UGPase clone UP2. The isolated fragment was filled with the Klenow fragment of DNA polymerase I. The fragment was then ligated to *Sac*I linkers and cloned in the antisense orientation downstream of the patatin promoter into the *Sac*I site of pBIN440.7 to make pBIN440.7-UP2, the binary plant transformation vector. The patatin promoter has been shown previously to direct expression of transgenes in a tuber-specific fashion (Jefferson *et al.*, 1990). The sense-UGPase binary vector consisted of the 1.6 kb *Nco*I fragment of the functional UGPase clone pTUG (Katsube *et al.*, 1991) which was end-filled (Klenow fragment of DNA polymerase I), ligated to *Sac*I linkers and cloned in the sense orientation into the *Sac*I site of pBIN440.7 to make pBIN440.7-sense-U4a. The binary vectors were introduced by triparental mating into *Agrobacterium tumefaciens* strain T37 which contained the modified nopaline-type avirulent helper plasmid pTiT37-SE (gift of the Monsanto Company). Potato tuber disks were transformed, regenerated, and PCR screened for transformation according to the methods of Spychalla and Bevan (1993).

Gene Regulation of UGPase mRNA and Coordinated Expression of Starch Synthesizing Transcripts

The antisense transcript was detected in tubers of seven independent transgenic lines (A5, A10, A11, A15, A17, A25 and A26) at harvest (Fig. 7.11) but not after six weeks in storage at 4°C (Fig. 7.12). An apparent reduction of sense UGPase RNA compared with the three wild-type plants or the three putative UGPase over-expressing lines (S6, S8 and S30) was demonstrated by transformants A10, A17, A25 and A26. Transformant A26 had the highest antisense RNA to sense RNA ratio of the clones tested.

Using probes for sucrose synthase, B and S subunits of AGPase, starch branching enzyme (SBE), forms I and II of granule-bound starch synthase (GBSS) and patatin, Northern blot experiments revealed a coordinated expression

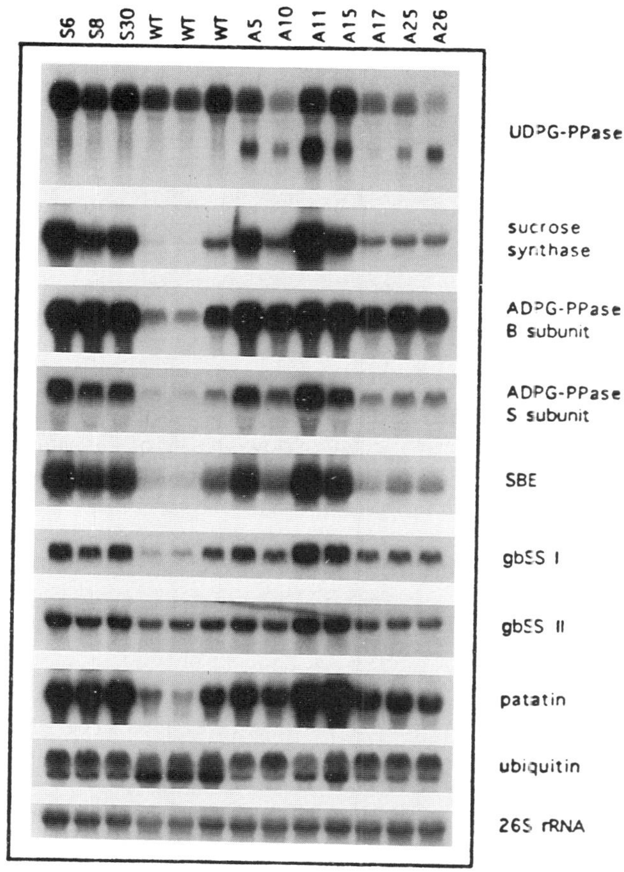

Figure 7.11. UGPase (UDPG-PPase), sink associated and control RNAs in wild-type (WT) and transgenic potato tubers at harvest. Each sample contained 25 μg of total RNA from an individual tuber. S6, S8 and S30 were transformed with a UGPase over-expression construct. A5, A10, A11, A15, A17, A25 and A26 were transformed with an antisense UGPase construct. (Spychalla *et al.*, 1994).

of the various RNA levels including UGPase and patatin, the major tuber protein (Fig. 7.11). Sucrose synthase demonstrated the widest variation of all the probes at harvest. Ubiquitin and 26S rRNA probes produced uniform signals which verified the uniformity of the sample loads.

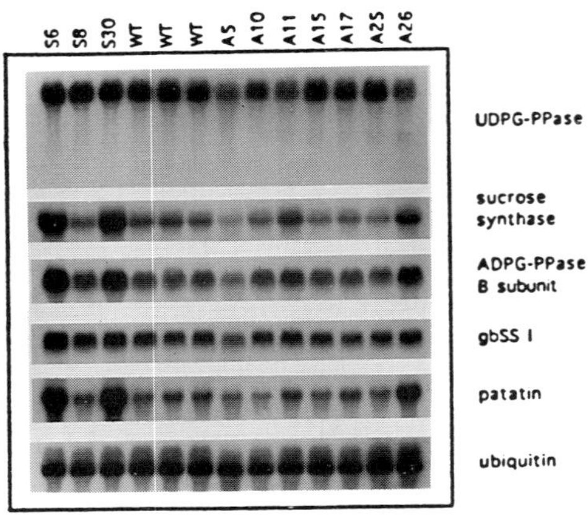

Figure 7.12. UGPase (UDPG-PPase), sink-associated and control RNAs in wild-type (WT) and transgenic potato tubers after six weeks of storage at 4°C. Lane headings are the same as described in Fig. 7.11. (Spychalla *et al.*, 1994.)

Antisense Inhibition of UGPase and Effect upon Sucrose Accumulation in Developing and Stored Potatoes

UGPase activity of both wild-type and transgenic potato tubers was measured at harvest and throughout storage at 4°C and 12°C for 6 weeks (Fig. 7.13). All of the antisense transgenic lines demonstrated reduced UGPase enzyme activity relative to the wild-type. This reduction (maximum 30% in transformant A26) was maintained throughout the 6 weeks of storage at both temperatures. No relationship could be consistently discerned between UGPase activity and hexose-phosphate or UDPG content between the transformants and wild-type clones (Spychalla *et al.*, 1994). Freshly harvested potatoes from plants expressing antisense RNA for UGPase, with activity suppressed to 4-5% of wild-type activity, demonstrated no significant difference in fresh weight, dry weight, starch, hexose-phosphate or UDPG content compared with control tubers (Zrenner *et al.*, 1993). Short-term storage of these potatoes (i.e. 18 days) resulted in no difference in starch and sugar content between control or transformed clones at either 4°C and 12°C (Zrenner *et al.*, unpublished data). Longer-term storage (i.e. 3-6 weeks) of the potato clones represented in Fig. 7.14 revealed that the sucrose content of the tubers was significantly correlated with UGPase activity (Fig. 7.14, panels B and C). There was a significant relationship between the mol hexose/mol sucrose ratio and UGPase activity among the transformants, particularly at the colder temperature (i.e. 4°C) (Fig. 7.14, panels E and F). Creation of transgenic plants, altered for a single gene product, permits experiments where metabolite pools can be determined allowing for the

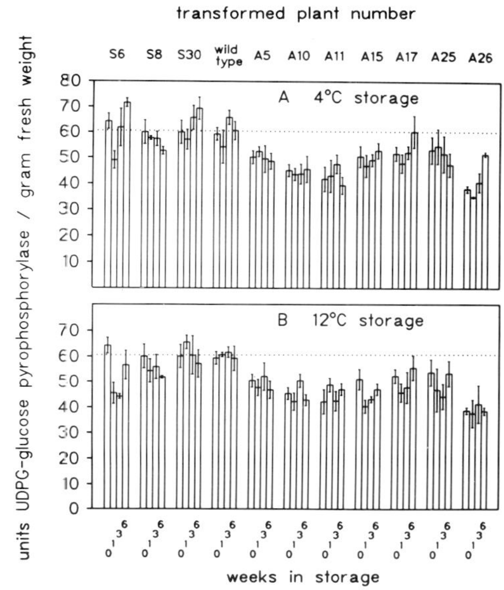

Figure 7.13. UGPase enzyme activity in wild-type (WT) and transgenic potato tubers at harvest and after 1, 3 and 6 weeks of storage at 4°C (A) and 12°C (B). Values are means from three individual tubers ± S.E. (Spychalla *et al.*, 1994.)

calculation of a metabolic control coefficient (C) for a targeted enzymatic step. C values are a measure of the contribution that a single step makes to control the metabolic flux through the pathway in question (Kacser and Porteous, 1987). The C values relating UGPase activity to sucrose content (Fig. 7.14, panels B and C) were determined according to Kacser and Porteous (1987). The C values expressed the percent change in sucrose content relative to the percent change in UGPase activity. Control coefficients were calculated over the range of UGPase activities from 35 to 75 IU/g fresh weight. Values for C of 0.40 and 0.76 (3 weeks at 4° and 12°C) and 0.48 and 0.58 (6 weeks of storage at 4° and 12°C) were determined. The coefficient values indicated an appreciable degree of control upon sucrose concentration in the post-harvest situation.

Conclusions

Variation in enzyme concentration does not appear to play a primary role in the developmental regulation of UGPase in the potato tuber. During field growth, 20 genetically diverse potato clones each demonstrated a high UGPase activity, varying from 40 to 55 IU/g fresh weight of tuber (unpublished data). Given the high levels of UGPase in developing potatoes along with its kinetic properties (low K_m values for the co-substrates UDPG and PPi; high V_{max}, and the K_{eq}

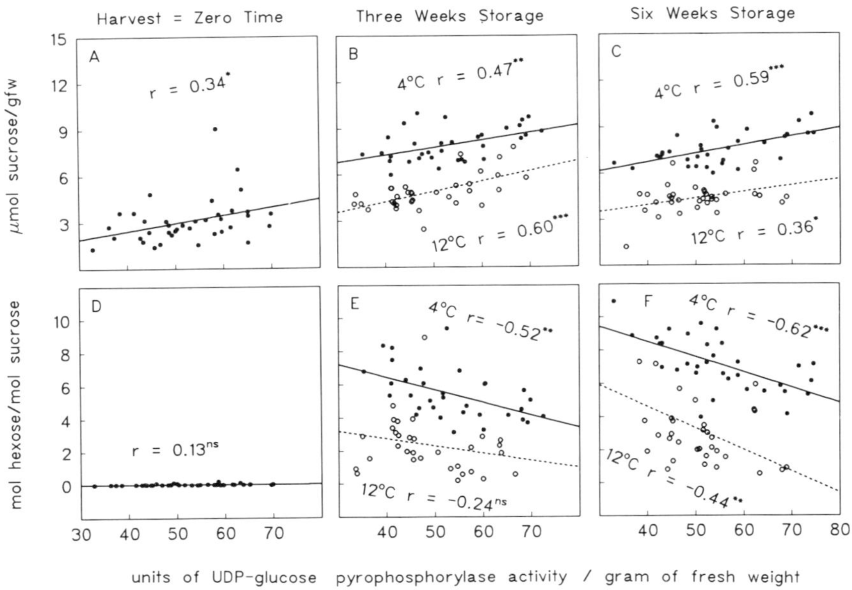

Figure 7.14. Plots of sucrose (panels A-C) and hexose:sucrose ratio (panels D-F) versus UGPase activity of potato tubers at harvest and after 1, 3 and 6 weeks storage at 4°C (•) and 12°C (o). The data points are values from individual tubers. Data were collected from three tubers of ten transformants and three wild-type tubers. The correlation coefficient (*r*) and the associated *P* value (* < 0.05, ** < 0.01, and *** < 0.001) are given for each regression. (Spychalla *et al.*, 1994.)

of the UGPase reaction in plants, which lies in the direction of Glc- 1-P synthesis (Turnquist and Hansen, 1973)), the conversion of UDPG to support starch synthesis and glycolytic reactions must occur rapidly in developing potato tubers (Sowokinos *et al.*, 1993). This interpretation is supported by the finding that potato plants expressing antisense RNA for UGPase with enzyme levels from 4 to 5% of wild-type showed no change in growth and development in relation to yield or starch content (Zrenner *et al.*, 1993).

Kinetic analysis suggests that UGPase may exert a greater influence on carbohydrate metabolism when the potato is in the sucroneogenic mode during storage, compared to tubers developing in the field. Kinetics in the direction of UDPG synthesis demonstrated a 'negative cooperative effect' in relation to the co-substrates Glc-1-P and UTP. It was suggested that enzymes with this property may 'cushion' the effect that environmental change would have on a metabolic process (Conway and Koshland, 1968). For example, when potatoes are subjected to cold stress and starch breakdown and hexose-phosphates increase in the cytosol, a decrease in the affinity of potato UGPase for Glc-1-P may help to slow gluconeogenesis and aid in shuttling carbons towards glycolysis and energy forming reactions (Fig. 7.4, reaction 1).

Altered sucrose levels in genetically engineered potato tubers along with biphasic kinetic patterns suggest that potato tuber UGPase may contribute greater control to the process of potato sweetening than would normally be expected for a near-equilibrium reaction. For example, phosphoglucoseisomerase (PGI) catalyzes a near-equilibrium reaction and has been thought to exist 'in excess' in plants and, therefore, is a poor candidate for regulation (Rolleston, 1972). However, the establishment of metabolic control coefficients for *Clarkia xantiana* mutants deficient in PGI, Neuhaus *et al.* (1989), demonstrated that small changes in PGI can indeed exert control over starch and sucrose metabolism. The genetic manipulation of glycolytic enzymes has supported views presented by Kacser (1987) that control is a property of the pathway as a whole and is shared by each of its component steps. It was clear that relatively small genetic changes in UGPase activity resulted in striking alterations in sucroneogenesis and the hexose:sucrose ratio in stored potato tubers (Fig. 7.14). Zrenner *et al.* (1993) demonstrated a remarkable increase in the amount of UGPase-specific RNA at 4°C compared with 8°C or 25°C and concluded that this implied a role for this enzyme in the process of 'cold-sweetening'. This supported the results of earlier studies that showed a significant correlation between glucose content and UGPase activity among genetically diverse potato clones (Sowokinos, 1990a). The question could be asked, did the reduced UGPase activity in potato tuber transformed with an antisense construct restrict the availability of substrates for SPS and thus regulate its catalytic activity? No discernible change in either UDPG or Fru-6-P content was noted when UGPase was reduced 30% (Spychalla *et al.*, 1994). One would expect, however, that with sufficient restriction of UGPase, the cytosolic levels of UDPG would be significantly reduced. This could markedly reduce the catalytic activity of SPS which requires high concentrations of this co-substrate to maintain a significant reaction rate (Fig. 7.4, reaction 2). Alternative interpretations may be that the concentration of an unidentified effector influencing the fine regulation of SPS may be affected, or that another enzyme(s) in the potato tuber had altered its expression in response to a reduction in UGPase. Such a coordinated expression of tuber carbohydrate metabolizing enzymes may indeed be possible as evidenced by the co-expression of sink enzyme RNA levels (Fig. 7.11).

UGPases purified from different cultivars possess different kinetic properties when measured in the direction of UDPG synthesis (Nakano *et al.*, 1989; Sowokinos *et al.*, 1993). Kazuta *et al.* (1991) identified five lysyl residues (Lys-263, Lys-329, Lys-367, Lys-409 and Lys-410) which are located at or near the UDPG(UTP)-binding site of potato tuber UGPase formed by the U4a cDNA. Modification of these residues with uridine di- and tri-phosphopyridoxals followed by reduction with sodium borohydride led to an inactivation of the enzyme. Site-directed mutagenesis experiments indicated that the lysyl residue at position 367 was found to be indispensable for the function of potato UGPase

(Katsube *et al.*, 1991). Coding of the UP2 cDNA substituted the amino acids Asn and Glu for two lysyl residues at positions 82 and 445, respectively (Spychalla *et al.*, 1994). Although these two lysyl substitutions are not at the active site, certain changes in three-dimensional configuration might indirectly lead to an alteration in kinetic properties.

Preliminary evidence with non-denaturing PAGE and kinetic analysis of tuber UGPase purified from cv. Norchip potatoes, suggested that this enzyme may exist as a dimer *in vitro*. This introduced the possibility that multimeric forms of the basic 53 kDa subunit might exist *in vivo*. Association of the basic monomeric subunit into higher polymeric forms during development and storage of potato tubers may permit this enzyme to respond to its environment and alter its catalytic properties. Additional research is necessary to determine whether: (i) multimeric forms of UGPase exist *in vivo* in potato tubers during growth and/or storage, (ii) additional genes are involved in affecting the association-dissociation of UGPase subunits, (iii) isoforms of UGPase exist in potato tubers, (iv) the kinetic and physicochemical properties of UGPase are similar between high and low sugar potato clones, and (v) temperature affects the biphasic kinetics of the UGPase reaction. In addition longer-term storage experiments (i.e. 6-8 months), with corresponding metabolite analysis, are necessary before a more thorough assessment can be made concerning the role of this enzyme in influencing the phenomenon of 'cold sweetening' in potato tubers.

Acknowledgements

The author wishes to thank P. McClean, A. Brovokov and G. Secor for sharing their data on the structure and organization of the UGPase gene. Appreciation is also extended to J. Spychalla who conducted the transformation, mRNA analysis and metabolite determination of potato tubers described in this chapter.

References

ap Rees, T. and Morrell, S. (1990) Carbohydrate metabolism in developing potatoes. *American Potato Journal* 67:835-847.

Barker, J. (1968) Studies in the respiratory and carbohydrate metabolism of plant tissues XXV. Changes in rate of CO2 output and in content of various phosphate compounds in potatoes. *New Phytology* 67:495-503.

Borovkov, A., McClean, P. and Secor G. (1993) Molecular cloning of the UDP-glucose pyrophosphorylase gene of potato. *Proceeedings of the 3rd International Symposium on the Molecular Biology of the Potato*. Santa Cruz, CA, p. 24 (Abstract).

Bruneau, J.M., Worrell, A.C., Cambou, B., Lando, D. and Voelker, T.A. (1991) Sucrose phosphate synthase, a key enzyme for sucrose biosynthesis in plants. *Plant Physiology* 96:473-478.

Claassen, P.A.M., Budde, M.A.W., de Ruyter, H.J., van Calker, M.H. and van Es, A. (1991) Potential role of pyrophosphate:fructose 6-phosphate phosphotransferase in carbohydrate metabolism of cold stored tubers of *Solanum tuberosum* cv Bintje. *Plant Physiology* 95:1243-1249.

Conway, A. and Koshland, D.E., Jr (1968) Negative cooperativity in enzyme action: the binding of diphosphopyridine nucleotide to glyceraldehyde 3-phosphate dehydrogenase. *Biochemistry* 7:4011-4022.

Dancer, J.E. and ap Rees, T. (1989) Phosphoribosyl pyrophosphate and the measurement of inorganic pyrophosphate in plant tissues. *Planta* 177:261-264.

Diamond, R.L., Farnsworth, P.A. and Loomis, W.F. (1976) Isolation and characterization of mutations affecting UDPG pyrophosphorylase activity in *Dictyostelium discoideum. Developmental Biology* 60:169-181.

Franke, J. and Sussman, M. (1971) Synthesis of uridine diphosphate glucose pyrophosphorylase during the development of *Dictyostelium discoideum. Journal of Biological Chemistry* 246:6381-6388.

Ginsburg, V. (1964) Sugar nucleotides and the synthesis of carbohydrates. *Advances in Enzymology* 26:35-88.

Goodwin, T.W. and Mercer, E.I. (1983) *Introduction to Plant Biochemistry.* Pergamon Press, Oxford, pp. 227-272.

Gustafson, G. and Gardner, J.E. (1972) Uridine diphosphate glucose pyrophosphorylase from *Sorghum vulgare*: purification and kinetic properties. *Journal of Biological Chemistry* 247:1387-1397.

Gustafson, G. and Wright, B.E. (1971) UDP-glucose pyrophosphorylase from the cellular slime mold. *Federation Proceedings, American Society of Experimental Biology* 30:1069.

Guy, C.L., Huber, J.L.A. and Huber, S.C. (1992) Sucrose phosphate synthase and sucrose accumulation at low temperature. *Plant Physiology* 100:502-508.

Hardy, P.J. and Norton, G. (1968) Sugar absorption and compartmentation in potato tuber slices. *New Phytology* 67:139-143.

Holdsworth, M.J., Grierson, C., Schuch, W. and Bevan, M. (1992) DNA-binding properties of cloned TATA-binding protein from potato tubers. *Plant Molecular Biology* 19:455-464.

Hondo, T., Hara, A. and Funaguma, T. (1983) The purification and some properties of the UDP-glucose pyrophosphorylase from the pollen of *Typhya latifolia* Linne. *Plant and Cell Physiology* 24:61-69.

Huber, S.C. and Huber, J.L.A. (1990) Regulation of spinach leaf sucrose-phosphate synthase by multisite phosphorylation. *Current Topics in Plant Biochemistry and Physiology* 9:329-343.

Isherwood, F.A. (1976) Mechanism of starch-sugar interconversion in *Solanum tuberosum. Phytochemistry* 15:33-41.

Jefferson, R., Goldsbrough, A. and Bevan, M. (1990) Transcriptional regulation of a patatin-1-gene in potato. *Plant Molecular Biology* 14:995-1006.

Jelitto, T., Sonnewald, U., Willmitzer, L., Hajirezeai, M. and Stitt, M. (1992) Inorganic pyrophosphate content and metabolites in potato and tobacco plants expressing *E. coli* pyrophosphatase in their cytosol. *Planta* 188:238-244.

Johnson, R. and Ryan, C.A. (1990) Wound-inducible potato inhibitor II genes: enhancement of expression by sucrose. *Plant Molecular Biology* 14:527-536.

Kacser, H. (1987) Control of metabolism. In: Davis, D.D. (ed.), *The Biochemistry of Plants*, Vol 12. Academic Press, New York, pp. 39-67.

Kacser, H. and Porteous, J.W. (1987) Control of metabolism: what do we have to measure? *Trends in Biochemical Science* 12:5-14.

Katsube, T., Kazuta, Y., Mori, H., Nakano, K., Tanizawa, K. and Fukui, T. (1990) UDP-glucose pyrophosphorylase from potato tuber: cDNA cloning and sequencing. *Journal of Biochemistry* 108:321-326.

Katsube, T., Kazuta, Y., Tanizawa, K. and Fukui, T. (1991) Expression in *Escherichia coli* of UDP-glucose pyrophosphorylase cDNA from potato tuber and functional assessment of the five lysyl residues located at the substrate-binding site. *Biochemistry* 30:8546-8551.

Kazuta, Y., Omura, Y., Tagaya, M., Nakano, K. and Fukui, T. (1991) Identification of lysyl residues located at the substrate-binding site in UDP-glucose pyrophosphorylase from potato tuber: affinity labeling with uridine di- and triphosphopyridoxals. *Biochemistry* 30:8541-8545.

Konishi, Y., Tanizawa, K., Muroya, S. and Fukui, T. (1991) Molecular cloning, nucleotide sequencing and affinity labeling of bovine liver UDP-glucose pyrophosphorylase. *Journal of Biochemistry* 114:61-68.

Kossmann, J., Visser, R.G.F., Müller-Röber, B., Willmitzer, L. and Sonnewald, U. (1991) Cloning and expression analysis of a potato cDNA than encodes branching enzyme: evidence for co-expression of starch biosynthetic genes. *Molecular and General Genetics* 230:39-44.

Lecharny, A. and Nguyen-Quoc, B. (1991) Sucrose synthase, ß-fructosidase and sucrose metabolism in leaves. *Plant Sciences Today* 16-18:107-108.

Mares, D.J., Sowokinos, J.R. and Hawker, J.S. (1985) Carbohydrate metabolism in developing potato tubers. In: Li, P.H. (ed.), *Potato Physiology*. Academic Press, New York, pp. 279-327.

Mohanty, B. and ap Rees, T. (1992) Demonstration and measurement of inorganic pyrophosphate in potato tubers. *Potato Research* 35:195-198.

Morrell, S. and ap Rees, T. (1986) Sugar metabolism in developing tubers of *Solanum tuberosum*. *Phytochemistry* 25:1579-1585.

Müller, B.T., Kossmann, J., Hannah, L.C., Willmitzer, L. and Sonnewald, U. (1990) One of two different ADP-glucose pyrophosphorylase genes from potato responds to elevated levels of sucrose. *Molecular and General Genetics* 224:136-146.

Murata, T. (1972) Sucrose phosphate synthase from various plant origins. *Agricultural Biological Chemistry* 36:1877-1884

Murphy, G.J.P. and Ward, E.S. (1990) In Howe, C.J. and Ward, E.S. (eds), *Nucleic Acid Sequencing: A Practical Approach*, IRL Press, Oxford, pp. 99-115.

Nakae, T. (1971) Multiple molecular forms of uridine diphosphate glucose pyrophosphorylase from *Salmonella typhimurium*. III. Interconversion of the various forms. *Journal of Biological Chemistry* 246:4404-4411.

Nakae, T. and Nikaido, H. (1971) Multiple molecular forms of uridine diphosphate glucose pyrophosphorylase from *Salmonella typhimurium*. I.

Catalytic properties of various forms. *Journal of Biological Chemistry* 246:4386-4396.

Nakano, K., Omura, Y., Tagaya, M. and Fukui, T. (1989) UDP-glucose pyrophosphorylase from potato tuber: purification and characterization. *Journal of Biochemistry* 106:528-532.

Neuhaus, H.E., Kruckeberg, A. L., Feil, R., and Stitt, M. (1989) Reduced-activity mutants of phosphoglucose isomerase in the cytosol and chloroplast of *Clarkia xantiana* II. Study of the mechanisms which regulate photosynthate partitioning. *Planta* 178:110-112.

Oparka, K.J. and Prior, D.A.M. (1987) 14C sucrose efflux from the perimedulla of growing potato tubers. *Plant Cell Environment* 10:667-675.

Peng, H.L. and Chang H.Y. (1993) Cloning of a human liver UDP-glucose pyrophosphorylase cDNA by complementation of the bacterial galU mutation. *Federation of European Biochemical Sopcieties Letters* 329:163-168.

Ragheb, J.A. and Dottin, R.P. (1987) Structure and sequence of UDP-glucose pyrophosphorylase gene of *Dictyostelium discoideum*. *Nucleic Acids Research* 15:3891-3906.

Roberts, J.K.M. (1990) Observation of uridine triphosphate:glucose-1-phosphate uridylyltransferase activity in maize root tips by saturation transfer 31P-NMR. Estimation of cytoplasmic PPi. *Biochimica et Biophysica Acta* 1051:29-36.

Rocha-Sosa, M., Sonnewald, U., Frommer, W., Stratmann, M., Schell, J. and Willmitzer, L. (1989) Both developmental and metabolic signals activate the promoter of a class I patatin gene. *EMBO Journal* 8:23-29.

Rolleston, F.S. (1972) A theoretical study to the use of measured intermediates in the study of the control of intermediary metabolism. *Current Topics in Cellular Regulation* 5:47-75.

Salanoubat, M. and Belliard, G. (1989) The steady-state level of potato sucrose synthase mRNA is dependent on wounding, anaerobiosis and sucrose concentration. *Gene* 84:181-185.

Sicher, R.C. (1986) Sucrose biosynthesis in photosynthetic tissue: rate-controlling factors and metabolic pathway. *Physiologia Plantarum* 67:118-121

Sowokinos, J.R. (1990a) Stress-induced alterations in carbohydrate metabolism. In Vayda, M. and Park, W. (eds), *Molecular and Cellular Biology of the Potato*. CAB International, Wallingford, UK, pp. 137-158.

Sowokinos, J.R. (1990b) Effect of stress and senescence on carbon partitioning in stored potatoes. *American Potato Journal* 67:873-881.

Sowokinos, J.R., Spychalla, J.P. and Desborough, S.L. (1993) Pyrophosphorylases in *Solanum tuberosum* IV. Purification, tissue localization, and physicochemical properties of UDP-glucose pyrophosphorylase. *Plant Physiology* 101:1073-1080.

Spychalla, J.P. and Bevan, M.W. (1993) *Agrobacterium*-mediated transformation of potato stem and tuber tissue, regeneration and PCR screening for transformation. In Lindsay, K. (ed), *Plant Tissue Culture Manual: Fundamentals and Applications*, Vol. B11, Kluwer Academic Publishers, Dordrecht, pp. 1-18.

Spychalla, J.P., Scheffler, B.E., Sowokinos, J.R. and Bevan, M.W. (1994) Cloning, antisense RNA inhibition, and the coordinated expression of UDP-glucose pyrophosphorylase with starch biosynthetic genes in potato tubers. *Journal of Plant Physiology* (In press).

Sung, S.S., Xu, D-P. and Black, C.C. (1989) Identification of actively filling sucrose sinks. *Plant Physiology* 89:1117-1121.

Tovey, K.C. and Roberts, R.M. (1970) Studies of the pyrophosphorylases of wheat. *Plant Physiology* 46:406-411.

Tsai, C.Y., Salamini, F. and Nelson O.E. (1970) Enzymes of carbohydrate metabolism in the developing endosperm of maize. *Plant Physiology* 46:299-306.

Tsuboi, K.K., Fukunaga, K. and Petricciani, J.C. (1969) Purification and specific kinetic properties of erythrocyte uridine diphosphate glucose pyrophosphorylase. *Journal of Biological Chemistry* 244:1008-1015.

Turner, J.F. (1969) Physiology of pea fruits VI. Changes in uridine diphosphate glucose pyrophosphorylase and adenosine diphosphate glucose pyrophosphorylase in the developing seed. *Australian Journal of Biological Science* 22:1145-1151.

Turnquist, R.L. and Hansen, R.G. (1973) Uridine diphosphoryl glucose pyrophosphorylase. In: Boyer, P.D. (ed.), *The Enzymes*, 3rd edn, Vol. VIII. Academic Press, New York, pp. 51-71.

Vella, J. and Copeland, L. (1990) UDP-glucose pyrophosphorylase from the plant fraction of nitrogen-fixing soybean nodules. *Physiologia Plantarum* 78:140-146.

Viola, R., Davies, H.V. and Chudeck, A.R. (1991) Pathways of starch and sucrose biosynthesis in developing tubers of potato (*Solanum tuberosum* L.) and seeds of faba bean (*Vicia faba* L.): elucidation by ^{13}C-nuclear-magnetic-resonance spectroscopy. *Planta* 183:202-208.

Weiner, H., Stitt, M. and Heldt, H.W. (1987) Subcellular compartmentation of pyrophosphate and alkaline pyrophosphatase in leaves. *Biochimica et Biophysica Acta* 893:13-21.

Wenzler, H., Mignery, G., Fisher, L. and Park, W.D. (1989) Analysis of a chimeric class-1 patatin-GUS gene in transgenic potato plants: high level of expression in tubers and sucrose-inducible expression in cultured leaf and stem explants. *Plant Molecular Biology* 12:41-50.

Wright, B.E. (1968) An analysis of metabolism underlying differentiation in *Dictyostelium discoideum*. *Journal of Cellular Physiology* 72:145-160.

Wright, B.E. (1970) The use of kinetic models to analyze differentiation. *Behavioral Science* 15:37-45.

Zrenner, R., Willmitzer, L. and Sonnewald, U. (1993) Analysis of the expression of potato uridinediphosphate-glucose pyrophosphorylase and its inhibition by antisense RNA. *Planta* 190:247-252.

Chapter Eight:

Post-Harvest Potato Tuber Glycerolipid Metabolism

James P. Spychalla
Spychalla Farms, Inc., Antigo, WI, USA.

The post-harvest potato tuber undergoes a number of physiological transitions and is exposed to a number of environmental pressures. At harvest the potato tuber shifts from a sink to a source tissue; cell division and enlargement cease; peridermal and wound suberization increase; and the tuber experiences a pulse of increased respiration. In storage, the tuber is then subject to low temperatures, desiccation, microbial pathogens, physical compression, non-ideal atmospheres, sprout-inhibiting xenobiotics and aging (Burton, 1989). Throughout this post-harvest period, one dynamic aspect of the tuber is its membrane glycerolipid metabolism. In the starch-storing tuber, lipids account for less than 0.1% of the tuber's fresh weight (Lepage, 1964; Galliard, 1968) and thus lipid metabolism *per se* might appear to be of little importance. However, the significance of tuber glycerolipid metabolism and membrane composition is realized in other aspects of metabolism. Membrane lipid composition influences the tuber's carbohydrate metabolism, tolerance to low temperatures and vigour as vegetative seed, all of which are of economic importance to growers and processors.

Briefly, plant glycerolipid synthesis originates in the plastid, where fatty acid synthase produces 16:0-ACP which may then be elongated to 18:0-ACP (Figure 8.1) (see legends to Tables 8.1 and 8.2 and Figure 8.1 for abbreviations) (Shimakata and Stumpf, 1982). The 18:0-ACP (but not 16:0-ACP) can be directly desaturated in the plastid stroma to form 18:1-ACP (McKeon and Stumpf, 1982; Shanklin and Somerville, 1991). The fatty acyl-ACP molecules may remain in the plastid or be converted to fatty acyl-CoA equivalents localized external to the plastid. In both the plastid and the endoplasmic reticulum, the acyl chains are then transesterified to the glycerol moiety of a nascent glycerolipid. Further desaturations and polar head group modifications may occur after the fatty acids have been incorporated into glycerolipids. These final glycerolipid synthetic reactions which occur in the plastid are termed the 'prokaryotic' pathway, while those in the endoplasmic reticulum are termed the 'eukaryotic' pathway (Roughan and Slack, 1982; Browse and Somerville, 1991).

The Molecular and Cellular Biology of the Potato, Second Edition
Edited by W.R. Belknap, M.E. Vayda and W.D. Park

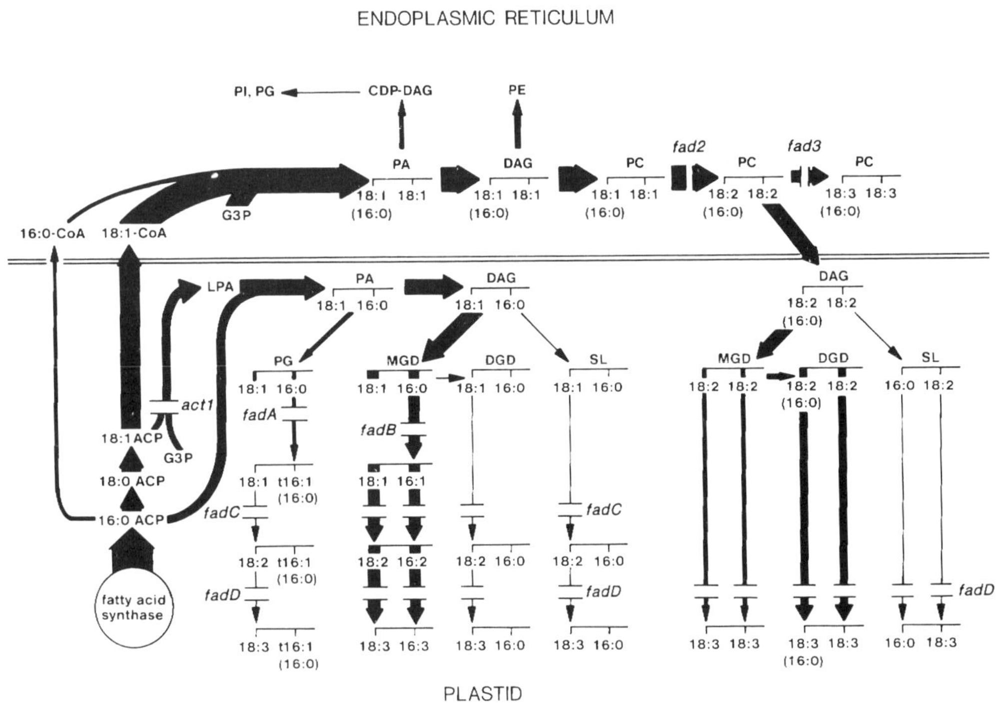

Figure 8.1. The major pathways of glycerolipid synthesis in leaf cells of *Arabidopsis*. Reprinted, with permission, from Somerville and Browse, 1991. The width of the arrows indicates the flux through the various steps. The breaks in the pathway represent the enzymatic deficiencies in various mutants of *Arabidopsis*. Abbreviations: ACP, acyl-carrier protein; CDP-DAG, cytidine diphosphate-DAG; CoA, coenzyme A; DAG, diacylglycerol; DGD, digalactosyldiacylglycerol; G3P, glycerol 3-phosphate; LPA, lysophosphatidic acid; MGD, monogalactosyldiacylglycerol; PA, phosphatidic acid; PC, phosphatidylcholine; PE, phosphatidylethanolamine; PG, phosphatidylglycerol; PI, phosphatidylinositol; SL, sulfolipid.

Plant glycerolipid synthesis has been reviewed by several authors (Harwood, 1988; Browse and Somerville, 1991; Ohlrogge *et al.*, 1991; Somerville and Browse, 1991; Slabas and Fawcett, 1992), and the enzymatic and free radical based mechanisms of glycerolipid catabolism have been discussed (Harwood and Russell, 1984; Kindl, 1984; Halliwell and Gutteridge, 1985; Lesham *et al.*, 1986; Thompson, 1988). Therefore this chapter will focus exclusively on potato tuber-specific aspects of membrane lipid metabolism, their physiological consequences and the prospects for transgenic improvement of the potato crop.

Potato Tuber Membranes and Their Glycerolipid Compositions

The glycerolipid and fatty acid compositions of the potato tuber and its membrane fractions have been fairly well characterized (Tables 8.1 and 8.2). As stated earlier, lipids represent less than 0.1% of the fresh weight of a potato tuber (Lepage, 1964; Galliard, 1968). Greater than 90% of tuber lipids are in forms containing esterified fatty acids, of which greater than 70% are polyunsaturated fatty acids, i.e. 18:2 and 18:3 (Galliard, 1968). The data in Tables 8.1 and 8.2 should be viewed with the understanding that it is difficult to obtain pure membrane preparations and membrane preparations isolated free from lipolytic activity. For example, PE is found in all cytoplasmic membranes isolated so far, but is considered a negative marker for purified intact chloroplasts (Dubacq and Kader, 1978). Thus the PE detected in both of the amyloplast preparations listed in Table 8.1, likely indicates cross-contamination. Nonetheless, the data in Table 8.1 are still quite informative. For example, it can be seen that the mitochondria contain only trace amounts of MGD and DGD. On the other hand, appreciable amounts of both MGD and DGD can be found in the amyloplast membranes. Pun *et al.* (1980) estimated that amyloplast membranes consist of 7.5% MGD and 5.6% DGD (% total lipids). By comparison, Fishwick and Wright (1980) obtained estimates of 13.7% MGD and 45% DGD (% of polar lipids) in amyloplast membranes. The same authors have also determined that amyloplast membranes contribute 67% of the MGD, 100% of the DGD, 94% of the SL and 89% of the PG in a tuber (Fishwick and Wright, 1980). Another limitation of the data in Table 8.1 is the fact that specific tuber membrane glycerolipid compositions are quite variable. Time-, temperature- and cultivar-dependent differences have all been documented in tuber glycerolipid compositions (Berkeley and Galliard, 1974; Jarvis *et al.*, 1974).

Plant cell membranes, including potato, have a capacity to share or mix their glycerolipids *in vitro* (Demandre *et al.*, 1978) and *in vivo* (Browse and Somerville, 1991) thus masking their origins. However, through positional analysis of their component fatty acyl chains, their origins can be discerned. Specifically, glycerolipids synthesized in the plastid contain C16 fatty acids in the *sn*-2 position, while those containing C18 in the *sn*-2 position are diagnostic

Table 8.1. Glycerolipid content of potato tubers and tuber membrane fractions[1].

Glycerolipid source	PC	PE	DPG	PA	PI	PS	PG	MGD	DGD	SL	TG	FFA	Reference
	mol percent of total lipids												
Cell Wall	3.3	3.1	-	-	3.0	-	-	4.6	1.9	-	9.4	9.1	Pun *et al.*,1980
Plasmalemma	12.6	17.5	1.1	-	7.2	-	-	17.1	12.5	-	-	-	Demandre,1976[2]
Plasmalemma	21.1	8.0	2.2	-	3.6	-	-	9.2	11.6	-	1.4	3.2	Pun *et al.*,1980
Plasmalemma[3]	37.0	41.0	2.0	-	18.0	-	-		2.0		-	-	Harwood and Russell,1984
Mitochondria (total)	36.0	25.6	11.2	-	6.4	-	-	1.9	4. 3	-	1.2	2.1	Pun *et al.*,1980
Mitochondria (total)	44.3	26.3	16.8	-		12.7		-	-	-	-	-	Meunier and Mazliak,1972
Mitochondria (outer)	52.6	25.0	12.1	-		10.3		-	-	-	-	-	Meunier and Mazliak,1972
Mitochondria (inner)	27.0	29.0	19.5	-		24.5		-	-	-	-	-	Meunier and Mazliak,1972
Mitochondria (outer)[3]	54.0	30.0	-	-	11.0	-	5.0		trace		-	-	Harwood and Russell,1984
Mitochondria (inner)[3]	41.0	37.0	15.0	-	5.0	-	3.0		trace		-	-	Harwood and Russell,1984
Microsomes	39.0	14.4	0.6	1.2	6.2	0.6	-	-	-		-	-	Ben Abdelkader, 1972[2]
Microsomes	19.5	13.0	1.0	-	5.9	-	-	10.9	15.3	-	3.2	4.7	Pun*et al.*, 1980
Microsomes[3]	44.0	32.0	1.0	-	15.0	2.0	2.0		6.0		-	-	Harwood and Russell,1984
Peroxisomes	16.3	41.0	-	-	10.9	-	-	10.0	trace	trace	-	-	Tchang, 1977[2]
Peroxisomes	7.3	2.9	0.4	-	1.0	-	2.3	7.5	8.1	-	6.4	8.3	Pun *et al.*,1980
Amyloplasts	11.9	2.5	-	-	1.9	-	-	7.5	5.6	-	11.9	12.5	Pun *et al.*,1980
Mature tuber	18.9	14.6	-	-	12.0	-	-	7.2	7.1	-	-	-	Lepage, 1964
Stored tuber	25.8	12.6	0.6	0.3	6.3	0.6	0.9	5.7	14. 2	1.3	15 . 4	-	Galliard, 1968
	mol percent of polar lipids												
Amyloplasts	16.4	5.0	16.8	-	5.3	-	2.4	13.7	45.0	5.3	5.6	12.5	Fishwick and Wright, 1980
Developing tuber	24.7	21.1	-	-	11.2	-	-	7.4	18.4	-	-	-	Berkeley and Galliard, 1974
Stored tuber	20.4	22.4	-	-	12.2	-	-	8.4	17.3	-	-	-	Berkeley and Galliard, 1974

[1]Missing values indicate that a lipid content was not reported. Abbreviations: PC, phosphatidylcholine; PE, phosphatidylethanolamine; DPG, diphosphatidylglycerol; PA, phosphatidic acid; PI, phosphatidylinositol; PS, phosphatidylserine; PG, phosphatidylglycerol; MGD, monogalactosyldiacylglycerol; DGD, digalactosyldiacylglycerol; SL, sulfolipid; TG, triacylglycerol; and FFA, free fatty acid. Boxes indicate combined values.

[2]PhD thesis cited in Mazliak (1983).

[3]Expressed as the percent of total acyl lipids.

Table 8.2. Fatty acid content of potato tubers and tuber glycerolipid fractions[1].

Fatty acid source and reference	mol percent						
	16:0	18:0	18:1	18:2	18:3	18:3/18:2	DB[2]
Tuber (Galliard, 1968)							
Total FA	19.7	2.8	3.5	59.8	15.2	0.25	7.5
PC	17.0	2.8	4.7	67.1	8.4	0.12	8.3
PE	18.7	3.3	5.8	64.3	7.9	0.12	7.2
DPG	9.7	6.6	9.7	57.8	16.2	0.28	10.7
PI	43.4	2.8	2.2	43.4	8.2	0.19	2.4
PG	47.7	4.8	6.0	38.0	3.5	0.09	1.8
MGD	0.8	nd	0.9	58.0	40.3	0.69	297.2
DGD	14.1	6.3	7.7	46.5	25.4	0.55	8.7
Stored tuber (Spychalla, 1989)							
Total FA 9°C	20.9	6.1	1.7	44.2	27.0	0.61	6.3
Tolal FA 3°C	21.5	5.6	1.6	39.1	32.2	0.83	6.5
Amyloplasts (Fishwick and Wright, 1980)							
PC	21.2	3.6	1.4	58.0	15.5	0.27	6.6
PE	21.0	3.4	1.6	58.5	14.5	0.25	6.7
PI	38.4	3.8	1.4	40.7	14.5	0.36	3.0
PG	47.4	3.9	2.5	33.4	11.8	0.35	2.0
MGD	1.6	0.7	0.6	73.1	23.5	0.32	94.5
DGD	6.3	5.1	1.5	68.9	18.0	0.26	16.9
Mitochondria (Pun *et al.*, 1980)							
Total FA	19.4	4.7	2.4	59.6	12.1	0.20	6.5
PC	17.5	4.0	4.8	60.3	13.3	0.22	7.7
PE	14.8	2.9	5.1	66.3	10.9	0.16	9.6
DPG	5.1	3.0	6.2	66.4	18.7	0.28	24.1
PI	37.2	10.1	3.4	37.3	12.0	0.32	2.4
Microsomes (Pun *et al.*, 1980)							
Total FA	24.7	6.8	2.0	49.6	15.4	0.31	4.7
PC	25.4	3.8	1.4	59.0	10.4	0.18	5.1
PE	27.3	5.2	1.1	58.4	8.0	0.14	4.4
DPG	30.0	8.9	0.8	45.3	15.0	0.33	3.5
PI	40.2	4.7	1.9	46.0	7.2	0.16	2.6
MGD	7.3	1.8	1.1	62.0	27.8	0.45	22.9
DGD	20.6	4.5	2.3	59.8	12.8	0.21	6.4

[1]Abbreviations: nd, not detected; DBI, double bond index; FA fatty acid; PC, phosphatidylcholine; PE, phosphatidylethanolamine; DPG, diphosphatidylglycerol; PI, phosphatidylinositol; PG, phosphatidylglycerol; MGD, monogalactosyldiacylglycerol; and DGD, digalactosyldiacylglycerol.

[2]DBI = [mol% 18:1 + (2 x mol% 18:2) + (3 x mol% 18:3)]/ [mol% 16:0 + mol% 18:0]

of synthesis in the endoplasmic reticulum (Figure 8.1) (Roughan and Slack, 1982). An analysis of tuber PC and PI found only C16/C18 or C18/C18 (*sn*-1/*sn*-2) molecular species (Demandre *et al.*, 1986; Justin *et al.*, 1989). The absence of C18/C16 molecular species indicates synthesis exclusively via the tuber's eukaryotic pathway. In a non-positional analysis of tuber MGD, only C18/C18 molecular species and a trace of C16/C18 were found (Demandre *et al.*, 1985). Other non-positional studies have found 0.8% 16:0 in tuber MGD and 1.6% 16:0 in amyloplast MGD (Table 8.2). The low levels of C16 fatty acids in the amyloplastic MGD would indicate that endoplasmic reticulum PC is the primary contributor to MGD synthesis even in the plastid (Figure 8.1). Curiously, the tuber's DGD which is synthesized from MGD (Figure 8.1) contains 14.1% 16:0, which is several-fold more 16:0 than that contained in MGD (Table 8.2). It is likely that the C18/C16 (*sn*-1/*sn*-2) species of MGD synthesized in the plastid is preferentially converted to DGD. In the case of *Arabidopsis* root, it is estimated that the endoplasmic reticulum 18:2 desaturase accounts for over 80% of the 18:3 present (Browse *et al.*, 1993). Given that *Arabidopsis* root and potato tubers are both non-chlorophyllous tissues, perhaps the eukaryotic desaturation pathway also predominates in tubers.

The synthesis of glycerolipids is eventually followed by their catabolism. Through radiolabeling studies, half-lives of total phospholipids from stored potato tubers were estimated to be: whole tubers at 11.2 days; mitochondria at 10.4 days; microsomes at 32 days; and the phospholipids of a cytosolic supernatant at 5.2 days (Ben Abdelkader and Mazliak, 1971). In addition, over the course of a 10 day labeling study, whole tuber 18:0 was found to be metabolically stable, while 18:1 had a rapidly decreasing specific activity, and the polyunsaturates 18:2 and 18:3 had specific activities which increased. These long half-lives suggest that tuber membrane phospholipid composition does not adjust quickly in response to environmental changes. An example would be the 4 months necessary for the tuber's increasing 18:3/18:2 fatty acid ratio to plateau when tubers are placed into post-harvest storage at 3°C or 9°C (Figure 8.2A).

Fatty acid analysis of potato tuber glycerolipids has shown that 16:0, 18:2 and 18:3 are the predominant fatty acids in all tuber glycerolipid species (Table 8.2). The tuber's DPG, MGD and DGD have high degrees of fatty acid unsaturation while PG has the lowest as measured by the double bond index (DBI, defined in legend to Table 8.2). A fatty acid which is conspicuous by its absence in Table 8.1 is 16:3, which is a product of the prokaryotic pathway (Figure 8.1). Some angiosperms (including the potato and *Arabidopsis*) have substantial amounts of 16:3 in their leaf lipids (16:3 plants) while others do not (18:3 plants) (Jamieson and Reid, 1971). These authors found that potato contained 8.6% of its leaf fatty acids as 16:3. In tomato, a related Solanaceous species, practically all leaf 16:3 is found in MGD (Demandre *et al.*, 1985) which is the most abundant chloroplast thylakoid membrane lipid (Gounaris and Barber, 1983). Thylakoid membranes are normally absent from a tuber, but upon greening, thylakoid-like granal stacks can be induced to form in tuber plastids (Sandelius and Liljenberg, 1982). Thus potato fatty acid composition depends upon the tissue and the physiological state of that tissue. There is also considerable time-, temperature- and cultivar-dependent variation in tuber fatty

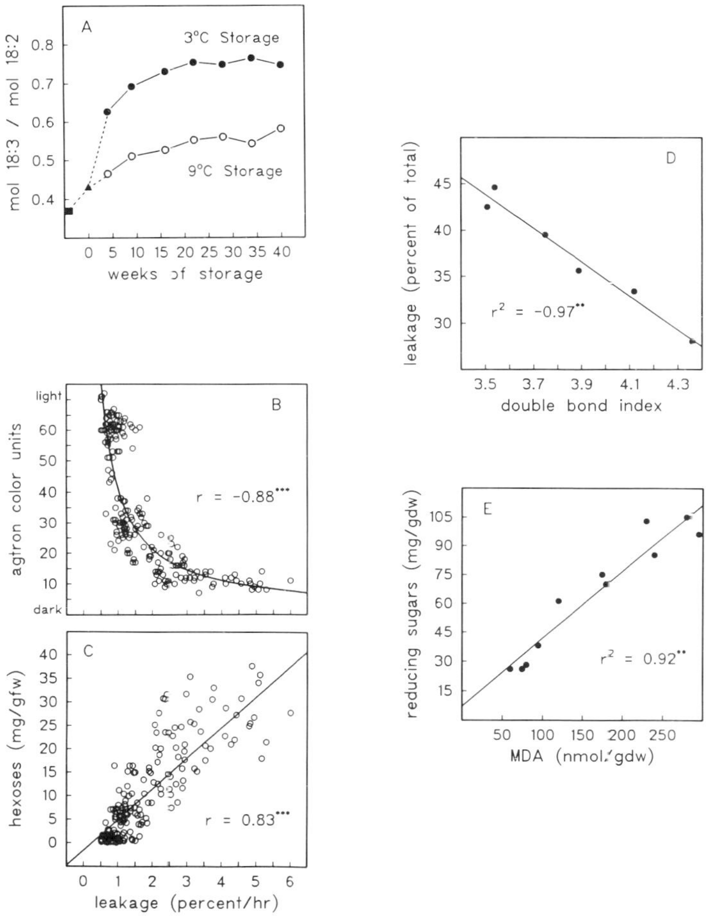

Figure 8.2. (A) Changes during storage in total linolenic acid (18:3) versus linoleic acid (18:2) content. Correlations of (B) potato chip (crisp) color measured as agtron units (i.e. units of relative light reflectance) versus rates of tuber membrane electrolyte leakage; (C) tuber hexose content (glucose + fructose) versus rates of tuber membrane electrolyte leakage; (D) tuber membrane electrolyte leakage versus tuber membrane glycerolipid double bond index (DBI); (D) tuber reducing sugar content versus tuber malondialdehyde (MDA) content. The data in panels A-E were replotted and reprinted with permission from Spychalla and Desborough (1990) (panels A-C), Knowles and Knowles (1989) (panel D) and Kumar and Knowles (1993) (panel E).

acid composition, particularly in the post-harvest situation (Figure 8.2A) (Lepage, 1964; Galliard, 1973; Jarvis *et al.*, 1974; Galliard *et al.*, 1975; Spychalla and Desborough, 1990a).

Post-Harvest Alterations to Glycerolipids and Membrane Function

Cellular compartmentation depends upon the integrity of membranes. Several instances of reduced post-harvest tuber membrane integrity have been documented. During post-harvest storage at 4°C, amyloplast membranes have been observed by electron microscopy to disintegrate (Ohad *et al.*, 1971). However, other reports have found that the amyloplast membrane remains intact, but develops an apparent deteriorated condition when stored at 5°C (Isherwood, 1976) or 9°C (Sowokinos *et al.*, 1987). Reduced potato tuber membrane integrity as measured by increased membrane electrolyte leakage has been implicated in the accumulation of sugars during low temperature storage (Figure 8.2B and C) (Isherwood and Kennedy, 1975; Isherwood, 1976; Workman *et al.*, 1976; Spychalla and Desborough, 1990a), extended storage (Isherwood, 1976) and stem-end sweetening of tubers (Shekar *et al.*, 1979). The free sugar content of a tuber is important because if the reducing sugars (glucose and fructose) accumulate to greater than 1-2 mg/g fresh weight, the suitability of the tubers for processed potato chips (crisps) and french fries will be reduced (Burton, 1989). In the above instances, a change in either the physical status or the chemical composition of the tuber membranes must account for the increased electrolyte leakage.

One of the most consistent and dramatic chemical changes to tuber membrane lipids induced by low temperature storage is an altered glycerolipid composition. During low temperature storage (4 or 5°C), an average 13% decline in PC, 27% increase in MGD and 29% increase in DGD occurred compared with tubers stored at warm temperatures (13 or 18°C) (Jarvis *et al.*, 1974; Galliard *et al.*, 1975). The MGD and DGD content of tuber lipids also increases with time. Tubers stored at 6°C were found to increase MGD and DGD through 2 months of storage at 6°C (Liljenberg *et al.*, 1978). The significance of these changes is unknown but may relate to the maintenance of membrane fluidity. In addition, the component fatty acids of these and other tuber glycerolipids become more highly unsaturated with time during low temperature storage.

The introduction of double bonds into membrane fatty acids dramatically alters the gel to liquid-crystalline phase transition temperature (T_C) of a membrane. For example, the T_C of a synthetic membrane composed exclusively of distearyl-PC is 54°C. The introduction of a single double bond into each acyl chain produces a dioleoyl-PC membrane which now has a T_C of -20°C (Cullis and Hope, 1985). Such increases in membrane lipid unsaturation during low temperature exposure have been interpreted as a beneficial acclimation response of biological membranes to maintain a constant fluidity (Sinensky, 1974) and presumably normal membrane function.

Analysis of mutant *Arabidopsis* lines has shown that normal plant function and development at low temperatures requires polyunsaturated fatty acids (Hughly and Somerville, 1992; Miquel *et al.*, 1993). Many plant tissues including potato increase the degree of unsaturation of their membrane fatty acids in response to low temperatures (Mazliak, 1979). Indeed, the observation that tubers induce oleoyl-PC desaturase activity (Bonnerot and Mazliak, 1984) and a higher degree of glycerolipid fatty acid unsaturation over time and during low temperature storage has been consistently demonstrated (Cherif and Ben Abdelkader, 1970; Cherif, 1973; Galliard *et al.*, 1975; Spychalla and Desborough, 1990a). An example of the kinetics of the low temperature dependent increase in fatty acid unsaturation (18:3/18:2 ratio) can be seen in Figure 8.2A. While the 18:3/18:2 ratio increased (Figure 8.2A) the sum of the two fatty acids remained constant throughout the storage experiment (Spychalla and Desborough, 1990a).

In an extended storage experiment, Knowles and Knowles (1989) found a high correlation ($r2 = -0.97$) between the DBI and membrane electrolyte leakage of cv. Russet Burbank tubers stored for 32 months at 4°C (Figure 8.2D). This increase in unsaturation is one chemical change which appears to confer resistance to increased rates of membrane electrolyte leakage in the stored tuber. In a ten month storage experiment, cultivars with the highest degree of membrane fatty acid desaturation were found to have the lowest rates of membrane electrolyte leakage and the lowest reducing sugar contents during long term cold storage (Spychalla and Desborough, 1990a). The degree of amyloplast membrane desaturation appears to be of particular importance in regards to low temperature induced sweetening. Tubers from the cv. Norchip are susceptible to low temperature sweetening while tubers from the numbered selection ND860-2 are resistant. Amyloplast membranes isolated from tubers of ND860-2 stored at 4°C are characterized by a higher fluidity, a lower T_C (27°C versus 35°C) and less degradation of unsaturated fatty acids compared to amyloplast membranes from the cv. Norchip (O'Donoghue *et al.*, 1993). Thus in potato tubers and their amyloplast membranes in particular, it appears that increased membrane fatty acid desaturation is necessary to mitigate the effects of low temperature storage.

In the case of potato leaves, a comparison between leaves of a species which acclimates to low temperatures (*S. commersonii*) and leaves of a potato species which does not acclimate (*S. tuberosum*) was performed. Increased PE, phospholipid:sterol ratio and 18:2 were all associated with cold acclimation in the *S. commersonii* (Weiss *et al.*, 1993). Analogous cold acclimation data from potato tuber is very limited. In two studies of PE during storage, the results were equivocal. The cv. Majestic had an approximately 13% higher PE content after 24 weeks of storage at 5°C than at 20°C, while the cv. Bintje PE responded with a 3% decrease in PE content (Galliard *et al.*, 1975). In a second study, the cv. Majestic again had 13% higher PE content after storage for 26 days at 4°C than at 18°C, the cv. Record responded with a 19% increase in PE content while the cultivars Desirée and Pentland Ivory both showed a 4% decrease in PE (Jarvis *et al.*, 1974). The observation that increased 18:2 was associated with the low temperature acclimating *S. commersonii* leaves appears inconsistent with the beneficial effects ascribed to a high tuber fatty acid DBI or 18:3/18:2 ratio in the previous paragraph. These mixed results and the lack of data on phospholipid:sterol ratios during storage of tubers prevents any analogies being drawn to cold acclimation in potato leaves.

Another aspect of post-harvest quality regards the performance of tubers as vegetative 'seed' in the subsequent growing season. Again, membrane glycerolipid composition and membrane integrity are important. In a study of 7-month-old seed tubers versus 19-month-old seed tubers, a component of vigorous seedling establishment was attributed to the maintenance of a high DBI and a high phospholipid content in the mother tuber (Knowles and Knowles, 1990).

Long-Term Storage and Membrane Degradation

All plant tissues age and eventually senesce. Tubers aged 7-10 months in storage are characterized by increased reducing sugar contents or as having undergone 'senescent sweetening' (Burton, 1989). Senescent sweetening may be associated with a deterioration in tuber membrane function. One of the most damaging phenomena that can occur to a senescing biological membrane is peroxidation of its component fatty acyl chains, resulting in a lowered membrane fluidity (Halliwell and Gutteridge, 1985). A decomposition product of peroxidized fatty acids is malondialdehyde (MDA) (Frankel and Neff, 1983). Estimates of MDA content in crude extracts can be based upon its reactivity with thiobarbituic acid (TBA) (Halliwell and Gutteridge, 1985). TBA-based measurements of MDA in potato indicated a cultivar-dependent 2-3-fold increase in tuber MDA production during the first 7 months of storage at 8°C in six cultivars (Lojkowska and Holubowska, 1989). Such increases in MDA- or TBA-reactive substances are commonly interpreted as an indication of increased glycerolipid oxidation. In another storage experiment, MDA content was found to be correlated ($r^2 = 0.94$) with the reducing sugar content of tubers stored for 35 months at 4°C (Figure 8.2E) (Kumar and Knowles, 1993). The correlation between MDA and reducing sugar content again indicates an interaction between tuber membrane composition and post-harvest carbohydrate metabolism. The same authors also studied several tuber antioxidants and present a strong argument that a gradual build-up of free radicals in stored tubers leads to peroxidative damage of tuber membrane lipids (Kumar and Knowles, 1993).

Plants possess a number of enzymatic and low molecular weight antioxidants which protect against the cytotoxic free radical species which can initiate membrane oxidation. Tuber antioxidants include superoxide dismutase (SOD), catalase (CAT), peroxidase (POD) and the membrane bilayer localized antioxidant α-tocopherol. These antioxidants are inducible under conditions of oxidative stress. SOD, CAT and α-tocopherol were all found to increase approximately 3-fold during 40 weeks of storage at both 3°C and 9°C (Spychalla and Desborough, 1990b). In a similar experiment the levels of SOD, CAT and POD were found to be approximately 2-fold higher in tubers stored for 20 months versus tubers stored for 8 months (Kumar and Knowles, 1993). It appears that during extended storage when tubers experience increased free radical mediated damage to membrane glycerolipids, they respond with an induction of antioxidants. There are also cultivar differences in antioxidant content. It is noteworthy that the cultivars Monona and Norchip which have low rates of membrane electrolyte leakage and low reducing sugar contents relative to the cultivars Red Pontiac and Russet Norgold, also have higher levels of SOD and CAT (Spychalla and Desborough, 1990b). Perhaps tuber membrane composition and function can be influenced by the antioxidative capacity of the stored tuber.

Transgenic Routes to Enhanced Chilling Tolerance

Transgenic manipulation of membrane glycerolipid composition has already succeeded in enhancing the chilling tolerance of tobacco, another related Solanaceous species. Murata *et al.* (1992) transformed tobacco with a cDNA for glycerol-3-phosphate acyltransferase from *Arabidopsis*. The *Arabidopsis* glycerol-3-phosphate acyltransferase preferentially transesterifies *cis*-unsaturated rather than saturated fatty acids to the *sn*-1 position of glycerolipids. This result was observed in the transgenic tobacco, which had a 12% increase in *cis*-unsaturated fatty acids in the PG fraction of leaf lipids. The degree of PG unsaturation may be one important factor in the determination of chilling sensitivity (Matsuo *et al.*, 1992). The enhanced chilling tolerance of the transgenic tobacco was characterized by reduced damage to photosynthetic capacity after a four hour treatment at 1°C, and by reduced chlorosis after 10 days of growth at 1°C. A similar transgenic experiment was performed by Wolter *et al.* (1992). They transformed *Arabidopsis* with the *Escherichia coli* glycerol-3-phosphate acyltransferase, which prefers saturated fatty acids as substrate. The transgenic *Arabidopsis* were characterized by more highly saturated fatty acids in all glycerolipid species especially PG (60% increase) and by an increased chilling sensitivity.

In another transgenic experiment, Polashock *et al.* (1993) overexpressed a yeast fatty acid desaturase in tobacco which increased the palmitoleic acid (16:1, Δ-9, *cis*) concentration, particularily in chloroplast glycerolipids. The transgenic tobacco had 42% higher photosynthetic rates at 6°C than wild-type, as well as 23% less chlorophyll loss, 4 days after a 1 h exposure to -10°C. Chilling tolerance has also been enhanced by the transgene-mediated introduction of dienoic fatty acids (18:2) into the membrane lipids of a cyanobacterium (Wada *et al.*, 1990). Wild-type *Anacystis nidulans* does not synthesize dienoic fatty acids, but the transgenic expression of the *desA* gene from the chilling-resistant cyanobacterium *Synechocystis* PCC6803 resulted in a dienoic fatty acid content of 11% of membrane lipids. This increase in glycerolipid unsaturation allowed the transgenic *A. nidulans* to maintain near full photosynthetic capacity at 5°C, while wild-type *A. nidulans* experienced a 50% drop in photosynthetic activity. These transgenic experiments have moved beyond correlative findings and have definitively shown that the degree of membrane lipid unsaturation can influence chilling resistance in plants.

The overexpression of chloroplastic SOD in tobacco has also been shown to impart enhanced chilling tolerance. The transgenic tobacco plants maintained higher photosynthetic rates at low temperatures and experienced reduced rates of membrane electrolyte leakage when exposed to methyl viologen (Gupta *et al.*, 1993). These effects were likely due to a reduction of activated oxygen-mediated damage to cellular membranes. In summary, these five transgenic experiments indicate that it may be possible to enhance potato tuber metabolism at low temperatures through the modification of membrane glycerolipids and/or cellular antioxidants.

Summary and Prospects

This chapter discussed several facets of post-harvest potato tuber glycerolipid metabolism and its consequences on tuber biochemistry and physiology. It now seems that transgenic improvement of the potato crop may be realized at the level of membrane glycerolipid metabolism by several routes.

1. *Increased degree of membrane glycerolipid unsaturation.* References to the importance of high levels of membrane glycerolipid unsaturation during low temperature exposure were made throughout this chapter. Several transgenic routes to increased glycerolipid unsaturation are plausible, e.g. via overexpression of the yeast -Δ9 fatty acid desaturase, the desaturase activities corresponding to the *Arabidopsis fadB*, *fadC*, *fadD*, *fad2* and *fad3* mutations (Figure 8.1), or the overexpression of a glycerol-3-phosphate acyl transferase with a substrate preference for unsaturated fatty acids. These experiments are now certainly feasible because several of the structural genes corresponding to the *Arabidopsis* desaturation mutations have been cloned, i.e. *fad2* (Okuley *et al.*, 1994), *fad3* (Arondel *et al.*, 1992; Yadav *et al.*, 1993) and *fadD* (Yadav *et al.*, 1993).
2. *Increased antioxidative capacity.* Increases in tuber antioxidative capacities via the overexpression of SOD, CAT and POD enzymes may mitigate free radical damage to glycerolipids during low temperature and long-term storage.
3. *Increased phospholipid/sterol ratio.* This seems to be one apparent means by which plant tissues (Yoshida, 1984), including potato leaves (Weiss *et al.*, 1993), acclimate to the cold. One apparent rate-limiting enzyme of phospholipid biosynthesis in soybean is CDP-choline:1,2-diacylglycerol choline phosphotransferase (CPT) (Cook and Cheesbrough, 1993). Tubers contain CPT activity (Jolliot *et al.*, 1982) and perhaps overexpression of CPT in tubers may increase their phospholipid:sterol ratio and subsequent cold tolerance.

Hopefully in the near future transgenic potato tubers with altered glycerolipid composition will be productive tissues for metabolic studies, in addition to possessing enhanced quality characteristics from long-term cold storage.

References

Arondel, V., Lemieux, B., Hwang, I., Gibson, S., Goodman, H.M. and Somerville C. (1992) Map-based cloning of a gene controlling omega-3 fatty acid desaturation in *Arabidopsis*. *Science* 258:1353-1355.

Ben Abdelkader, A. (1972) Biogenèse des lipides membranaires pendant la 'survie' de tranches de tubercules de pomme de terre. PhD Thesis, University of Paris. 184 pp.

Ben Abdelkader, A. and Mazliak. P. (1971) Renouvellement des lipides dans diverses fractions cellulaires de parenchyme de pomme de terre ou d'inflorescence de chou-fleur. *Physiologie Végétale* 9:227-240.

Berkeley, H.D. and Galliard, T. (1974) Lipids of potato tubers. III. Effect of growth and storage on lipid content of the potato tuber. *Journal of the Science of Food and Agriculture* 25:861-867.

Bonnerot, C. and Mazliak, P. (1984) Induction of the oleoyl-phosphatidylcholine desaturase activity during the storage of plant organs. A comparison between potato and Jerusalem artichoke tubers. *Plant Science Letters* 35:5-10.

Browse, J. and Somerville, C. (1991) Glycerolipid synthesis: biochemistry and regulation. *Annual Review of Plant Physiology and Plant Molecular Biology* 42:467-506.

Browse, J., McConn, M., James, D.J., Jr. and Miquel, M. (1993) Mutants of *Arabidopsis* deficient in the synthesis of α–linolenate. Biochemical and genetic characterization of the endoplasmic reticulum linoleoyl desaturase. *Journal of Biological Chemistry* 268:16345-16351.

Burton, W.G. (1989) *The Potato*, 3rd edn. Longman Scientific and Technical, Essex, 742 pp.

Cherif, A. (1973) Métabolisme des lipides dans le tubercule de pomme de terre (*Solanum tuberosum* L.). 1. Evolution des lipides au cours de la conservation des tubercules. *Potato Research* 16:126-147.

Cherif, A. and Ben Abdelkader, A.B. (1970) Analyse quantitative des acides gras présents dans différentes régions du tubercule de pomme de terre; Variation au cours de la conservation à 10°C. *Potato Research* 13:284-295.

Cook, N.E. and Cheesbrough, T.M. (1993) Developmental regulation of CDP-choline:1,2-diacylglycerol choline phosphotransferase in young soybean seeds. *National Plant Lipid Cooperative. Plant Lipid Symposium*, Minneapolis, MN, p. 24 (abstract).

Cullis, P.R. and Hope, M.J. (1985) Physical properties and functional roles of lipids in membranes. In: Vance, D.E. and Vance, J.E. (eds), *Biochemistry of Lipids and Membranes*. The Benjamin/Cummings Publishing Company, Inc., Menlo Park, pp. 25-72.

Demandre, C. (1976) Les membranes cytoplasmiques de tubercule de pomme de terre (*Solanum tuberosum*, L.). PhD Thesis, University of Paris. 184 pp.

Demandre, C., Douady, D., Kader, J.C. and Mazliak, P. (1978) What is the origin of the lipids of potato tuber plasmalemma? *Plant Science Letters* 11:275-283.

Demandre, C., Tremolieres, A., Justin, A.M. and Mazliak, P. (1985) Analysis of molecular species of plant polar lipids by high-performance and gas liquid chromatography. *Phytochemistry* 24:481-485.

Demandre, C., Tremolieres, A., Justin, A.M. and Mazliak, P. (1986) Oleate desaturation in six phosphatidylcholine molecular species from potato tuber microsomes. *Biochimica et Biophysica Acta* 877:380-386.

Dubacq, R. and Kader, J.C. (1978) Free flow electrophoresis of chloroplasts. *Plant Physiology* 61:465-468.

Fishwick, M.J. and Wright, A.J. (1980) Isolation and characterization of amyloplast envelope membranes from *Solanum tuberosum*. *Phytochemistry* 19:55-59.

Frankel, E.N. and Neff, W.E. (1983) Formation of malonaldehyde from lipid oxidation products. *Biochimica et Biophysica Acta* 754:264-270.

Galliard, T. (1968) Aspects of lipid metabolism in higher plants. I. Identification and quantitative determination of the lipids in potato tubers. *Phytochemistry* 7:1907-1914.

Galliard, T. (1973) Lipids of potato tubers. 1. Lipid and fatty acid composition of tubers from different varieties of potato. *Journal of the Science of Food and Agriculture* 24:617-622 .

Galliard, T., Berkeley, H.D. and Matthew, J.A. (1975) Lipids of potato tubers. Effects of storage temperature on total, polar and sterol lipid content and fatty acid composition of potato tubers. *Journal of the Science of Food and Agriculture* 26:1163-1170.

Gounaris, K. and Barber, J. (1983) Monogalactosyldiacylglycerol: the most abundant polar lipid in nature. *Trends in Biochemical Sciences* 8:378-381.

Gupta, A.S., Heinen, J.L., Holaday, A.S., Burke, J.J. and Allen, R.D. (1993) Increased resistance to oxidative stress in transgenic plants that overexpress chloroplastic Cu/Zn superoxide dismutase. *Proceedings of the National Academy of Sciences USA* 90:1629-1633.

Halliwell, B. and Gutteridge, J.M.C. (1985) *Free Radicals in Biology and Medicine*. Clarendon Press, Oxford, 346 pp.

Harwood, J.L. (1988) Fatty acid metabolism. *Annual Review of Plant Physiology and Plant Molecular Biology* 39:101-138.

Harwood, J.L. and Russell, N.J. (1984) *Lipids in Plants and Microbes*. George Allen and Unwin Ltd., London, 162 pp.

Hugly, S. and Sommerville, C. (1992) A role for membrane lipid polyunsaturation in chloroplast biogenesis at low temperature. *Plant Physiology* 99:197-202.

Isherwood, F.A. (1976) Mechanism of starch-sugar interconversion in *Solanum tuberosum*. *Phytochemistry* 15:33-41.

Isherwood, F.A. and Kennedy, M.G.H. (1975) The composition of the expressed sap from cold stored potatoes. *Phytochemistry* 14:83-84.

Jamieson, G.R. and Reid, E.H. (1971) The occurrence of hexadeca-7,10,13-trienoic acid in the leaf lipids of angiosperms. *Phytochemistry* 10:1837-1843.

Jarvis, M.C., Dalziel, J. and Duncan, H.J. (1974) Variations in free sugars and lipids in different potato varieties during low-temperature storage. *Journal of the Science of Food and Agriculture* 25:1405-1409.

Jolliot, A., Justin, A.M., Bimont, E. and Mazliak, P. (1982) Regulation by lipids of plant microsomal enzymes. III. Phospholipid dependence of the cytidine-diphospho-choline phosphotransferase of potato microsomes. *Plant Physiology* 70:206-210.

Justin, A.M., Demandre, C. and Mazliak, P. (1989) Molecular species synthesized by phosphatidylinositol synthases from potato tuber, pea leaf and soya bean. *Biochimica et Biophysica Acta* 1005:51-55.

Kindl, H. (1984) Lipid degredation in higher plants. In: Numa, S. (ed.), *Fatty Acid Metabolism and Its Regulation*. Elsevier, Amsterdam, pp. 181-204.

Knowles, L.O. and Knowles, N.R. (1990) Changes in fatty acid composition of phospholipids from different ages of potato seed-tubers during sprouting. *Annals of Botany* 65:217-223.

Knowles, N.R. and Knowles, L.O. (1989) Correlations between electrolyte leakage and degree of saturation of polar lipids from aged potato (*Solanum tuberosum* L.) tuber tissue. *Annals of Botany* 63:331-338.

Kumar, G.N.M. and Knowles, N.R. (1993) Changes in lipid peroxidation and lipolytic and free-radical scavenging enzyme activities during aging and sprouting of potato (*Solanum tuberosum*) seed-tubers. *Plant Physiology* 102:115-124.

Lepage, M. (1964) The lipid components of white potato tubers (*Solanum tuberosum* L.). *Lipids* 3:477-481.

Lesham, Y.Y., Halevy, A.H. and Frenkel, C. (1986) *Processes and Control of Plant Senescence*. Elsevier Science Publishers B.V., Amsterdam, 215 pp.

Liljenberg, C, Sandelius, A.S. and Selstam, E. (1978) Effect of storage in darkness and in light on the content of membrane lipids of potato tubers. *Physiologia Plantarum* 43:154-159.

Lojkowska, E. and Holubowska, M. (1989) Changes of the lipid catabolism in potato tubers from cultivars differing in susceptibility to autolysis during the storage. *Potato Research* 32:463-470.

Matsuo, T., Ide, S. and Shitida, M. (1992) Correlation between chilling sensitivity of plant tissues and fatty acid composition of phosphatidylglycerols. *Phytochemistry* 31:2289-2293.

Mazliak, P. (1979) Temperature regulation of plant fatty acyl desaturases. In Lyons, J.M., Graham, D. and Raison, J.K. (eds), *Low Temperature Stress in Crop Plants. The Role of the Membrane*. Academic Press, New York, pp. 391-404.

Mazliak, P. (1983) Plant membrane lipids: changes and alterations during aging and senesence. In: Lieberman, M. (ed.), *NATO Advanced Study on Post-Harvest Physiology and Crop Preservation*. Plenum Press, New York, pp. 123-140.

McKeon, T.A. and Stumpf, P.K. (1982) Purification and characterization of the stearoyl-acyl carrier protein desaturase and the acyl-acyl carrier protein thioesterase from maturing seeds of safflower. *Journal of Biological Chemistry* 257:12141-12147.

Meunier, D. and Mazliak, P. (1972) Différence de composition lipidique entre les deux membranes des mitochondries de pomme de terre. *Comptes Rendus Hebdomadaires des Séances de L'Académie des Sciences*, Série D: *Sciences Naturelles* 275:213-216.

Miquel, M., James, D., Jr, Dooner, H. and Browse, J. (1993) *Arabidopsis* requires polyunsaturated lipids for low temperature survival. *Proceedings of the National Academy of Sciences USA* 90:6208-6212.

Murata, N., Ishizaki-Nishizawa, O., Higashi, S., Hayashi, H., Tasaka, Y. and Nishida, I. (1992) Genetically engineered alteration in the chilling sensitivity of plants. *Nature* 356:710-713.

O'Donoghue, E.P., Marangoni, A.G. and Yada, R.Y. (1993) Role of the amyloplast membrane in low temperature sweetening of potato tubers. *Plant Physiology* 102:599 (abstract).

Ohad, I., Friedberg, I., Ne'eman, Z. and Schramm, M. (1971) Biogenesis and degradation of starch. I. The fate of the amyloplast membranes during maturation and storage of potato tubers. *Plant Physiology* 47:465-477.

Ohlrogge, J.B., Browse, J. and Somerville, C. (1991) The genetics of plant lipids. *Biochimica et Biophysica Acta* 1082:1-26.

Okuley, J., Lightner, J., Feldmann, K., Lark, E. and Browse, J. (1994) *Arabidoposis FAD2* gene encodes the enzyme for polyunsaturated lipid synthesis. *The Plant Cell* 6:147-158.

Polashock, J., Lipson, D., Berkowitz, G., Chin, C.K. and Martin, C. (1993) Expression of yeast delta-9 fatty acid desaturase activity enhances chilling resistance in tobacco. *Plant Physiology* 102:921 (abstract).

Pun, W.H., Khan, A.A., Chung, I., Haydar, M. and Hadziyev, D. (1980) Lipid distribution in potato tubers. *Potato Research* 23:57-74.

Roughan, P.G. and Slack, C.R. (1982) Cellular organization of glycerolipid metabolism. *Annual Review of Plant Physiology* 33:97-123.

Sandelius, A.S. and Liljenberg, C. (1982) Light-induced changes in the lipid composition and ultrastructure of plastids from potato tubers. *Physiologia Plantarum* 56:266-272.

Shanklin, J. and Somerville, C.R. (1991) Stearoyl-acyl-carrier protein desaturase from higher plants is structurally unrelated to the animal and fungal homologs. *Proceedings of the National Academy of Sciences USA* 88:2510-2514.

Shekar, V.C., Iritani, W.M. and Magnuson, J. (1979) Starch-sugar interconversion in *Solanum tuberosum* L. II. Influence of membrane permeability and fluidity. *American Potato Journal* 56:225-235.

Shimakata, T. and Stumpf, P.K. (1982) Isolation and function of spinch leaf ß-ketoacyl-[acyl-carrier-protein] synthases. *Proceedings of the National Academy of Sciences USA* 79:5808-5812.

Sinensky, M. (1974) Homeoviscous adaptation - a homeostatic process that regulates the viscosity of membrane lipids in *Escherichia coli*. *Proceedings of the National Academy of Sciences USA* 71:522-525.

Slabas, A.R. and Fawcett, T (1992) The biochemistry and molecular biology of plant lipid biosynthesis. *Plant Molecular Biology* 19:169-191.

Somerville, C. and Browse, J. (1991) Plant lipids: metabolism, mutants, and membranes. *Science* 252:80-87.

Sowokinos, J.R., Orr, P.H., Knoper, J.A. and Varns, J.L. (1987) Influence of potato storage and handling stress on sugars, chip quality and integrity of the starch (amyloplast) membrane. *American Potato Journal* 64:213-226.

Spychalla, J.P. (1989) The contribution of endogenous antioxidants and fatty acids to the processing quality of stored tubers of *Solanum tuberosum.* Ph.D. thesis, University of Minnesota. 142 pp.

Spychalla, J.P. and Desborough, S.L. (1990a) Fatty acids, membrane permeability, and sugars of stored potato tubers. *Plant Physiology* 94:1207-1213.

Spychalla, J.P. and Desborough, S.L. (1990b) Superoxide dismutase, catalase, and α-tocopherol content of stored potato tubers. *Plant Physiology* 94:1214-1218.

Tchang, F. (1977) Biosynthèse des lipides contenus dans les peroxysomes du tubercule de pomme de terre (*Solanum tuberosum*, L.). PhD Thesis, University of Paris. 105 pp.

Thompson, J.E. (1988) The molecular basis for membrane deterioration during senescence. In: Nooden, L.D. and Leopold, A.C. (eds), *Senescence and Aging in Plants.* Academic Press, San Diego, CA, pp. 51-83.

Wada, H., Gombos, Z. and Murata, N. (1990) Enhancement of chilling tolerance of a cyanobacterium by genetic manipulation of fatty acid desaturation. *Nature* 347:200-203.

Weiss, L.S., Whitaker, B.D. and Palta, J.P. (1993) Temporal changes in plasma membrane lipids during cold acclimation of potato species differing in acclimation capacity. *Plant Physiology* 102: 463 (abstract).

Wolter, F.P., Schmidt, R. and Heinz, E. (1992) Chilling sensitivity of *Arabidopsis thaliana* with genetically engineered membrane lipids. *EMBO Journal* 11:4685-4692.

Workman, M., Kerschner, E. and Harrison, M. (1976) The effect of storage factors on membrane permeability and sugar content of potatoes and decay by *Erwinia carotovora* var. *atroseptica* and *Fusarium rosseum* var. *sambucinum. American Potato Journal* 53:191-204.

Yadav, N.S., Wierzbicki, A., Aegerter, M., Caster, C.S., Pérez-Grau, L., Kinney, A.J., Hitz, W.D., Booth, J.R., Jr, Schweiger, B., Stecca, K.L., Allen, S.M., Blackwell, M., Reiter, R.S., Carlson, T.J., Russell, S.H., Feldman, K.A., Pierce, J. and Browse, J. (1993) Cloning of higher plant omega-3 fatty acid desaturases. *Plant Physiology* 103:467-476.

Yoshida, S. (1984) Chemical and biophysical changes in the plasma membrane during cold acclimation of mulberry bark cells (*Morus bombycis* Koidz. cv Goroji). *Plant Physiology* 76:257-265.

Chapter Nine:

Functionality of Semi-Artificial Transit Peptides Encoded by Gene Constructs Derived from the Potato Gene for Granule-Bound Starch Synthase

Feike R. van der Leij, Jaap S. Bosscher[†] and Will J. Feenstra[†]
Department of Biotechnology, Swiss Federal Institute of Technology, Zürich, Switzerland. [†]*Department of Genetics, University of Groningen, Haren, The Netherlands.*

Granule-bound starch synthase (GBSS), also called the waxy protein, is the enzyme in potato which is solely responsible for the accumulation of amylose in the starch granules of plastids (van der Leij *et al.*, 1991a). This nuclear encoded protein is synthesized as a precursor (pre-GBSS) with an N-terminal transit peptide (TP) necessary for routing to, and transport into, the plastids. All nuclear encoded proteins which are located inside the plastids share as part of the precursor sequence a TP, which is cleaved during or after translocation (for reviews see Keegstra *et al.*, 1989; Archer and Keegstra, 1990; de Boer and Weisbeek, 1991). The TP of the waxy protein from maize has been successfully employed to study the transport of *Escherichia coli* ß-glucuronidase (GUS) into chloroplasts and amyloplasts, both *in vitro* (Klösgen *et al.*, 1989) and *in vivo* (Klösgen and Weil, 1991).

The analysis of an amylose-free (*amf*) potato mutant revealed a point deletion in the portion of the GBSS gene encoding the TP (van der Leij *et al.*, 1991b). A GBSS-like protein with a slightly higher relative molecular weight (M_r) than the wild-type GBSS is detectable in membrane preparations of *amf* tubers. It has been proposed that this protein is a pre-GBSS with an insufficient TP, formed as a result of reinitiation of translation of *amf* mRNA (van der Leij *et al.*, 1991b). In order to investigate whether modified TPs can be used to verify the proposed reinitiation of translation, and to enable a deletion analysis of TP sequences, we constructed a series of semi-artificial GBSS genes which were introduced in potato. Although not applicable to address the reinitiation problem properly, some of the semi-artificial constructs provided experimental data, useful for a better understanding of structural requirements of TPs.

Although the sequences of numerous nuclear encoded precursors of plastid proteins of a variety of plant species are known, the present knowledge of

The Molecular and Cellular Biology of the Potato, Second Edition
Edited by W.R. Belknap, M.E. Vayda and W.D. Park

structural features of TPs is limited. This is due to both the lack of consensus of primary structures and lengths of TPs and the limited knowledge of TP secondary structures. In general TPs consist of a relatively high amount of serine, threonine and alanine residues, basic residues, and almost no acidic residues. Higher plant TPs with known cleavage sites vary in length from 25 (*Lol p*Ib preprotein, Singh *et al.*, 1991) to 97 amino acid residues (precursor of a glutamate synthase, Sakakibara *et al.*, 1991). Based on predicted cleavage sites of TP sequences derived from DNA sequences, the minimum and maximum number of residues might diverge even more (Archer and Keegstra, 1990; de Boer and Weisbeek, 1991; Friemann *et al.*, 1992).

By analyzing a representative set of non-homologous TPs, von Heijne *et al.* (1989) were able to characterize the terminal domains of TPs. The N-terminal domain, i.e. the first ten residues of TPs, have relatively low amounts of charged residues, as well as few glycine and proline residues. For example, the frequency of appearance of an arginine residue between position 2 and position 11, when compared with its general frequency in the mature proteins, is 11%, and those for glycine and proline are 26%, and 29%, respectively. The first two residues of the N-terminus, methionine and alanine, are highly conserved. It is thought that the methionine residue is removed from the preprotein after translation (Flinta *et al.*, 1986; von Heijne *et al.*, 1989).

The C-terminus of a TP is normally rich in arginine residues, and therefore relatively hydrophilic, and carries a poorly conserved cleavage recognition sequence (Gavel and von Heijne, 1990). Smeekens *et al.* (1986) showed that removal of the last seven residues from the TP of pre-ferredoxin from *Silene pratensis* abolishes transport completely.

The central domain is probably predominantly involved in binding to the membrane (Reiss *et al.*, 1989; Perry *et al.*, 1991), and large alterations in this region do not necessarily affect function (Reiss *et al.*, 1989).

Von Heijne and Nishikawa (1991) proposed a random coil model for the part of TPs between position 2 of the N-terminus and position -3 relative to the cleavage site. Such a random coil state of TPs could result in an unfolded conformation of this targeting peptide, whether or not in association with cytoplasmic factors (Archer and Keegstra, 1991; de Boer and Weisbeek, 1991; von Heijne and Nishikawa, 1991). Recently, the first empirical information about transit peptide secondary structure was obtained (Pilon *et al.*, 1992), indicating a flexible structure which can change its conformation depending on its environment.

Here, we report the effect of the substitution of two residues in the N-terminal region of the wild-type TP of potato GBSS, and we report the study of semi-artificial TPs, ranging in length from 40 to 99 amino acid residues.

```
                   BamHI                       SphI     BamHI            EspI SphI          BglII    BglII    EspI

92-mer      5'-GATCCACT   GCTATTTCATCAACTGCATGCGCTGGATCCCCTATTAAGGTGGGCTCAGCATGCTTCGGAGATGGAAGATCTTGGAG-TCTAGC-3'
91-mer          3'-GTGA   CGATAAAGTAGTTGACGTACGCGACCTAGGGGATAATTCCACCCGAGTCGTACGAAGCCTCTACCTTCTAGAACCTC-AGATCGAGT-5'
pTRV151 MAS   G  S  T     A  I  S  S  T  A  C  A  G  S  P  I  K  V  G  S  A  C  F  G  D  G  R  S  W  R  S  S SATIVC
CAB     MAA      S  T  M  A  I  S  S  T  A  M  A  G  T  P  I  K  V  G  S        F  G  E  G  R  I  T
        **       *  *     *  *  *  *  *  *  .  *  *  .  *  *  *  *  *  *        *  *  .  *  *
```

Figure 9.1. Sequence of the artificial oligonucleotide, coding for an amino acid sequence based on the transit peptide (TP) of a CAB protein from maize. After annealing of the 91-mer to the 92-mer the DNA fragment has sticky ends, compatible and, after ligation, redigestible with *Bam*HI and *Esp*I (*Cel*II). Restriction sites are indicated. The resulting transit peptide sequence of pTRV151 is compared with the maize CAB TP sequence. Asterisks are identical residues, dots are conservative substitutions. The sequence SATIVC of pTRV151 resembles the C-terminus of the wild-type potato GBSS TP.

Construction of Semi-Artificial Transit Peptides

Primers for polymerase chain reactions (PCRs) were synthesized with restriction sites (Fig. 9.2A) to facilitate cloning of PCR fragments. To prevent the use of primers with multiple annealing sites all primers were checked by a matrix scan for their unicity in the GBSS gene and the pUC18 sequence.

The oligomer as shown in Fig. 9.1 was devised on the basis of the sequence of one of the shortest transit peptides known (Keegstra *et al.*, 1989), i.e., the transit peptide of a chlorophyll a/b binding protein (CAB) from maize (Sullivan *et al.*, 1989, see Fig. 9.1). The codon usage was adapted to the preferences known for the GBSS gene of potato (van der Leij *et al.*, 1991b), internal methionine triplets were removed, and artificial restriction sites were introduced to allow the removal of parts of the sequence after cloning in front of the mature GBSS coding sequence. The reversed-phase HPLC-purified 91-mer and 92-mer were obtained from Eurosequence (Groningen, The Netherlands). Annealing of the 91-mer and 92-mer was accomplished by mixing the preparations at 93°C and slowly cooling down to 40°C, resulting in a double-stranded DNA molecule with sticky ends, compatible for ligation as *Bam*HI-*Esp*I fragments (Figs 9.1, and 9.2B). The double-stranded oligomer was phosphorylated prior to ligation.

To allow easy manipulation and eventual sequencing, the reorganized GBSS genes were first constructed in pUC18 (Yanisch-Perron *et al.*, 1985), using as host the *Escherichia coli* strain JM83 (Vieira and Messing, 1982) and its r^-m^+ derivative TB1 (Johnston *et al.*, 1986). An exemplification of the construction of the pTRV plasmids is shown in Fig. 9.2B (TRV: transit reorganization vector). After construction in pUC, all the gene constructs were cloned as *Sal*I-*Kpn*I fragments in the plant transformation vector pBIN19 (Bevan, 1984), and the pTRV numbers were raised by 100; for example, pTRV181 is the pBIN analog of pTRV81 (Fig. 9.2B).

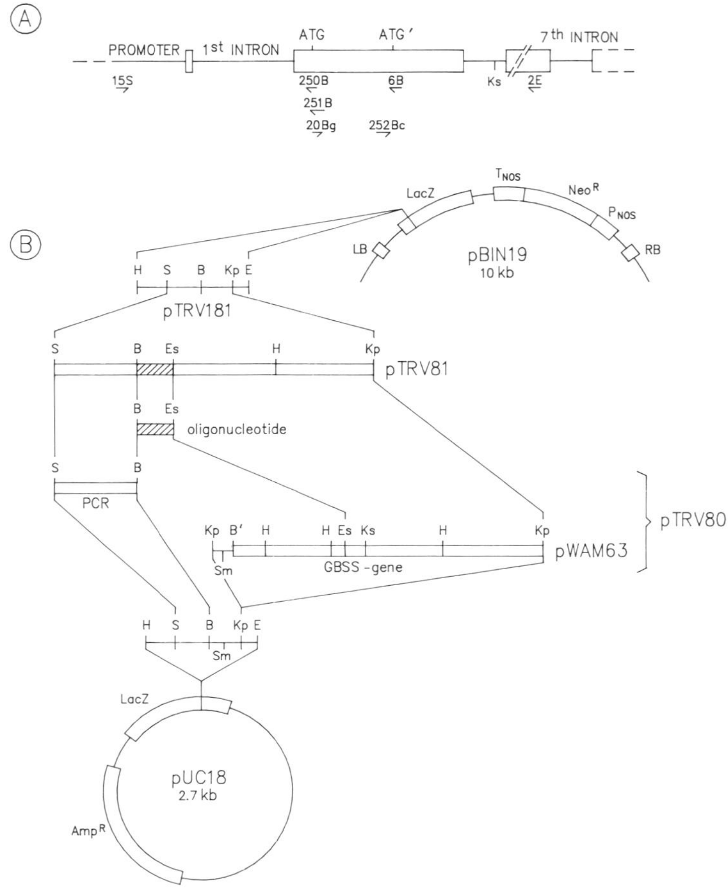

Figure 9.2. Construction of pTRV plasmids. (A) Locations and directions of the primers. The potato GBSS gene is partially represented with boxes for exons. The restriction sites introduced in the primers are indicated. Abbreviations as in B. B) Scheme of the construction of pTRV181 as exemplification of the construction of the pTRV vectors. The promoter of the potato granule-bound starch synthase (GBSS) gene was amplified by a polymerase chain reaction (PCR) (Saiki *et al.*, 1988) and inserted in pWAM63 to create pTRV80. The vector pWAM63 consists of pUC18 with the genomic GBSS sequence as *Kpn*I fragment, including all regulatory sequences. The oligonucleotide, which is shown in Fig. 9.1, was cloned in pTRV80 as *Bam*HI-*Spe*I fragment, replacing the *Bam*HI-*Esp*I fragment with the 5' *Kpn*I site, to form pTRV81. The total gene construct was placed into the binary plasmid pBIN19 as *Sal*I-*Kpn*I fragment. For clarity, only part of the circular plasmid pBIN19 is shown with the multiple cloning site in reverse. The fragments are not drawn to scale. LB = left border, RB = right border; LacZ = ß-galactosidase gene; B' = *Bam*HI/*Bgl*II fusion site.

The backbone plasmid pWAM63 was obtained after ligation of the GBSS gene-containing *Kpn*I fragment from pWAM60, which is the pUC analog of pWAM100 (van der Leij *et al.*, 1991a). Upstream of the GBSS gene in pWAM63 different PCR-fragments were cloned in the *Sal*I and *Bam*HI sites, all starting with primer 15S (5'-ATACTGTCGACGAGCATTTCCC-3'), which anneals at positions -425--404 of the GBSS gene. In this manner, a promoter fragment, including the TACAAAT- and CTAT-boxes as well as all three G-box-like sequences (van der Leij *et al.*, 1991b), and the untranslated leader containing the first intron were cloned in the right orientation. The positions and directions of the primers used for PCR are shown in Fig. 9.2A.

Primer 15 was used in combination with primer 6B (5'-TTGGTCT-CGGATCCGGATGCCAT-3', positions 177-200) (van der Leij *et al.*, 1992) for the construction of pTRV80 (Fig. 9.2B). After digestion of pTRV80 with *Bam*HI and *Esp*I the described oligonucleotide (Fig. 9.1) was inserted to form pTRV81 (Fig. 9.2B). Sequencing of the part of DNA with the inserted oligonucleotide revealed that two out of four candidate plasmids contained the proper insert.

For the construction of pTRV151 (Fig. 9.3) pTRV50 was made correspondingly to pTRV80, using primer 250B (5'-GGATCCGATGTGT-GGTCTACAAAAA-3', positions -21-4) instead of primer 6B. A PCR fragment starting with primer 252Bc (5'-GGCTCTGATCAAGAACTAAT-ACTAAGG-3', positions 140-166) and overlapping the unique *Ksp*I site (Fig. 9.2A,B) by the use of primer 2E (5'-ATTGAATTCAAGTAGCAA-3', positions 1149-1166) as the second primer, was amplified from pTRV181 (Fig. 9.2B, Fig. 9.3), and cloned as *Bcl*I-*Ksp*I fragment in pTRV50, which was digested with *Bam*HI and *Ksp*I, to form pTRV51 (not shown).

pTRV68 was made from pTRV60 (not shown), which was constructed as pTRV80, but with the use of primer 251B (5'-GTGAGATCTGGATCC-GCTTGCCATGTGATGTGTGG-3', positions -11-24) instead of primer 6B. Subsequent cloning of a *Bgl*II and *Ksp*I digest of a fragment amplified from a wild-type template with primer 20Bg (5'-ATGGCAAGATCTACAGCTTCAC-3', positions 1-22) and primer 2E resulted in pTRV68. Plasmids pTRV51 and pTRV68 were constructed entirely in duplicate to reduce the chance of false results based on sequence alterations due to PCR. Deletion of the *Bam*HI fragment of pTRV181 resulted in pTRV183. pTRV184 was made by deleting the *Bgl*II fragment (Figs 9.1 and 9.4). Since pBIN19 carries a *Bgl*II site the latter construct was first made in pUC as pTRV84.

Plant transformation and analysis of transformants

The diploid *Solanum tuberosum* L. homozygous *amf* lines 1029-32 and 1031-29 ($2n = 2x = 24$) (Jacobsen *et al.*, 1989) were grown *in vitro* on solid MS medium (Murashige and Skoog, 1962) as described previously (van der Leij *et al.*, 1991a).

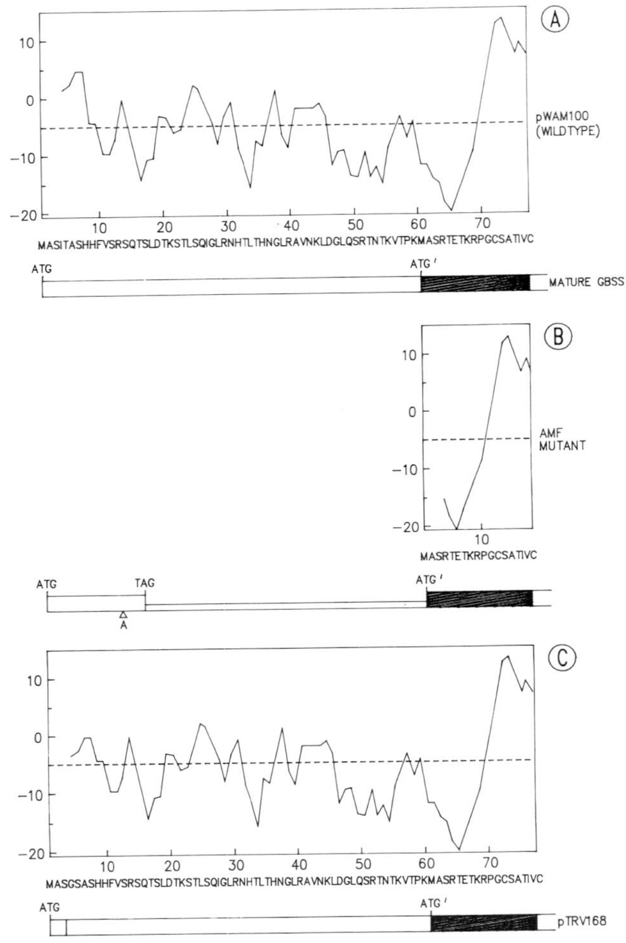

Figure 9.3. Transit peptide amino acid sequences with their corresponding hydrophobicity plots and gene structures. The plots of hydropathic indices are according to Kyte and Doolittle (1982), with scores above the dotted line for hydrophobic, and scores beneath the dotted line for hydrophilic residues. The interval used for the calculations was nine amino acid residues. (A) The potato wild-type GBSS TP as encoded by pWAM100. The translational start (ATG), and the proposed second start in the mutant (ATG'), are indicated. The black part starts at ATG' and ends at the processing site, where the sequence of the mature protein begins. (B) The mutant sequence of the *amf* allele carries a frame-shift mutation, which is indicated by Δ A, leading to termination of translation at the TAG stop-sequence. The untranslated region is shown as a small horizontal bar, followed by the proposed mutant GBSS with a non-functioning TP, as indicated in black. (C) The TP of pTRV168 starts at the normal ATG and differs from

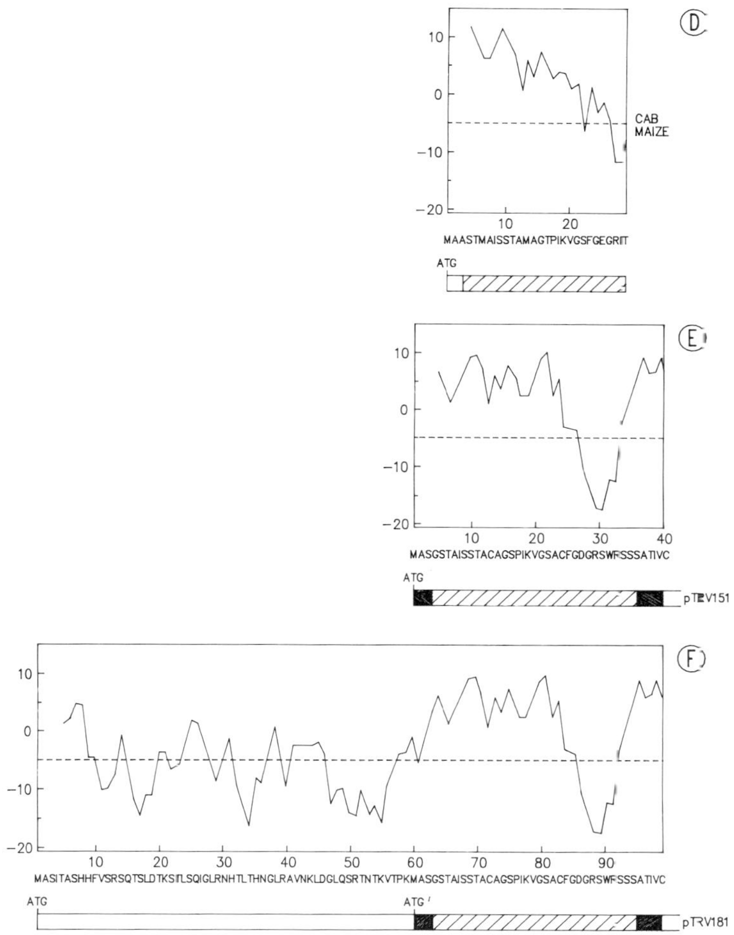

the wild-type TP by the substitution of isoleucine-threonine by glycine-serine at position 4-5, leading to a slightly lower hydrophobicity at the N-terminus. (D) The 29 amino acid TP sequence of a maize chlorophyll a/b binding protein (CAB). The processing site of this TP is not known and the TP could be longer than expected before (Sullivan *et al.*, 1989). The hatched bar represents the part used as the basis for the oligonucleotide as shown in Fig. 9.1. (E) The plasmid pTRV151 is derived from pTRV181 (F), and starts with the mutant sequence surrounding the first ATG. The resulting TP-like sequence is 40 amino acid residues long. (F) A TP of 99 amino acid residues is encoded by pTRV181. The coding sequence starts with the first 62 amino acid residues of the wild-type pre-GBSS, followed by 31 residues encoded by the designed oligonucleotide (positions 63-93), and continuing the total GBSS sequence from position 74 (panel 9.3A), which in this construct is at position 94.

Triparental mating and stem segment transformation with *Agrobacterium rhizogenes* were carried out as described previously (van der Leij *et al.*, 1991a; Visser *et al.*, 1989). Primary hairy roots and kanamycin resistant (Km^R) hairy roots were screened for their starch composition as described (van der Leij *et al.*, 1991a). To enhance starch synthesis the roots were placed in the dark for 24-48 h prior to analysis.

Computer analysis

From the PC/Gene software package (Intelligenetics, Geneva) version 6.26, we used four different methods (Garnier *et al.*, 1978; Chou and Fasman, 1979; Novotny and Auffray, 1984; Gascuel and Golmard, 1988) to predict the secondary structures of the transit peptides. The method of Chou and Fasman (1979) gives a prediction of the propensities of subsequences to form ß-turns, the other programs also calculate the propensities of subsequences to form other secondary structures. Only when all the programs used predicted a same sort of structure were the predictions considered to be reliable. For a general impression of the hydropathic distributions the method of Kyte and Doolittle (1982) was used. For a detailed analysis of the N-terminal part of the TPs, different hydrophobicity scales were used (see von Heijne, 1987 and references therein; Roseman, 1988). Hydropathic moment analyses were done with the University of Wisconsin Genetics Computer Group package (Devereux *et al.*, 1984), using the method of Eisenberg *et al.* (1984).

Analysis of the Functionality of pTRV151, pTRV168 and pTRV181

After introduction of pTRV181 complementation was observed with expected frequency (Table 9.1), indicating normal transcriptional activity of the promoter sequences and proper functioning of the sequence based on the CAB TP within the wild-type TP. The protein encoded by pTRV181 carries the first 59 amino acid residues of the wild-type GBSS TP and then continues with the same sequence as the pTRV151 preprotein (Fig. 9.3E and F).

We elongated the coding sequence of the presumed *amf* GBSS (Fig. 9.3B) by replacing part of it with the described oligonucleotide sequence (Fig. 9.3D), and made the proposed site of reinitiation (ATG', Fig. 9.3) the start of translation. As noticed earlier (van der Leij *et al.*, 1991b), the presence of the basic arginine residue in position 4 of the sequence MASRT, originating from the GBSS gene, would be exceptional for a TP sequence. Therefore, our strategy of inserting an artificial sequence included the removal of the arginine residue. Given the limitations of our cloning strategy this removal resulted in either the introduction of an aspartic acid, a tryptophan, or a glycine residue in the N-terminus. According to the absolute frequencies of these residues in TPs (von

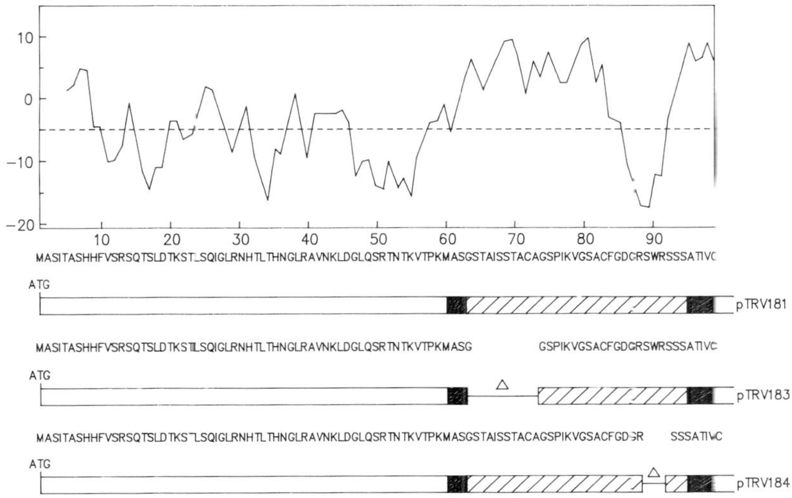

Figure 9.4. Hydrophobicity plot, amino acid sequence and gene structure of the pTRV181 TP, drawn as in Fig. 9.3F. The deletion clones pTRV183 and pTRV184 were constructed as described. Deleted parts are indicated by a single line with a Δ and are drawn to scale compared with the hydrophobicity plot of pTRV181. Predicted turns (>) and a non-amphiphilic ß-structure (/\/\/) are indicated.

Heijne *et al.*, 1989), replacing arginine by glycine was the most obvious choice. By creating a *Bam*HI site at the corresponding position in the DNA sequence the arginine-threonine residues at positions 4 and 5 were replaced by glycine-serine (pTRV151, Fig. 9.3E). However, this construct did not lead to the formation of an active GBSS (Table 9.1, pTRV151).

Introduction of pTRV168 did not result in complementation (Table 9.1). The only difference between pTRV168 and the wild-type sequence is the substitution of glycine-serine for isoleucine-threonine at positions 4 and 5 of the TP (Figs 9.3A and C).

Transit Peptide Deletion Analysis

With the ability of pTRV181 to complement the *amf* mutation, we had a tool to investigate a number of necessary conditions required for a functional TP. Therefore, the plasmids pTRV183 and pTRV184 were introduced in the *amf* mutant and primary hairy roots were screened (Table 9.1). The results of transformation with pTRV183 show that deletion of the hydrophobic part encoded by the *Bam*HI fragment completely abolished complementation. Similarly, no complementation was observed when a small portion of the

Table 9.1. Amylose in hairy roots after the introduction of pTRV constructs in the amylose-free mutant.

	pTRV number						
	151a	151b	168a	168b	181	183	184
Number of red-staining KmR hairy roots	8	nd[1]	6	nd	16	nd	nd
Number of blue-staining KmR hairy roots[2]	0	nd	0	nd	21	nd	nd
Number of red-staining primary hairy roots	31	51	85	63	52	67	101
Number of blue-staining primary hairy roots	0	0	0	0	13	0	0

Blue-staining with iodine: + amylose; red-staining: - amylose (a and b are constructs which were made and introduced in duplicate).

[1]nd = not determined

[2]KmR = kanamycin-resistant

hydrophilic dip before the assumed processing site was removed by *Bgl*II digestion (pTRV184).

Secondary Structure Predictions of Known GBSS TPs

Different statistical methods were applied to the GBSS TPs known from the sequenced genes from maize (Klösgen *et al.*, 1986), barley (Rohde *et al.*, 1988), wheat (Clark *et al.*, 1991), potato (van der Leij *et al.*, 1991b), pea (Dry *et al.*, 1992) and two genes from rice (Wang *et al.*, 1990; Okagaki, 1992), to obtain predictions of secondary structures. At the primary sequence level these TPs show little similarity. The results of the statistical analyses indicate an irregular structure for the main part of each TP, with randomly distributed subsequences with bending propensity. All programs predict for all wild-type GBSS TPs a ß-strand at positions -5 to -2 relative to the (putative) cleavage sites. This possible ß-strand was not affected in any of the deletion clones derived from pTRV181. The same methods were applied on the pre-CAB sequence which we had used as guide-line for the design of the constructs pTRV151 and pTRV181. A ß-strand behind a hydrophilic dip with arginine residues (Fig. 9.3D) is predicted just behind the cleavage site proposed by Sullivan *et al.* (1989). The position of this ß-strand in the maize sequence is in the pTRV151 and pTRV181 constructs taken by the potato cleavage site (Fig. 9.3E and F).

Substitution of Two Residues in the N-terminal Domain of the GBSS TP Abolishes Correct Transport into the Plastid

Among the constructs we made to screen for the functionality of the encoded TPs, one construct consisted of the wild-type GBSS gene, with the exception of altered triplets 4 and 5. This construct, pTRV168, encodes a pre-GBSS with the N-terminus MASGS instead of MASIT. Its transcription is driven by the same promoter as used for the other pTRV constructs, of which pTRV131 showed normal expression in root tips. From the results we obtained by transforming the mutant with pTRV168 (Table 9.1) it is clear that a glycine-serine sequence at positions 4 and 5 in the GBSS TP is fatal to appropriate function. This finding is of importance for the understanding of the primary and secondary structural requirements of TPs.

Although it has been previously reported that strong structure-breaking amino acids like glycine and proline are not common within the first part of TPs (von Heijne *et al.*, 1989), several TPs with glycines or prolines at positions 2, 3, 4 or 5 have been reported (for an overview, see von Heijne *et al.*, 1991). Moreover, the presence of glycine-serine is known for positions 6-7 of the different TPs of pea GBSSI and pea GBSSII (Dry *et al.*, 1992) and for a spinach root acyl carrier protein (Schmid and Ohlrogge, 1990). A glycine at position 6 is known for the TP of starch phosphorylase type L from potato (Nakano *et al.*, 1989). TPs of preproteins from gymnosperms are known with sequences MASGAA, MATAVGCS (Jansson and Gustafsson, 1991) and MATGAGAGA (Yamamoto *et al.*, 1988).

The dominant character of glycine is its structure-breaking or turn-forming capacity (Levitt, 1978; von Heijne, 1987), which is not diminished by conformational stability in a terminal sequence. The exchange of the threonine residue for serine may in itself be considered as harmless to the functioning of the TP, regarding the fact that both amino acids are hydroxylated, and that up to six serine residues can be present in this region of transit peptides (Keegstra *et al.*, 1989). However, serine residues also have turn-forming propensities, and, when next to glycine in the pTRV168 preprotein, enhance the probability to bend.

A comparison of the predictions of secondary structures revealed that extra bending propensity is indeed introduced in the pTRV168 protein relative to the wild-type pre-GBSS. According to the method of Chou and Fasman (1979) a turn would be formed by the tetrapeptide ASGS in the pTRV168 TP. However, the exact secondary structure of the first part of the TPs is not predictable. All three computing programs used predict α-helices for different subsequences of the wild-type N-terminus, which are diminished in the pTRV168 sequence. To date there are no proteins with known secondary and tertiary structure which share significant parts of the sequence ASITASHHF or ASGSASHHF (data not shown).

In view of the possibility that functional TPs can have some bending propensity in proximity to the N-terminal residue, other factors could also be involved in the non-functioning of the pTRV168 preprotein. Such bending has been calculated for the TP of a precursor of the small subunit of ribulose 1,5-bisphosphate carboxylase/oxygenase from rice, with the sequence MAPSVMAS (Xie and Wu, 1988) (data not shown). Fig. 9.3C shows a slightly reduced hydrophobicity of the N-terminus of the pTRV168 gene product when compared with the N-terminus of wild-type pre-GBSS (Fig. 9.3A). The sequence surrounding position 11 of the wild-type TP of potato GBSS is already less hydrophobic than in most other wild-type TPs, and the alteration introduced in the pTRV168 sequence might be fatal, purely for reasons of hydropathy. There are TPs known with less average hydrophobicities calculated for this part, but these are less likely to bend in close proximity to the N-terminus than was found for the pTRV168 TP. Therefore, a combined effect on the secondary structure of the N-terminus might be the reason for failure of the pTRV168 TP to transport GBSS into the plastid correctly. Whether this effect is most severe on binding, recognition or translocation remains unknown.

These data demonstrate that TPs can be very sensitive to small alterations in the N-terminal domain. Further site-directed mutagenesis studies, e.g., substitution of glycine for isoleucine and alanine for threonine (MASGAA is found for a TP from *Pinus sylvestris*) or only glycine for isoleucine, together with structural and *in vitro* analyses, could shed more light on this intriguing result.

A Short TP Based on a CAB TP and Starting with MASGS Cannot Transport GBSS into the Plastids

We have introduced a semi-artificial TP sequence, based on the TP of a CAB preprotein from maize, in place of a part of both the sequence expected to encode an *amf* pre-GBSS (pTRV151, Fig. 9.3E), and the wild-type pre-GBSS (pTRV181, Fig. 9.3F). The proper functioning of the pTRV181 TP indicated that the extension, which is almost at the C-terminus of the TP, did not interfere with accurate transport across the plastid membranes. However, the pTRV151 TP has the same glycine-serine residues relative to the N-terminus as the non-functioning pTRV168 TP. Therefore, it is likely that the sequence glycine-serine (see above) is responsible for the non-functioning of the TP encoded by pTRV151. However, it cannot be excluded that the CAB-derived sequence also would require more length, or more positive charge to successfully transport GBSS (which is larger and more negatively charged than CAB) across the negatively charged membranes into the plastid. In addition to the encoded TPs, pTRV168 and pTRV151 also differ in that the latter has the sequence CCCAAG immediately upstream of the initiation codon, where pTRV168 has the wildtype sequence CATCAC. In view of a higher matching score for the pTRV151 sequence when compared with the consensus sequence formulated by Lütcke *et al.* (1987), this difference seems to be of minor importance. Both translational starts are the first AUG sequences in the resulting mRNAs.

A Functional Semi-artificial TP of 99 Residues is Sensitive to the Deletion of Different Parts of its Sequence

In the pTRV181 sequence the normally used TP cleavage site, which appears to be conserved at the secondary structural level for all known waxy proteins, remained unaltered. Since there are no other subsequences with a predicted ß-structure in the preceding sequence, we assume that processing of the preprotein is at the normally used cleavage site. Given this assumption, the construct pTRV181 (Figs 9.3C and 9.4) encodes a TP of 99 amino acid residues, consisting of the first 62 residues and the C-terminal six residues of the wild-type GBSS TP, intervened by the protein sequence as derived from the TP of a CAB precursor from maize (Figs 9.1 and 9.3D). This semi-artificial TP precedes the mature form of GBSS, which is transported *in vivo* into the amyloplasts of columella cells of root tips, where synthesis and accumulation of amylose take place (Table 9.1). The constructs pTRV183 and pTRV184 each carry a deletion within the TP sequence (Fig. 9.4), both of which abolish complementiation by these constructs (Table 9.1).

The resemblance in the distribution of hydropathic residues between the GBSS precursors of four plant species has been discussed earlier (van der Leij *et al.*, 1991b), and similar patterns have been found for other TPs (e.g., Singh *et al.*, 1991; Pilon *et al.*, 1992, and our own unpublished results). The most striking feature of these distributions is the strongly hydrophilic region prior to an increase in hydrophobicity at the region surrounding the processing site. Removal of three residues from this hydrophilic part, as in the case for pTRV184, reduces the hydrophilicity, and does not result in an active GBSS. Of course, the pTRV184 TP could also be non-functional due to conformational alterations. At present, nothing is known about the secondary structure of this part of other TPs. Speculations about possible amphiphilic ß-strand formation (von Heijne *et al.*, 1989) cannot be clearly confirmed by our analyses (data not shown), which included extensive comparisons between hydropathic moment plots derived by the method of Eisenberg *et al.* (1984).

The same kind of explanations as given for the sensitivity to deletions in the C-terminal part can be given for the finding that a central hydrophobic part cannot be deleted without consequences (pTRV183, Fig. 9.4, Table 9.1). Firstly, the overall hydrophobicity of the TP of pTRV183 could be too low. Secondly, drastic alterations in secondary structure could be the consequence of the deletion. The latter explanation is supported by the finding that with the deletion of the hydrophobic part also the subsequence with the strongest bending propensity of the pTRV181 TP is removed.

Conclusions

One of the objectives of this study was to investigate the applicability of designed sequences for the analysis of a particular case of second site initiation of translation. We have to conclude that the constructs described here are not useful for addressing this question properly. However, the complementation system used in this study, which combines the advantages of the rapid development of hairy roots and the easily applicable iodine staining of starch, is a valuable tool for studying the efficient transport of GBSS into the plastids and for studying the consequences of different sequence alterations.

In this study we showed that small alterations in TP structure can have severe effects, and, on the other hand, that semi-artificial TP sequences made by replacement with a designed sequence can be functional. In view of the limited knowledge of structural aspects involved in TP functionality, no unequivocal explanations can be given yet for the results presented here. The finding that replacing isoleucine-threonine with glycine-serine is fatal for GBSS TP functioning is in our view most interesting, and additional studies of this case would be desirable.

References

Archer, E.K. and Keegstra, K. (1990) Current views on chloroplast protein import and hypotheses on the origin of the transport mechanism. *Journal of Bioenergetics and Biomembranes* 22:789-810.

Bevan, M. (1984) Binary *Agrobacterium* vectors for plant transformation. *Nucleic Acids Research* 12:8711-8721.

Chou, P.Y. and Fasman, G.D. (1979) Prediction of ß-turns. *Biophysical Journal* 26:367-384.

Clark, J.R., Robertson, M. and Ainsworth, C.C. (1991) Nucleotide sequence of a wheat (*Triticum aestivum* L.) cDNA clone encoding the *waxy* protein. *Plant Molecular Biology* 16:1099-1101.

de Boer, A.D. and Weisbeek, P.J. (1991) Chloroplast topogenesis: protein import, sorting and assembly. *Biochimica et Biophysica Acta* 1017:221-253.

Devereux, J., Haeberli, P. and Smithies, O. (1984) A comprehensive set of sequence analysis programs for the VAX. *Nucleic Acids Research* 12:387-395.

Dry, I., Smith, A., Edwards, A., Bhattacharyya, M., Dunn, P. and Martin, C. (1992) Characterisation of cDNAs encoding two isoforms of granule-bound starch synthase which show differential expression in developing storage organs of pea and potato. *The Plant Journal* 2:193-202.

Eisenberg, D., Weiss, R.M. and Terwilliger, T.C. (1984) The hydrophobic moment detects periodicity in protein hydrophobicity. *Proceedings of the National Academy of Sciences USA* 81:140-144.

Flinta, C., Persson, B., Jörnvall, H. and von Heijne, G. (1986) Sequence determinants of cytosolic N-terminal protein processing. *European Journal of Biochemistry* 154:193-196.

Friemann, A., Brinkmann, K. and Hachtel, W. (1992) Sequence of a cDNA encoding nitrite reductase from the tree *Betula pendula* and identification of conserved protein regions. *Molecular and General Genetics* 231:411-416.

Garnier, J., Osguthorpe, D.J. and Robson, B. (1978) Analysis of accuracy and implications of simple methods for predicting the secondary structure of globular proteins. *Journal of Molecular Biology* 120:97-120.

Gascuel, O. and Golmard, J.L. (1988) A simple method for predicting the secondary structure of globular proteins: implications and accuracy. *Computational Applied Bioscience* 4:357-365.

Gavel, Y. and von Heijne, G. (1990) A conserved cleavage-site motif in chloroplast transit peptides. *FEBS Letters* 261:455-458.

Jacobsen, E., Hovenkamp-Hermelink, J.H.M., Krijgsheld, H.T., Nijdam, H., Pijnacker, L.P., Witholt, B. and Feenstra, W.J. (1989) Phenotypic and genotypic characterization of an amylose-free starch mutant of the potato. *Euphytica* 44:43-48.

Jannson, S. and Gustafsson, P. (1991) Evolutionary conservation of the chlorophyll *a*/*b*-binding proteins: cDNAs encoding Type I, II and III LHC I polypeptides from the gymnosperm Scots pine. *Molecular and General Genetics* 229:67-76.

Johnston, T.C., Thompson, R.B. and Baldwin, T.O. (1986) Nucleotide sequence of the *LuxB* gene of *Vibrio harveyi* and the complete amino acid sequence of the beta subunit of bacterial luciferase. *Journal of Biological Chemistry* 261:4805-4811.

Keegstra, K., Olsen, L.J. and Theg, S.M. (1989) Chloroplastic precursors and their transport across the envelope membrane. *Annual Review of Plant Physiology and Plant Molecular Biology* 40:471-501.

Klösgen, R.B. and Weil, J.-H. (1991) Subcellular location and expression level of a chimeric protein consisting of the maize *waxy* transit peptide and the ß-glucuronidase of *Escherichia coli* in transgenic potato plants. *Molecular and General Genetics* 225:297-304.

Klösgen, R.B., Gierl, A., Schwarz-Sommer, Z. and Saedler, H. (1986) Molecular analysis of the *waxy* locus of *Zea mays*. *Molecular and General Genetics* 203:237-244.

Klösgen, R.B., Saedler, H. and Weil, J.-H. (1989) The amyloplast-targeting transit peptide of the *waxy* protein of maize also mediates protein transport *in vitro* into chloroplasts. *Molecular and General Genetics* 217:155-161.

Kyte, J. and Doolittle, R.F. (1982) A simple method for displaying the hydropathic character of a protein. *Journal of Molecular Biology* 157:105-132.

Levitt, M. (1978) Conformational preferences of amino acids in globular proteins. *Biochemistry* 17:4277-4285.

Lütcke, H.A., Chow, K.C., Mickel, F.S., Moss, K.A., Kern, H.F. and Scheele, G.A. (1987) Selection of AUG initiation codons differs in plants and animals. *EMBO Journal* 6:43-48.

Murashige, T. and Skoog, F. (1962) A revised medium for rapid growth and bioassays with tobacco tissue cultures. *Physiologica Plantarum* 15:473-497.

Nakano, K., Mori, H. and Fukui, T. (1989) Molecular cloning of cDNA encoding potato amyloplast α-glucan phosphorylase and the structure of its transit peptide. *Journal of Biochemistry* 106:691-695.

Novotny, J. and Auffray, C. (1984) A program for prediction of protein secondary structure from nucleotide sequence data: application to histocompatibillity antigens. *Nucleic Acids Research* 12:243-255.

Okagaki, R.J. (1992) Nucleotide sequence of a long cDNA from the rice *waxy* gene. *Plant Molecular Biology* 19:513-519.

Perry, S.E., Buvinger, W.E., Bennett, J. and Keegstra, K. (1991) Synthetic analogues of a transit peptide inhibit binding or translocation of chloroplastic precursor proteins. *Journal of Biological Chemistry* 266:11882-11889.

Pilon, M., Rietveld, A.G., Weisbeek, P.J. and de Kruijff, B. (1992) Secondary structure and folding of a functional chloroplast precursor protein. *Journal of Biological Chemistry* 267:19907-19913.

Reiss, B., Wasmann, C.C., Schell, J. and Bohnert, H.J. (1989) Effect of mutations on the binding and translocation functions of a chloroplast transit peptide. *Proceedings of the National Academy of Sciences USA* 86:886-890.

Rohde, W., Becker, D. and Salamini, F. (1988) Structural analysis of the *waxy* locus from *Hordeum vulgare*. *Nucleic Acids Research* 16:7185-7186.

Roseman, M.A. (1988) Hydrophilicity of polar amino acid side-chains is markedly reduced by flanking peptide bonds. *Journal of Molecular Biology* 200:513-522.

Saiki, R.K., Gelfland, D.H., Stoffel, S,. Sharf, S.J., Higuchi, R.G., Horn, G.T., Mullis, K.B. and Erlich, H.A. (1988) Primer-directed enzymatic amplification of DNA with a thermostable DNA polymerase. *Science* 239:487-491.

Sakakibara, H., Watanabe, M., Hase, T. and Sugiyama, T. (1991) Molecular cloning and characterization of complementary DNA encoding for ferredoxin-dependent glutamate synthase in maize leaf. *Journal of Molecular Biology* 266:2028-2035.

Schmid, K.M. and Ohlrogge, J.B. (1990) A root acyl carrier protein-II from spinach is also expressed in leaves and seeds. *Plant Molecular Biology* 15:765-778.

Singh, M.B., Hough, T., Theerakulpisut, P., Avjioglu, A., Davies, S., Smith, P.M., Taylor, P., Simpson, R.J., Ward, L.D., McCluskey, J., Puy, R. and Knox, R.B. (1991) Isolation of cDNA encoding a newly identified major allergenic protein of rye-grass pollen: Intracellular targeting to the amyloplast. *Proceedings of the National Academy of Sciences USA* 88:1384-1388.

Smeekens, S., Geerts, D., Bauerle, C. and Weisbeek, P. (1989) Essential function in chloroplast recognition of the ferredoxin transit peptide processing region. *Molecular and General Genetics* 216:178-182.

Sullivan, T.D., Christensen, A.H. and Quail, P.H. (1989) Isolation and characterization of a maize chlorophyll a/b binding protein that produces high levels of mRNA in the dark. *Molecular and General Genetics* 215:431-440.

van der Leij, F.R., Visser, R.G.F., Oosterhaven, K., van der Kop, D.A.M., Jacobsen, E. and Feenstra, W.J. (1991a) Complementation of the amylose-free starch mutant of potato (*Solanum tuberosum* L.) by the gene encoding granule-bound starch synthase. *Theoretical and Applied Genetics* 82:289-295

van der Leij, F.R., Visser R.G.F., Ponstein A.S., Jacobsen E. and Feenstra W.J. (1991b) Sequence of the structural gene for granule-bound starch synthase of potato (*Solanum tuberosum* L.) and evidence for a single point deletion in the *amf* allele. *Molecular and General Genetics* 228:240-248

van der Leij, F.R., Enequist, H., Terpstra, P. and Feenstra, W.J. . (1992) Chimerism as the basis for the occurrence of amylose synthesizing clones derived from an amylose-free potato mutant. *Plant Science* 81:135-139.

Vieira, J. and Messing, J. (1982) The pUC plasmids, an M13mp7-derived system for insertion mutagenesis and sequencing with synthetic universal primers. *Gene* 19:259-268.

Visser, R.G.F., Jacobsen, E., Witholt, B. and Feenstra, W.J. (1989) Efficient transformation of potato (*Solanum tuberosum*) with a binary vector in *Agrobacterium rhizogenes*. *Theoretical and Applied Genetics* 78:594-600.

von Heijne, G. (1987) *Sequence Analysis in Molecular BIology, Treasure Trove or Trivial Pursuit*. Academic Press, San Diego.

von Heijne, G., Hirai, T., Klösgen, R.B., Steppuhn, J., Bruce, B., Keegstra, K. and Herrmann, R. (1991) CHLPEP - A database of chloroplast transit peptides. *Plant Molecular Biology Reporter* 9:104-126

von Heijne, G., Steppuhn, J. and Herrmann, R.G. (1989) Domain structure of mitochondrial and chloroplast targeting peptides. *European Journal of Biochemistry* 180:535-545.

von Heijne, G. and Nishikawa, K. (1991) Chloroplast transit peptides: the perfect random coil? *FEBS Letters* 278:1-3.

Wang, Z., Wu, Z., Xing, Y., Zheng, F., Guo, X., Zang, W. and Hong M (1990) Nucleotide sequence of rice waxy gene. *Nucleic Acids Research* 18:5898.

Xie, Y. and Wu, R. (1988) Nucleotide sequence of a ribulose-1,5-bisphosphate carboxylase/oxygenase small subunit gene (*rbcS*) in rice. *Nucleic Acids Research* 16:7749.

Yamamoto, N., Kano-Murakami, Y., Matsuoka, M., Ohashi, Y. and Tanaka, Y. (1988) Nucleotide sequence of a full length cDNA clone of ribulose bisphosphate carboxylase small subunit gene from dark-grown pine (*Pinus tunbergii*) seedling. *Nucleic Acids Research* 16:11830.

Yanisch-Perron, C., Vieira, J. and Messing, J. (1985) Improved M13 phage cloning vectors and host strains: nucleotide sequences of M13mp18 and pUC19 vectors. *Gene* 33:103-119.

Chapter Ten:

Potato Alternative Oxidase: Detection of mRNA by PCR and Tissue-Specific Differences in the Protein Levels

Carrie Hiser and Lee McIntosh[†]
Michigan State University, DOE Plant Research Laboratory, and [†]Biochemistry Department, East Lansing, MI, USA.

In addition to the cyanide-sensitive cytochrome pathway of electron transport, the inner membrane of higher plant mitochondria possesses a cyanide-resistant alternative pathway. The alternative pathway consists of an alternative terminal oxidase which accepts electrons from the ubiquinone pool, reduces O_2 to H_2O, and does not conserve energy (for reviews, see Day *et al.*, 1980: Lance *et al.*, 1985). The alternative pathway has been predominantly investigated in the Araceae. In their inflorescences, these plants show high levels of alternative path respiration which raises the temperature of the spadices above ambient temperature, thereby volatilizing chemicals to attract insect pollinators (reviewed in Meeuse and Raskin, 1988). The function of the alternative pathway in non-thermogenic plants, which possess much lower levels of cyanide-resistant respiration, is unknown. However, with the development of monoclonal antibodies against the alternative oxidase (Elthon *et al.*, 1989) and isolation of the alternative oxidase structural gene (Rhoads and McIntosh, 1991), the alternative pathway can be studied at the molecular level in non-thermogenic plants.

The aging of potato tuber slices is a model system in which to study induction of the alternative pathway. The respiration of mitochondria isolated from fresh tuber tissue is almost entirely cyanide-sensitive, while mitochondria from aerobically aged slices exhibit significant levels of cyanide-resistant, alternative path respiration (Dizengremel and Lance, 1976; Theologis and Laties, 1978). The potato tuber alternative oxidase was shown to be an approximately 36 kDa protein which was not detected in mitochondria from fresh tissue, but was present in mitochondria from aged slices (Hiser and McIntosh, 1990). This data suggested *de novo* synthesis of the alternative oxidase during the aging process, although the mode of regulation of this process was not ascertained.

The Molecular and Cellular Biology of the Potato, Second Edition
Edited by W.R. Belknap, M.E. Vayda and W.D. Park

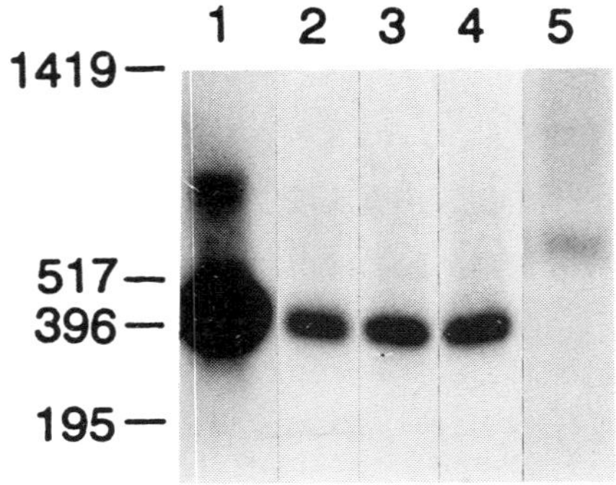

Figure 10.1. Detection of the potato alternative oxidase mRNA by PCR amplification. Total RNA was isolated (Yeh *et al.*, 1991) from fresh and aged slices of Russet-type tubers (Hiser and McIntosh, 1990), and 10 µg samples were reverse-transcribed and PCR-amplified (Kawasaki, 1990). Amplified samples from fresh slices (lane 2), and slices aged for 12 (lane 3) and 24 h (lane 4) were blotted and probed with the plasmid containing the potato alternative oxidase cDNA clone (Hiser, 1991). Controls for PCR amplification included 1 ng of plasmid carrying the potato alternative oxidase cDNA clone (lane 1), and 1 µg Russet Burbank genomic DNA (lane 5). Size markers in nucleotides are shown to the left.

Detection of Alternative Oxidase mRNA by PCR

To investigate further the mode of regulation of the potato alternative oxidase, we attempted to observe mRNA levels in tubers. The mRNA from the potato alternative oxidase gene appears to be of very low abundance. No reproducible signal could be detected on Northern blots of total RNA or mRNA isolated from tubers or leaves (data not shown). Reverse transcription of the RNA samples followed by PCR amplification was necessary to detect the message (Fig. 10.1). An approximately 390 bp amplification product was predicted from the location of binding of the PCR primers to the potato cDNA sequence (Hiser, 1991), and this band was visible on the ethidium bromide stained gel (data not shown) and the blot when the plasmid carrying the potato alternative oxidase cDNA clone was used as the PCR template (lane 1). In the reverse-transcribed, PCR-amplified RNA samples, bands of the predicted size were not seen on the stained gel although several non-specific amplification products were visible (data not shown). After hybridization of the blot, the predicted band appeared in fresh samples (lane 2) and in samples aged for 12 or 24 h (lanes 3 and 4). Because the PCR technique is very sensitive, small amounts of contaminating DNA in the RNA samples might have caused these signals. However, when genomic DNA was PCR-amplified for comparison (lane 5), the product was about 150 bp larger than the product from the reverse-transcribed RNA samples. This showed that the signals in lanes 2-4 were not caused by contaminating DNA. Therefore, a low level of alternative oxidase message seems to be present in both fresh and aged tuber tissue even though, in mitochondria isolated from fresh tubers, the alternative oxidase protein is not (or only faintly) detectable and respiration is almost entirely cyanide-sensitive (Hiser and McIntosh, 1990). Inherent limitations in the PCR technique precluded accurate quantification of the message levels in the samples, so differences in alternative oxidase mRNA levels between fresh and aged tissue may exist but were not detected here. Therefore, some degree of regulation of the alternative oxidase in potato tubers by alteration of mRNA levels should not be excluded. However, these results suggest that the synthesis of the alternative oxidase protein that occurs upon aging of tuber slices (Hiser and McIntosh, 1990) may be regulated in another way, perhaps at the level of translation.

Evidence for Intron(s) in the Potato Alternative Oxidase Gene

When genomic DNA was PCR-amplified, the product was approximately 150 bp longer than the product of PCR-amplified reverse-transcribed RNA (Fig. 10.1). This suggested the presence of an intron(s) in the potato gene located between the primer binding sites. Three introns have been reported in a genomic clone of the *Sauromatum guttatum* alternative oxidase (Rhoads and McIntosh, 1993; see Fig. 10.2). The data from potato could not distinguish between one intron or multiple introns totaling about 150 bp, nor could the exact location be ascertained. The potato intron(s), however, must occur within the region constituting exon 3 in the *S. guttatum* gene, since the primer binding sites lie near the ends of this region (Fig. 10.2). When analyzed by the SEQANAL secondary structure predictor program (Crofts, 1987), this region was predicted to contain two potential transmembrane helices on either side of a third helix which may span the membrane or may be a surface helix (Hiser, 1991; Rhoads and McIntosh, 1991). The amino acid sequence of this region is highly conserved between the dicots potato (Hiser and McIntosh, submitted), *Arabidopsis thaliana* (Kumar and Soll, 1992), soybean (Whelan *et al.*, 1993) and tobacco (Vanlerbergh and McIntosh, 1994); the monocots *S. guttatum* (Rhoads and McIntosh, 1991) and rice (Nickels and McIntosh, unpublished results); and even the yeast *Hansenula anomala* (Sakajo *et al.*, 1993). Although the sequences of genomic clones from plants other than *S. guttatum* are not yet available, our data show that the location of the introns will not necessarily be conserved, even within such a highly conserved gene.

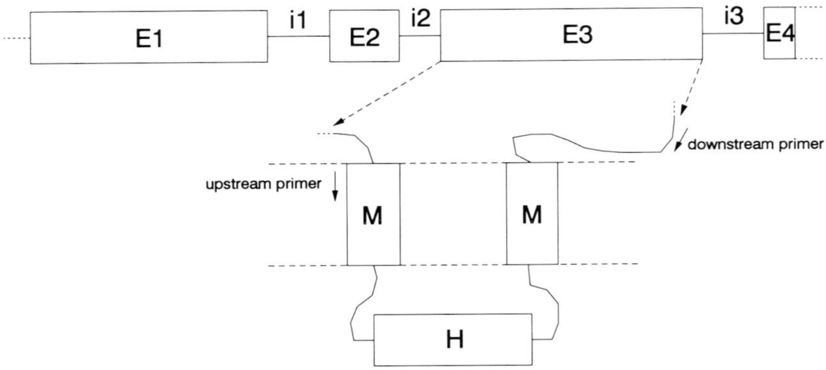

Figure 10.2. An intron(s) in the potato alternative oxidase gene is located in a highly conserved region corresponding to the third exon (E3) of the *Sauromatum guttatum* gene (Rhoads and McIntosh, 1993). This region is predicted to contain two α-helices that span the inner mitochondrial membrane (M) and a potential surface α-helix (H). The upstream and downstream PCR primers, which border the area in which the potato intron(s) is located, are denoted by arrows. E1-E4 are exons and i1-i3 are introns in the known structure of the *S. guttatum* gene.

Tissue-Specific Differences in Alternative Oxidase Proteins

The alternative oxidase from Russet-type potato tuber mitochondria was shown to be an approximately 36 kDa polypeptide (Hiser and McIntosh, 1990). On a Western blot probed with the monoclonal antibody against the alternative oxidase (Elthon *et al.*, 1989), this polypeptide was not detected in mitochondria from fresh tubers, but was present in mitochondria from tuber slices aged 24 h (Hiser and McIntosh, 1990). To investigate the alternative oxidase protein in the FL1607 potato line and in tissues other than tubers, a Western blot of mitochondria isolated from FL1607 leaves, roots and tubers was probed with the same monoclonal antibody (Fig. 10.3). In leaf mitochondria, the alternative oxidase appeared as a closely-spaced doublet of polypeptides of about 35.5 kDa and 36 kDa (lane L). This doublet was also present in root mitochondria (lane R), although at a lower level than in leaves. As in previous experiments that had used Russet-type tubers, only one alternative oxidase polypeptide of about 36 kDa was present in mitochondria isolated from FL1607 tuber slices aged 24 h (lane AT). This polypeptide was only faintly visible in mitochondria isolated from fresh FL1607 tuber tissue (lane FT). These differences in the potato alternative oxidase were therefore tissue-specific and not due to variations between different potato lines.

The reason for the presence of multiple alternative oxidase polypeptides (generally in the 35-37 kDa range) in several plant tissues, including potato leaves, remains unknown. Some plant tissues that possess alternative pathway

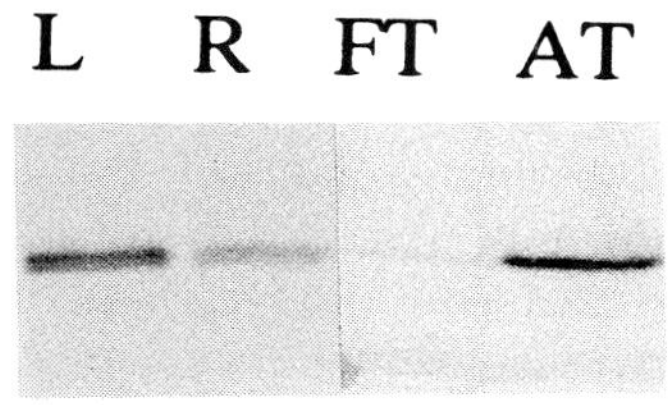

Figure 10.3. FL1607 potato alternative oxidase polypeptides. Mitochondria were isolated from tissue culture-grown FL1607 (Wenzler *et al.*, 1989) leaves and roots by the method of Boutry *et al.* (1984), and from fresh tissue or tissue aged for 24 h from greenhouse-grown FL1607 tubers as described (Hiser and McIntosh, 1990). Samples of 50 µg of mitochondrial protein from leaves (lane L) or roots (lane R) or 500 µg of mitochondrial protein from fresh tubers (lane FT) or from tuber slices aged for 24 h (lane AT) were subjected to SDS-PAGE and Western blotting (Hiser and McIntosh, 1990) and reacted with a monoclonal antibody against the *Sauromatum guttatum* alternative oxidase (Elthon *et al.*, 1989).

capacity possess only one alternative oxidase polypeptide (Elthon *et al.*, 1989; Obenland *et al.*, 1990), including aged potato tuber slices (Hiser and McIntosh, 1991). A single nuclear gene encodes the alternative oxidase structural protein in plants (Rhoads and McIntosh, 1993) and *Neurospora* (Lambowitz *et al.*, 1988), so multiple polypeptides are probably generated by post-translational modification(s) of a single precursor. Which mature polypeptide(s) encodes the functional oxidase is unknown. In several aroid species, a single polypeptide was reported in the tissues possessing low alternative pathway capacity while additional, lower molecular weight polypeptides were present in tissues with high alternative pathway capacity (Elthon and McIntosh, 1987). These observations suggest that lower molecular weight form(s) may be associated with alternative oxidase activity. However, in 3-9 day old soybean cotyledons, an increase in alternative path capacity is correlated with an increase in a lower molecular weight band in nongreen samples and with an increase in a higher molecular weight band in green samples (Obenland *et al.*, 1990). This suggests that different forms of the alternative oxidase may be active in different tissues. Clearly, differential processing/modification occurs in the different potato tissues, but its relationship to the activity of the alternative oxidase remains unclear. Since the alternative oxidase polypeptides in potato leaves were so closely spaced on the Western blots, any correspondence of one of the leaf polypeptides with the single tuber polypeptide could not be determined with certainty. Although the single tuber polypeptide must be active, determination of the amount of the active polypeptide(s) in potato leaves was not possible. One recent report suggested another mode of regulation of the alternative pathway: the alternative oxidase of some plants may exist as a mixture of a dimeric form (intermolecularly cross-linked by disulfide bonds) and a monomeric form, with the level of alternative pathway capacity correlating with the level of the monomeric form (Umbach and Siedow, 1993). This possibility remains to be investigated in potato.

Conclusions

Regulation of the alternative oxidase of potato appears to be complex and seems to vary between tissues. Earlier work suggested that, during the aging of tuber slices, increasing levels of alternative oxidase capacity were correlated with *de novo* synthesis of the alternative oxidase protein. The results presented here cannot exclude that the increased levels of tuber alternative oxidase protein during the aging process were due to increased transcription or mRNA stability, but suggest that another mode of regulation may occur, perhaps at the level of translation. Differences in alternative oxidase polypeptides between leaves, roots and tubers suggest that differential processing and/or modification occurs in different potato tissues. The potential for regulation of alternative oxidase activity in this manner is being investigated. These tissue-specific differences may assist in delineating the role(s) of the alternative pathway in various non-thermogenic plant tissues.

References

Boutry, M., Faber, A-M., Charbonnier, M. and Briquet, M. (1984) Microanalysis of plant mitochondrial protein synthesis products: detection of variant polypeptides associated with cytoplasmic male sterility. *Plant Molecular Biology* 3:445-452.

Crofts, A. (1987) SEQANAL program (University of Illinois).

Day, D.A., Arron, G.P. and Laties, G.G. (1980) Nature and control of respiratory pathways in plants: interaction of cyanide-resistant respiration with the cyanide-sensitive pathway. In: Davies, D.D. (ed.) *The Biochemistry of Plants*, Academic Press, New York, Vol. 2 pp. 179-241.

Dizengremel, P. and Lance, C. (1976) Control of changes in mitochondrial activities during aging of potato slices. *Plant Physiology* 58:147-151.

Elthon, T.E. and McIntosh, L. (1987) Identification of the alternative terminal oxidase of higher plant mitochondria. *Proceedings of the National Academy of Sciences USA* 84:8399-8403.

Elthon, T.E., Nickels, R. and McIntosh, L. (1989) Monoclonal antibodies to the alternative oxidase of higher plant mitochondria. *Plant Physiology* 89: 1311-1317.

Hiser, C. (1991) *Molecular and Developmental Aspects of Respiratory Complexes in Higher Plant Mitochondria*. PhD dissertation, Biochemistry Department, Michigan State University.

Hiser, C. and McIntosh, L. (1990) Alternative oxidase of potato is an integral membrane protein synthesized *de novo* during aging of tuber slices. *Plant Physiology* 93:312-318.

Kawasaki, E. (1990) Methods for RNA-PCR. In: *Perkin-Elmer Spring 1990 Biotechnology Catalog*, p. 19.

Kumar, A.M. and Soll, D. (1992) *Arabidopsis* alternative oxidase sustains *Escherichia coli* respiration. *Proceedings of the National Academy of Sciences USA* 89:10842-10846.

Lambowitz, A.M., Sabourin, J.R., Bertrand, H., Nickels, R. and McIntosh, L (1988) Immunological identification of the alternative oxidase of *Neurospora crassa* mitochondria. *Molecular and Cellular Biology* 9:1362-1364.

Lance, C., Chauveau, M. and Dizengremel, P. (1985) The cyanide-resistant pathway of higher plant mitochondria. In: Douce, R. and Day, D. (eds), *Higher Plant Cell Respiration*. Springer-Verlag, New York, pp. 202-247.

Meeuse, B.J.D. and Raskin, I. (1988) Sexual reproduction in the arum lily family, with emphasis on thermogenicity. *Sexual Plant Reproduction* 1:3-15.

Obenland, D., Dietham, R., Shibles, R. and Stewart, C. (1990) Relationship of alternative respiratory capacity and alternative oxidase amount during soybean seedling growth. *Plant and Cell Physiology* 31:897-901.

Rhoads, D.M. and McIntosh, L. (1991) Isolation and characterization of a cDNA clone encoding an alternative oxidase protein of *Sauromatum guttatum* (Schott). *Proceedings of the National Academy of Sciences USA* 88:2122-2126.

Rhoads, D.M. and McIntosh, L. (1993) The salicylic acid-inducible alternative oxidase gene *aox1* and genes encoding pathogenesis-related proteins share regions of sequence similarity in their promoters. *Plant Molecular Biology* 21:615-624.

Sakajo, S., Minagawa, N. and Yoshimoto, A. (1993) Characterization of the alternative oxidase protein in the yeast *Hansenula anomala*. *FEBS Letters* 318:310-312.

Theologis, A. and Laties, G.G. (1978) Relative contribution of cytochrome-mediated and cyanide-resistant electron transport in fresh and aged slices. *Plant Physiology* 62:232-237.

Umbach, A.L. and Siedow, J.N. (1993) Covalent and noncovalent dimers of the cyanide- resistant alternative oxidase protein in higher plant mitochondria and their relationship to enzyme activity. *Plant Physiology* 103:845-854

Vanlerbergh, G.C. and McIntosh, L. (1994) Mitochondrial electron transport regulation of nuclear gene expression. *Plant Physiology* 105:867-874.

Wenzler, H., Mignery, G., May, G. and Park, W. (1989) A rapid and efficient transformation method for the production of large numbers of transgenic potato plants. *Plant Science* 63:79-85.

Whelan, J., McIntosh, L. and Day, D.A. (1993) Sequencing of a soybean alternative oxidase cDNA clone. *Plant Physiology* 103:1481.

Yeh, K.-W., Juang, R.-H. and Su, J.-C. (1991) A rapid and efficient method for RNA isolation from plants with high carbohydrate content. *BRL Focus* 13:102-103.

Chapter Eleven:

Polyphenol Oxidase in Potato Tubers

P.W. Thygesen, I.B. Dry and S.P. Robinson
CSIRO Division of Horticulture, Adelaide 5001, Australia.

Polyphenol oxidase (PPO) is a copper-containing enzyme of higher plants which catalyzes the conversion of monophenols to *o*-diphenols and *o*-dihydroxyphenols to *o*-quinones (Vaughn *et al.*, 1988). It is responsible for the enzymic browning in plants and such damage is of economic and nutritional significance for fruit and vegetables (Vamos-Vigyazo, 1981). PPO is localized in plastids and, while membrane-associated, it is not an integral membrane protein (Vaughn *et al.*, 1988). The phenolic substrates of PPO are normally found in the vacuole, and their oxidation occurs only when there has been tissue injury and loss of subcellular compartmentation. The quinone products can oxidatively polymerize and react with various amino acid groups of proteins giving rise to black or brown pigment deposits. Blackspot bruising of potato tubers occurs when mechanical damage of potato tubers initiates enzymic browning resulting in the formation of red, brown and black pigments. The most abundant substrate for PPO in potato tubers is chlorogenic acid (Dao and Friedman, 1992) although tyrosine is an important determinant of the degree of browning observed (Matheis, 1987b; Corsini *et al.*, 1992). Despite being implicated in host defense against infection by pathogens (Mayer, 1987) and attack by pests in some instances (Steffens *et al.*, 1990), no *in vivo* function has yet been shown for PPO (Vaughn *et al.*, 1988).

The advent of molecular biology has allowed the critical appraisal of the functions and importance of many plant genes, and the manipulation of their expression. The reliable transformation of potato provides great scope for the improvement of this crop by genetic engineering (Vayda and Belknap, 1992). The first step for such an approach is the isolation and characterization of the genes of interest. To this end we have recently cloned a number of genes encoding PPO from potato and have investigated the expression of PPO in various potato tissues.

The Molecular and Cellular Biology of the Potato, Second Edition
Edited by W.R. Belknap, M.E. Vayda and W.D. Park

Table 11.1. PPO activity in potato tissues.

Tissue	PPO (U . g FW)1
Young leaves	17
Old leaves	7
Stems	2
Flowers	95
Roots	144
Stolons	221
Tuber buds	236
Mature tubers	45

Tissues were extracted in 25 mM MES, 10 mM ascorbate, pH 6.0 and PPO was determined by measuring oxygen uptake in 50 mM NaPi, pH 6.0 at 25°C using an oxygen electrode. The reaction was started by the addition of 4-methylcatechol to a final concentration of 2 mM.
[1]One unit of PPO activity is defined as 1 μmol O_2 consumed min^{-1}. Tuber buds weighed less than 0.1 g and mature tubers weighed 100 g.

PPO Activity in Potato

Hunt *et al.* (1993) reported that PPO protein was immunologically detectable in both young and old leaves using antisera to *Solanum berthaultii* trichome PPO, but they did not correlate this with activity measurements. To obtain some insight into the regulation of PPO expression in potatoes, PPO activity was determined in various potato tissues (Table 11.1). The activity of PPO from many species is enhanced or activated by the presence of anionic detergents (Mayer, 1987; Vaughn *et al.*, 1988). However, the addition of SDS to potato extracts resulted in a decrease in activity rather than an increase (data not shown). PPO activity was generally low in photosynthetic tissues but was higher in young leaves than in fully expanded leaves. There was significant PPO activity in tubers, flowers and roots but the highest activity was in stolons and tuber buds (Table 11.1).

The PPO activity of tubers at varying stages of development is shown in Fig. 11.1. Total PPO activity per tuber increased more or less linearly with increasing tuber size (Fig. 11.1) suggesting continued synthesis of the enzyme throughout tuber development. On a fresh weight basis, PPO activity was very high in small tubers, but initially decreased as the tubers increased in size and was then relatively constant for tubers larger than 10 g (Fig. 11.1). PPO activity per gram fresh weight was higher in the peel tissue of tubers than in the cortex, particularly in the later stages of tuber development (Fig. 11.1). Comparison of the total PPO activity in the peel and cortex of tubers revealed that although the specific activity of peel was 2-3 times higher than that of cortex tissue, the peel

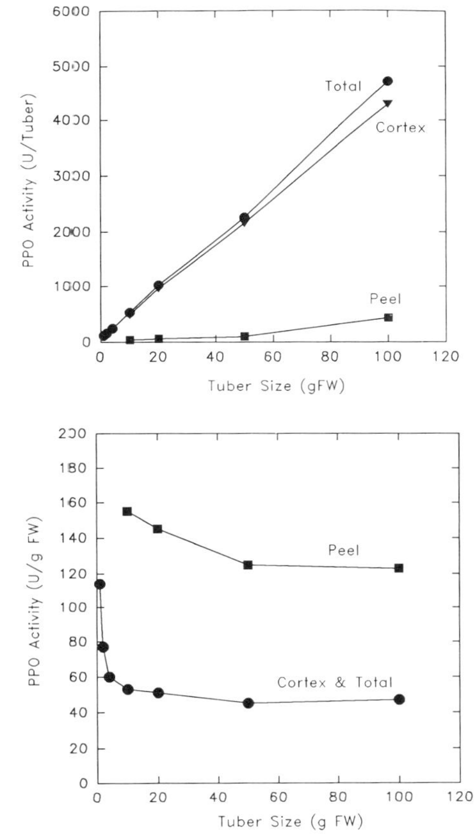

Figure 11.1. PPO activity in potato tubers. Tubers of varying size were collected from actively growing field potatoes. PPO activity was determined in the peel and cortex tissue as in Table 11.1 and total PPO was calculated. The upper panel shows PPO expressed on a per tuber basis and in the lower panel activity is expressed per gram fresh weight of each tissue.

only contributed 2-3% of the fresh weight and less than 10% of the total PPO activity (Fig. 11.1). These data suggest that there is differential expression of PPO activity in tubers and other potato tissues. In early tuber development there appears to be a rapid accumulation of PPO activity which then levels out and is maintained as the tubers continue to grow. The higher PPO levels observed in peel are consistent with a potential role in plant defense, as the peel will be the first point of contact for an invading pathogen or insect herbivore.

Sequence	Source
A **P** A P P **P** **D** L S S - - - - -	*Solanum tuberosum* tuber
S **P** I P P **P** **D** L K S C G V A H	*S. tuberosum* leaf [1]
S **P** I P P **P** **D** L K S X G V A H	*S. berthaultii* trichome [2]
A **P** I P P **P** **D** L K S X G T A -	Tomato trichome [2]
S **P** I S P **P** **D** F/L S K C G P P -	Broadbean leaf [3]
A **P** I Q A **P** **D** I S K C G T A T	Grape berry[4]

Figure 11.2. N-terminal amino acid sequences of PPO proteins. An alignment of the sequences of the mature PPO proteins from various tissues was made and absolutely conserved residues are shown in bold. Ambiguous residues are represented by X, and residues not determined by -. The sequence for *Solanum tuberosum* tuber was obtained from PPO purified from the cortex of mature tubers (see text). *S. berthaultii* is a wild relative of *S. tuberosum*. 1: deduced from the cDNA sequence in Hunt *et al.* (1993), 2: Yu *et al.* (1992), 3: Robinson and Dry (1992). 4: Dry and Robinson (1994).

Isolation of cDNA Clones Encoding PPO From Potato Tubers

PPO was purified 110-fold from the cortex tissue of 200-300 g tubers. The resulting preparations exhibited only two protein bands on denaturing SDS-PAGE gels with apparent molecular mass of 60 and 40 kDa (data not shown). The N-terminal amino acid sequences of the two proteins were obtained by microsequencing. The sequence of the 40 kDa protein was found to be identical to the tuber storage protein, patatin (Park *et al.*, 1983). The N-terminal sequence of the 60 kDa protein was compared with those of other PPOs (see Fig. 11.2). The sequence is very similar to that of PPO from potato leaf and trichomes of the wild potato relative *S. berthaultii* and tomato. There is also obvious homology to the N-terminal sequence of PPO from grape and broad bean. The nature and extent of the observed sequence conservation thus validated the identity of the 60 kDa protein from potato tubers as PPO.

A number of cDNA clones encoding PPO were isolated from young, developing potato tuber tissue (< 1 g, cv. Norchip) by 3'-rapid amplification of cDNA ends (3'-RACE; Frohman *et al.*, 1989) using oligonucleotide primers designed from conserved regions of known PPO sequences. Full-length clones were obtained using the 5'-RACE technique (Frohman *et al.*, 1989). We have identified a number of distinct cDNAs from potato tubers which can be grouped into three families, based on sequence homology, which we have designated POT32, POT33 and POT72. At this stage we have concentrated on characterizing representatives of the POT32 and POT33 families as these appear

Table 11.2. Homology between PPO sequences.

	POT32	
	% identity	% similarity
POT33	80	86
Potato leaf'[1]	71	80
Tomato leaf[2]	68	77
Tomato flower[3]	68	76
Broadbean leaf[4]	39	50
Grape berry[5]	38	54

The degree of homology to the potato tuber cDNA clone POT32 at the amino acid level was determined using the sequence analysis program PALIGN (PC/Gene, Intelligenetics, Inc.).
POT33 is another tuber cDNA clone. Similar residues are defined as those that can be functionally interchanged. Remaining sequences are from: [1]Hunt *et al.* (1993), [2]Newman *et al.* (1993), [3]Shahar *et al.* (1992), [4]Cary *et al.* (1992), [5]Dry and Robinson (1994).

to be the most highly expressed. The clones we have isolated are related to, but distinct from, those isolated from potato leaves by Hunt *et al.* (1993).

Each clone is ~2 kb in length and contains an open reading frame of ~1.8 kb which encodes a protein of ~ 67 kDa. The N-terminal sequence of PPO purified from potato tuber cortex shown in Fig. 11.2 is identical to the deduced amino acid sequence of clones from the POT32 family. Taking account of the known N-terminal sequence of potato tuber PPO, both POT32 and POT33 would encode mature proteins of ~58 kDa following the cleavage of a putative ~9.3 kDa plastid transit peptide. This predicted size is in good agreement with the observed size of the PPO protein purified from tuber cortex which was 60 kDa. The ~9.3 kDa sequence possesses many of the features characteristic of chloroplast transit sequences (Keegstra *et al.*, 1989) as do the corresponding regions in PPO clones from potato leaf (Hunt *et al.*, 1993), tomato flower (Shahar *et al.*, 1992; Newman *et al.*, 1993) broad bean leaf (Cary *et al.*, 1992) and grape berry (S.P. Robinson and I.B. Dry, unpublished results). Indeed, all the PPO genes characterized to date encode ~68 kDa proteins which possess plastid transit peptide sequences and are processed to mature proteins of ~60 kDa following import into the plastid. In the case of potato tubers the PPO protein is presumably imported into the amyloplasts of the tuber cells. Early estimates of the size of PPO varied widely (Vaughn *et al.*, 1988) and there were reports of two distinct PPOs in potato, a ~59 kDa and a ~45 kDa protein (Matheis, 1987a; Steffens *et al.*, 1990). However subsequent characterization of

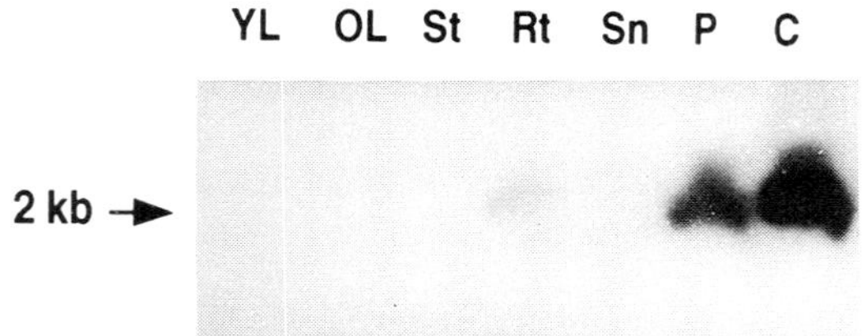

Figure 11.3. Expression of POT32 in potato tissues. Northern blot of 20 μg of total RNA from each tissue hybridized with PPO clone POT32. Blots were hybridized at 65°C and washed in 0.1 x SSC at 65°C. YL: young leaves; OL: mature leaves; St: stems; Rt: roots; Sn: stolons; P: peel; C: cortex.

PPO genes has found no evidence for a 45 kDa form (Yu *et al.*, 1992; Hunt *et al.*, 1993) and the differences in apparent molecular weight of PPO in potato most probably reflect artifacts related to the isolation protocol or the electrophoretic conditions used (Robinson and Dry, 1992).

The relative identity and similarity of the deduced amino acid sequences of PPO genes isolated from a number of different plant tissues is shown in Table 11.2. The potato tuber cDNAs POT32 and POT33 are more closely related to each other than to the potato leaf PPO. The PPOs of potato and tomato are very similar, as might be predicted given how closely the two species are related. The PPO sequences from broad bean and grape (both phylogenetically divergent from potato), exhibit a lesser degree of sequence similarity with potato PPO. The major area of homology occurs in two histidine-containing regions of the PPO protein proposed to form the copper-binding domains (data not shown). These regions have been identified by their homology to the known copper-binding sites of hemocyanins and tyrosinases (Cary *et al.*, 1992; Shahar *et al.*, 1992; Newman *et al.*, 1993; Dry and Robinson, 1994).

Expression Patterns of PPO Genes in Potato Tissues

POT32 was used to probe total RNA samples isolated from a range of potato tissues in order to determine its pattern of expression. There was considerable hybridization to both tuber cortex and peel RNA and some hybridization to roots, but no hybridization to RNA from leaves, stems or stolons (Fig. 11.3). Analogous experiments using POT33 as a probe also failed to elicit any hybridization to RNA from green tissues and very little hybridization to tuber cortex RNA (data not shown). These data suggest that the predominant PPO transcript in tuber cortex is from the POT32 gene family. This contention is supported by the identity of the N-terminal sequence of the PPO protein isolated from tuber cortex tissue with the sequence of POT32. These results also indicate that POT32 (and POT33) is preferentially, if not exclusively, expressed in tuber tissues and that they are not varietal variants of the cloned leaf genes (Hunt *et*

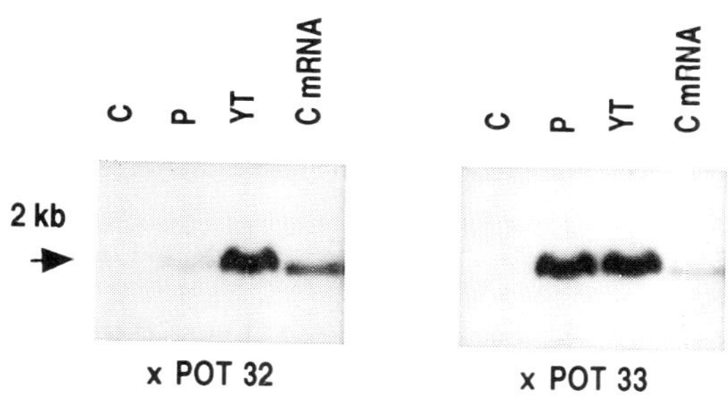

Figure 11.4. Expression of POT32 and POT33 in potato tuber tissues. Northern blots as in Fig. 11.3 with 20 μg total RNA or 1 μg mRNA. C: cortex of 10 g tubers; P: peel of 10 g tubers; YT: whole young tubers (<1 g); C mRNA: mRNA from cortex of 10 g tubers.

al., 1993). The absence of hybridization to RNA from stolons, the progenitor organ of tubers, despite the high PPO activity detected, suggests that there may be a stolon-specific PPO gene (which does not cross-hybridize with POT32) yet to be isolated. Alternatively, high levels of transcription of a PPO gene may only occur in the growing tip, which comprises a very small part of the stolon tissue used for RNA isolation. The PPO enzyme thus synthesized may remain functional in the rest of the stolon. It is also possible that the level of PPO transcript detected in a particular tissue is not correlated with the PPO activity due to factors such as the rate of mRNA turnover.

RNA isolated from peel and cortex tissues of 10 g potato tubers and from young tubers (<1 g) were probed with either POT32 or POT33; the results of this experiment are shown in Fig. 11.4. Both probes hybridized strongly to RNA from young tubers as might be predicted given the high PPO activity measured in these tubers. POT32 is obviously more highly expressed in cortex than POT33, but the most striking result is the very strong hybridization of POT33 to RNA from peel relative to the weak hybridization of POT32 to this tissue. This suggests that POT33 may be a more peel-specific form of PPO in tubers. Further experiments are being undertaken to elucidate the pattern of expression of POT32 and POT33 within the tuber and throughout tuber development. We are also investigating when and where the other PPO transcripts isolated from tuber tissues are expressed.

Given that there are seven PPO genes present in tomato (Newman *et al.*, 1993) it is perhaps not surprising that a number of PPO genes (at least six) are also present in the closely related potato plant, which is a tetraploid. We have isolated at least four distinct PPO genes in potato tubers and an additional two PPO genes have been described in potato leaves by Hunt *et al.* (1993). Despite the high levels of PPO activity in potato tissues such as flowers, roots and stolons (Table 11.1) it is difficult to detect correspondingly high levels of mRNA transcripts of the known genes in these tissues (flowers: Hunt *et al.*

1993; roots: Fig. 11.2, stolons: Fig. 11.3 and P.W.Thygesen and S.P. Robinson, unpublished results). This may result from expression of known PPO genes only at specific developmental stages or in specific tissue types or there may be additional PPO genes in potato which are expressed preferentially or specifically in these tissues. Tissue-specific expression of PPO genes has been reported in tomato (Shahar *et al.*, 1992; Newman *et al.*, 1993) and it has been postulated that control of this expression is mediated by different promoter elements found preceding these genes (Newman *et al.*, 1993). It will therefore be of interest to characterize the 5' regions of the PPO genes from potato tubers and to elucidate the signal transduction pathway(s) involved in their regulation.

The isolation and characterization of genes encoding PPO provides the basis for the creation of specific PPO mutants, either deficient in or over-expressing this activity. Analysis of such mutants may provide evidence to elucidate the *in vivo* function(s) of PPO and whether the manipulation of its expression' can alter the agronomic characteristics of the potato plant.

Note

The control of PPO activity in potatoes forms part of a patent application entitled 'Polyphenol oxidase genes CSIRO, PCT/AU92/00356'.

References

Cary, J.W., Lax, A.R. and Flurkey, W.H. (1992) Cloning and characterization of cDNAs coding for *Vicia faba* polyphenol oxidase. *Plant Molecular Biology* 20:245-253.

Corsini, D.L., Pavek, J.J. and Dean, B. (1992) Differences in free and protein-bound tyrosine among potato genotypes and their relationship to internal blackspot resistance. *American Potato Journal* 69:423-435.

Dao, L. and Friedman, M. (1992) Chlorogenic acid content of fresh and processed potatoes determined by ultraviolet spectrophotometry. *Journal of Agricultural Food and Chemistry* 40:2152-2156.

Dry, I.B. and Robinson, S.P. (1994) Molecular cloning and characterisation of grape berry polyphenol oxidase. *Plant Molecular Biology*, in press.

Frohman, M.A., Dush, M.K. and Martin, G.R. (1989) Rapid amplification of full-length cDNAs from rare transcripts: amplification using single gene-specific oligonucleotide primers. *Proceedings of the National Academy of Sciences USA* 85:8998-9002.

Hunt, M.D., Eannetta, N.T., Haifeng, Y., Newman, S.M. and Steffens, J.C. (1993) cDNA cloning and expression of potato polyphenol oxidase. *Plant Molecular Biology* 21:59-68.

Keegstra, K., Olsen, L.J. and Theg, S.M. (1989) Chloroplastic precursors and their transport across the envelope membranes. *Annual Review of Plant Physiology*. 40:471-501.

Matheis, G. (1987a) Polyphenol oxidase and enzymatic browning of potatoes (*Solanum tuberosum*) I. Properties of potato polyphenol oxidase. *Chemie Mikrobiologie Technologie Lebensmittel* 11:5-12.

Matheis, G. (1987b) Polyphenol oxidase and enzymatic browning of potatoes (*Solanum tuberosum*) II. Enzymatic browning and potato constituents. *Chemie Mikrobiologie Technologie Lebensmittel* 11:33-41.

Mayer, A.M. (1987) Polyphenol oxidases in plants - recent progress. *Phytochemistry* 26:11-20.

Newman, S.M., Eannetta, N.T., Yu, H., Prince, J.P., de Vicente, M.C., Tanksley, S.D. and Steffens, J.C. (1993) Organization of the tomato polyphenol oxidase gene family. *Plant Molecular Biology* 21:1035-1051.

Park, W.D., Blackwood, C., Mignery, G.A., Hermodson, M.A. and Lister, R.M. (1983) Analysis of the heterogeneity of the 40,000 molecular weight tuber glycoprotein of potatoes by immunological methods and NH_2-terminal sequence analysis. *Plant Physiology* 71:156-160.

Robinson, S.P. and Dry, I.B. (1992) Broad bean leaf polyphenol oxidase is a 60-kilodalton protein susceptible to proteolytic cleavage. *Plant Physiology* 99:317-323.

Shahar, T., Hennig, N., Gutfinger, T., Haleven, D. and Lifshcitz, E. (1992) The tomato 66.3-kD polyphenol oxidase gene:molecular identification and developmental expression. *The Plant Cell* 4:135-147.

Steffens, J.C., Kowalski, S.P. and Yu, H. (1990) Characterization of glandular trichome and plastid polyphenol oxidases of potato. In Vayda, M. and Park, W.D. (eds), *The Molecular and Cellular Biology of the Potato*. CAB International, Wallingford, UK,pp. 103-112.

Vamos-Vigyazo, L. (1981) Polyphenol oxidase and peroxidase in fruits and vegetables. *CRC Critical Reviews in Food Science and Nutrition* 15:49-127.

Vaughn, P.F.T., Eason, R., Paton, J.Y. and Ritchie, G.A. (1988) Polyphenol oxidase: the chloroplast enzyme with no established function. *Physiologica Plantarum* 72:659-665.

Vayda, M.E. and Belknap, W.R. (1992) The emergence of transgenic potatoes as commercial products and tools for basic science. *Transgenic Research* 1:149-163.

Yu, H., Kowalski, S.P. and Steffens, J.C. (1992) Comparison of polyphenol oxidase expression in glandular trichomes of *Solanum* and *Lycopersicon* species. *Plant Physiology* 100:1885-1890.

Chapter Twelve:

Gene Expression During Early Tuber Development

David J. Hannapel
Department of Horticulture, Iowa State University, Ames, IA, USA.

Despite many recent advances made in understanding the controling mechanisms of developmental biology, we still know very little about the fundamental basis of the regulation of plant growth. Plants represent a diverse group of organisms with unique biochemical and physiological processes (Goldberg, 1989). Despite their morphological simplicity, plants have developmental genetic programs that are as complex as animal systems. With the recent advent of such technology as gene transfer, expression in transgenic plants, and regeneration of whole plants from single cells, it has become possible to examine specific molecular events responsible for controling plant growth processes. One such developmental system that encompasses a number of important biological processes like carbon partitioning, source-sink relations and the control of growth correlation is potato tuberization. The process of tuberization is a complex one, involving a number of biochemical and morphological events including storage protein synthesis, sucrose transport, starch accumulation, and cell division and expansion. The mechanism of tuberization has been the subject of considerable investigation in the past several decades (Chapman, 1958; Gregory, 1965; Vreugdenhill and Struik, 1989). Environmental factors such as photoperiod, temperature and nitrogen nutrition have a significant influence in controling tuber formation and development (Gregory, 1965).

Under an optimum set of environmental conditions, the potato plant is programmed to synthesize a specific group of proteins and to accumulate large amounts of starch at the tip of an underground stem (stolon) to produce the storage organ (tuber). The fundamental mechanisms for initiating the biochemical events leading to tuberization and controling photosynthate partitioning in whole plants are important biological questions that remain unanswered. Specific genes coding for abundant proteins associated with the process of tuberization have been characterized and shown to be excellent biochemical markers for this developmental system (Lee *et al.*, 1983; Hannapel *et al.*, 1985; Park *et al.*, 1985). Recent advances in understanding potato gene expression have been made at the molecular level studying the major tuber proteins, patatin and the proteinase inhibitors. Both have known metabolic

The Molecular and Cellular Biology of the Potato, Second Edition
Edited by W.R. Belknap, M.E. Vayda and W.D. Park

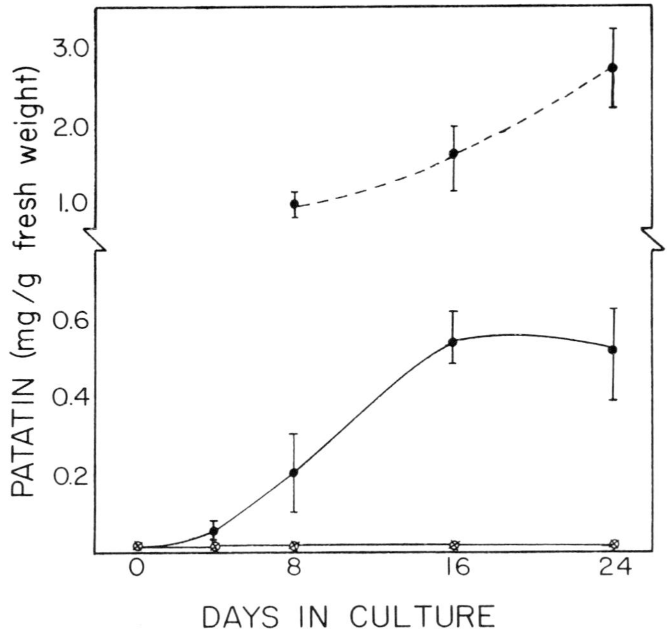

Figure 12.1. Patatin content of tubers (----) and stolon tips (——) cultured *in vitro* with 10 µM kinetin (closed circles), 3 µM GA3 plus kinetin (open circels), or no hormones (x). Vertical bars represent standard deviation. Tubers from stolon tips of the cv. 'Superior' first appeared on day 8. The stolon tips assayed were those that did not produce any tubers. All cultures were kept in the dark and grown on a Murashige and Skoog salt medium plus 6% sucrose with hormones or 2% sucrose without hormones.

activities and their functional roles in the potato plant have been discussed (Ryan, 1981; Racusen, 1984).

Inhibition by GA of Tuber Storage Protein Accumulation

As an example of how a specific tuber storage protein can work as a biochemical marker for the process of tuberization, consider the effect of gibberellic acid (GA) on tuberization. There are numerous examples in which GA is correlated with the inhibition of tuberization. GA has been shown to reduce tuber yield in whole plants and to prevent tuber formation in leaf bud cuttings, *in vitro* cultured stem pieces, and etiolated shoots (Harmey *et al.*, 1966; Garcia-Torres and Gomez-Campo, 1973; Kumar and Wareing, 1974; Menzel, 1980). The noninduced state in whole plants grown under long days has been correlated with high endogenous levels of GA in leaves, while short days result in the converse (Okazawa, 1960;

Railton and Wareing, 1973). GA affects carbohydrate distribution to favor stem and leaf growth in a response which appears to mimic the effects of high temperature (Menzel, 1980). In summary, while there are certainly other phytohormones involved in controling tuberization, the evidence implies that the level of GA is an important factor in the initiation and subsequent development of potato tubers.

GA_3 inhibition of the accumulation of patatin was demonstrated utilizing an *in vitro* cultured stolon system. When cultured stolon tips were placed on an inductive medium (6% sucrose, 10 μM kinetin), approximately 75% formed tubers. These tubers accumulated large amounts of patatin, reaching 3.0 mg/g fresh weight by 24 days (Fig. 12.1). The remainder of the cultured stolon tips on this medium developed as swollen stolons which contained approximately 500 μg/g fresh weight patatin. Stolons cultured on the 3 μM GA_3 plus 10 μM kinetin medium continued to develop as stolons and contained less than 10 μg/g fresh weight patatin. These results were comparable to those obtained on the medium with 2% sucrose and no hormones. These results indicate that patatin accumulation is readily controled by the same phytohormone that has such a profound effect on tuber development. GA also inhibits sucrose-induced mRNA accumulation for patatin and the proteinase inhibitors in leaves (D. Hannapel, unpublished data).

Differential Expression of Tuber Proteins

Even though patatin and the other major tuber proteins do not exhibit strict tuber-specific expression, their very close correlation with the early events of tuber development and their relative abundance in tubers distinguishes them as biochemical markers for the ongoing processes of tuberization. It is possible that substantial tuber protein accumulation in nontuber tissue reflects the induction of specific biochemical events of tuberization without the accustomed morphological response. To ascertain if such biochemical induction is possible in potato species incapable of attaining tuber morphology, I have studied the induced expression of two distinct potato tuber genes present in genomes of plants that do not bear tubers.

Diploid potato species of the series Etuberosa do not bear tubers and have been identified by geneticists as a reservoir of germplasm of potential value for breeding for disease resistance, quality and environmental adaptation (Hermsen, 1980). In an attempt to transfer nontuber potato germplasm to *Solanum tuberosum*, Ehlenfeldt and Hanneman (1984) identified a triploid hybrid ($2n = 3x = 36$), which produced only stolons, and a pentaploid hybrid ($2n = 5x = 60$), which produced tubers (see Table 12.1). The triploid hybrid (BC) resulted from a cross between colchicine-induced *Solanum brevidens* (a diploid species without tubers) and *S. chacoense* (a diploid tuber-bearing species). The pentaploid hybrid (BCT) resulted from a cross between the triploid (BC) hybrid and *S. tuberosum* 'Wis AG 231' ($2n = 4x = 48$). The triploid hybrid has a 2:1 ratio of genomes for

Table 12.1. Production of hybrids utilizing tuber-bearing and nontuber-bearing potato species.

Ratio of nontuber-bearing genomes to tuber-bearing genomes	Cross	Ploidy level	Morphology
	S. brevidens[1] (colchicine-treated) x *S. chacoense*[2] ↓		
2 : 1	*brevidens/chacoense* hybrid (BC)	3*x*	stolons
	BC x *S. tuberosum* 'Wis Ag 231' ↓		
2 : 3	BC/*S. tuberosum* hybrid (BCT)	5*x*	tubers

[1] *S. brevidens* is a nontuber-bearing diploid species

[2] *S. chacoense* is a tuber-bearing diploid species

nontuber-bearing plants to tuber-bearing plants. The pentaploid hybrid has a larger number of genomes from the tuber-bearing plants (a ratio of 2:3). A comparison of morphology and genome ratios in these crosses suggests that tuberization may be controled by a dosage or threshold effect (Ehlenfeldt and Hanneman, 1984).

Southern blots of genomic DNA probed with two tuber cDNAs were performed to determine if related sequences were present in the genomes of plants that do not produce tubers. Fig. 12.2 A and B show the results of the analysis of patatin genes and the cathepsin D 22 kDa proteinase inhibitor, respectively. Both the nontuber genomes contain sequences related to the tuber gene probes (Fig. 12.2, lanes 1 and 2), but the complexity of these related sequences is greatly reduced compared with the genomes for tuberizing plants (Fig. 12.2, lanes 3-6).

Although sequences related to the tuber protein genes are present in nontuber genomes, the status of their functionality must be assessed. Because the Etuberosa types do not produce stolons or tubers, it is impossible to study their induction in these organs. But the evaluation of tuber formation from axillary buds of petiole-leaf cuttings is a suitable model system for determining a plant's potential to form tubers (Ewing, 1985). When a single petiole-leaf cutting with an intact axillary bud and a 1-2 cm section of stem attached is taken from plants grown under short days and placed in moist sand, the axillary buds will grow out as small tubers. With similar cuttings taken from plants grown under long days, the axillary buds will grow out as shoots (Paiva *et al.*, 1982). Table 12.2 summarizes the results of an evaluation of the morphological capacity of the six species and hybrids examined in this study (plus cv. 'Superior') to form tubers or stolons from buds of petiole-leaf cuttings. As shown previously (Ewing, 1985),

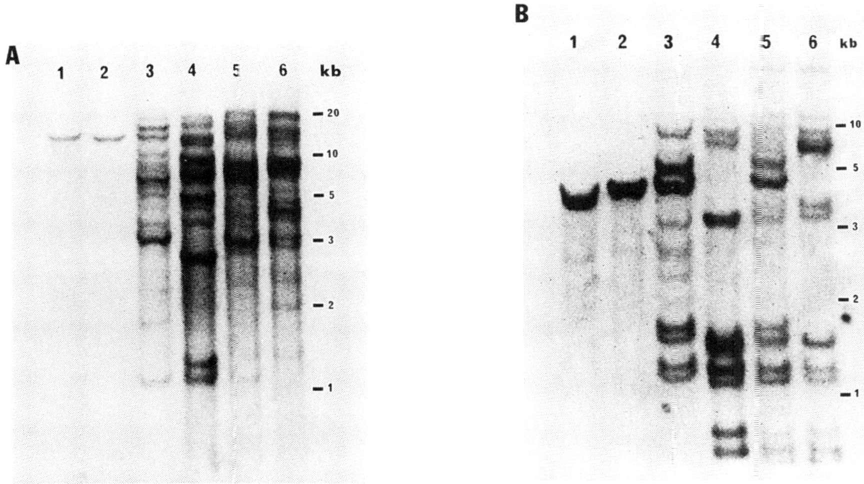

Figure 12.2. Southern blot analysis of *Eco*RI-digested genomic DNA from 1) *S. etuberosum*, 2) *S. brevidens*, 3) BC triploid hybrid, 4) *S. chacoense*, 5) BCT pentaploid hybrid, and 6) *S. tuberosum* 'Wis AG 231' hybridized to either ^{32}P-labeled patatin cDNA insert from pGM01 (A) or the cathepsin D inhibitor cDNA p740 (B). Ten micrograms of DNA from each sample was loaded per lane.

Table 12.2. Evaluation of tuber formation from axillary buds of petiole leaf cuttings from *Solanum* species and hybrids.

Species/hybrid	No. of cuttings formed	No. of tubers formed	No. of stolons formed	No. of shoots formed
S. etuberosum	18	0	0	18
S. brevidens	18	0	0	18
S. brevidens x *S. chacoense*	24	0	3	21
S. brevidens/chacoense x *S. tuberosum* 'Wis AG 231'	20	15	2	3
S. chacoense	18	0	15	3
S. tuberosum 'Wis AG 231'	18	17	0	1
S. tuberosum 'Superior'	10	10	0	0

Table 12.3. Starch accumulation in petioles of petiole-leaf cuttings with the axillary bud excised.

	mg starch / 15 mg dry weight		
Species/hybrid	0 day petioles[1]	28 day petioles	tuber
S. etuberosum	0.4 ± 0.1	3.5 ± 0.4	---
S. brevidens	0.9 ± 0.1	6.7 ± 1.5	---
S. brevidens x *S. chacoense*	1.9 ± 0.6	9.2 ± 0.6	---
S. brevidens/chacoense x *S. tuberosum* 'Wis AG 231'	0.3 ± 0.1	8.5 ± 0.4	---
S. chacoense	1.3 ± 0.2	9.7 ± 0.8	---
S. tuberosum 'Wis AG 231'	0.8 ± 0.04	10.0 ± 0.7	9.4 ± 0.3
S. tuberosum 'Superior'	0.4 ± 0.03	8.0 ± 0.2	9.0 ± 0.5

1 0 day petioles were analyzed directly from whole plants, while 28 day petioles were removed from whole plants with the axillary bud excised, incubated under an 8 h photoperiod for 28 day and then analyzed for starch content.

there seemed to be intermediate stages of organ development between tubers (strong induction) and shoots (no induction). This category is designated 'stolons'; under these conditions, axillary bud growth is characterized by swelling, lack of leaf development and reduction in chlorophyll content. As expected, the three nontuber types did not product axillary tubers on cuttings, and the BC hybrid produced very few stolons (Table 12.2). The three *S. tuberosum* types produced a very high proportion of axillary tubers. *S. chacoense*, a diploid tuber-bearing type, produced mostly stolons but no tubers. Combining these results with those from whole-plant morphology (Ehlenfeldt and Hanneman, 1984) demonstrated conclusively that the Etuberosa types lacked the capacity to form either stolons or tubers. By manipulating sink availability, however, the tuber proteins can be induced to accumulate without tuber formation along with large amounts of starch, in stems and petioles from a petiole-leaf cutting system (Paiva *et al.*, 1983). These petiole-leaf cuttings with the axillary bud removed can be used as a system for inducing expression of tuber protein genes in the absence of tuber morphology.

Because starch is known to accumulate synchronously with the induction of tuber proteins in these petioles, starch content was measured in both 0 and 28 day petioles. These results are summarized in Table 12.3. As expected, large amounts of starch accumulated in those petioles where tuber proteins also accumulated (BC, BCT, *S. chacoense* and 'Wis AG 231'), but substantial amounts of starch also accumulated in the petioles from the nontuber types, even

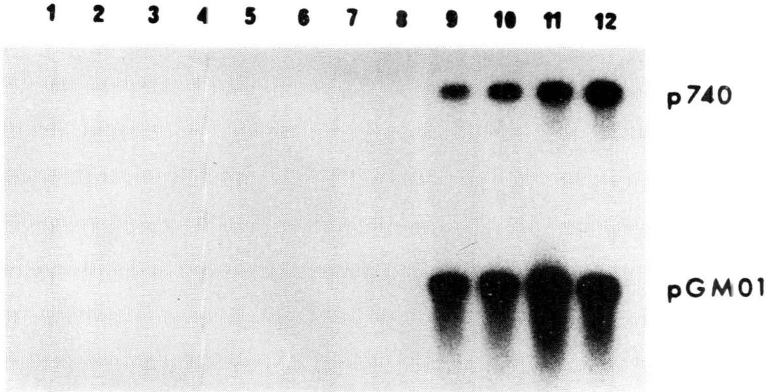

Figure 12.3. Northern blot analysis of polysomal RNA extracted from 0 day (lanes 1-6) or 28 day (lanes 7-12) petioles hybridized to either the ^{32}P-labeled patatin cDNA pGM01 or the cathepsin D inhibitor cDNA p740. RNA was extracted from petioles of *S. etuberosum* (lanes 1 and 7), *S. brevidens* (lanes 2 and 8), the BC triploid hybrid (lanes 3 and 9), the BCT pentaploid hybrid (lanes 4 and 10), *S. chacoense* (lanes 5 and 11) and *S. tuberosum* 'Wis AG 231' (lanes 6 and 12). Ten micrograms of polysomal RNA was loaded for the 0 day samples and the 28 day samples of *S. etuberosum* and *S. brevidens*. Three micrograms was loaded for the remaining 28 day samples.

though the two major tuber protein groups did not. *S. etuberosum* and *S. brevidens* 28 day petioles exhibited 9-fold and 7-fold increases, respectively, in starch levels during the 28 days of culture. This compares with a 12-fold increase for 'Wis AG 231' petioles.

To assess tuber gene expression at the translational level under the inductive conditions of the petiole-cutting system, polysomal mRNA was extracted from both 0 and 28 day petioles and analyzed by using Northern blot hybridization (Fig. 12.3). Denatured mRNAs were probed with either labeled pGM01 (patatin) or p740 (cathepsin D inhibitor) inserts. No signal was detected on the Northern blots with either probe against polysomal mRNA extracted from either 0 day or 28 day petioles of *S. etuberosum* and *S. brevidens* (Fig. 12.3, lanes 1, 7 and 2, 8). The cathepsin D inhibitor gene is wound-inducible in *S. brevidens* and *S. etuberosum* leaves (Hansen and Hannapel, 1992) but no patatin transcripts have been detected in RNA isolated from either of these two nontuber-bearing plant species. Substantial amounts of patatin mRNA (approximately 1500 nucleotides in length) accumulated by 28 day to a similar level in all four stolen or tuber-producing types (Fig. 12.3, lanes 9-12). The highest accumulation of the cathepsin D inhibitor transcript (approximately 1000 nucleotides in length) occurred in mRNA from induced petioles of *S. chacoense* and AG 231 (Fig. 12.3, lanes 11 and 12), with lower cathepsin D inhibitor transcript levels in mRNA from the two hybrids (Fig. 12.3, lanes 9 and 10). These reductions in cathepsin D inhibitor transcript amounts were evident in all replications and were verified using a related but different cDNA probe encoding a cathepsin D inhibitor type tuber protein (data not shown). The same patterns of expression can be reproduced in leaf explants cultured on 300 mM sucrose (Hannapel, 1990). Again, these data indicate that tuber storage proteins appear to be good biochemical markers of the process of tuberization. Nontuber-bearing potato species contain sequences homologous to the tuber protein cDNAs. If these sequences represent functional genes, they cannot be induced under conditions favorable for the induction of their tuber-bearing counterparts. However, there is much more to tuberization than storage protein synthesis and starch accumulation. Recent studies using antisense constructs have shown that a 90% reduction of patatin accumulation (Höfgen and Willmitzer, 1992) and a 95% reduction of starch accumulation (Müller-Röber *et al.*, 1992) in tubers had very little effect on tuber morphology.

Characterization of the Early Events of Potato Tuber Development

Gregory (1965) has considered the development of potato tubers in three general stages: i) tuber initiation, characterized by differentiation and the development of tuber primordium without evidence of any visible swelling; ii) tuber enlargement, characterized by accumulation of storage proteins and starch, rapid cell division and visible swelling of the stolon tip; and iii) tuber maturation, when the organ passes into the dormant period. To date, most studies of the tuberization process have focused on expression of the storage proteins and activity of starch synthesis enzymes. Both of these are representative of the biochemical events of the tuber enlargement stage.

Both morphological and biochemical analytical approaches have been taken to examine the mechanisms associated with tuber initiation and development. Duncan and Ewing (1984) used single-node potato leaf cuttings to study anatomical changes associated with tuber formation from axillary buds. They showed that starch deposition, cell mitosis and cell enlargement were the earliest detectable changes in anatomy associated with tuber initiation. Because of its uniformity, the axillary bud system is an excellent model to examine early events of tuber morphology. By manipulating sink availability potato plants can be induced to form tubers from stem axillary buds (Ewing, 1985; Hannapel, 1991; Wheeler *et al.*, 1988). When petiole-leaf stem cuttings taken from plants grown under short days are placed in moist sand, the axillary buds will grow out as small tubers. With similar cuttings taken from plants grown under long days, the axillary buds grow out as shoots (Paiva *et al.*, 1982; Wheeler *et al.*, 1988). Since stem cuttings are reproducible and easy to manipulate, they have been widely used to study the induction of tuberization. Initial developmental changes in short-day axillary buds have been detected as follows: day 1, starch accumulation and cell division; day 2, patatin accumulation; day 4, cell expansion; and day 6, obvious tuber formation (Duncan and Ewing, 1984; Hannapel, 1991).

These are several advantages in using the axillary bud tuber system:

1. *Axillary tuber development closely mimics underground stolon tuber development.* Starch and total protein accumulation in axillary tubers is equivalent on a fresh weight basis to normal tubers (Paiva *et al.*, 1982). The protein banding patterns resolved by both nondenaturing and SDS PAGE for both tuber types is virtually identical (Paiva *et al.*, 1982). The two protein types are also immunologically identical based on Ouchterlony double immunodiffusion (Paiva *et al.*, 1982).
2. *The development of axillary tubers is both rapid and synchronous.* Increases in bud fresh weight can be detected after 3-4 days of culture (see Fig. 12.4). Axillary tubers reach maturity by 2-3 weeks and most of the buds develop uniformly (Paiva *et al.*, 1982). Such synchrony of development is impossible to control with tuberizing underground stolons (Vreugdenhill and Struik, 1989).
3. *Patatin accumulation has been well characterized during axillary tuber development.* Patatin accumulation in developing axillary tubers has been quantified by immunodetection techniques and compared with underground stolon tubers (Paiva *et al.*, 1982). After 10 days in culture, patatin makes up only 3% of the soluble protein present in axillary tubers. At day 42, patatin content has increased to 40% (Paiva *et al.*, 1982). This percent increase in patatin is comparable to that in field-grown tubers (Paiva *et al.*, 1983). Aside from the storage proteins, very little is known about the proteins present in tubers during the early stages of development. One reason for this is that, in developing tubers, there is a rapid accumulation of the storage proteins which make up about 80% of the soluble protein present in a 3-10 g field-grown tuber. The developmental synchrony of axillary tubers makes it possible to focus on the early stages of tuberization before many important gene products are diluted by the large-scale synthesis of the tuber storage proteins.

In recent studies (Hannapel, 1991), the early events associated with tuber initiation in developing axillary tubers have been monitored. These studies have produced some interesting results:

1. Accompanying the rapid increases in both bud fresh weight and starch content (Fig. 12.4) were substantial increases in both total RNA and protein. Total protein accumulation (on a protein/fresh weight basis) is highest before (day 4, Table 12.4) and immediately after (day 6, Table 12.4) the first visible signs of swelling in developing axillary bud tubers. The initial increase in total protein (day 4, Table 12.4) occurs before any substantial increase in the developing bud's fresh weight (see Fig. 12.4). Obviously, the slight decrease in total protein after day 6 is due to the rapid accumulation of starch during this tuber enlargement stage. In addition, total RNA increases by almost 2-fold beginning on day 4 and reaches its highest level in day 6 buds (Table 12.4).

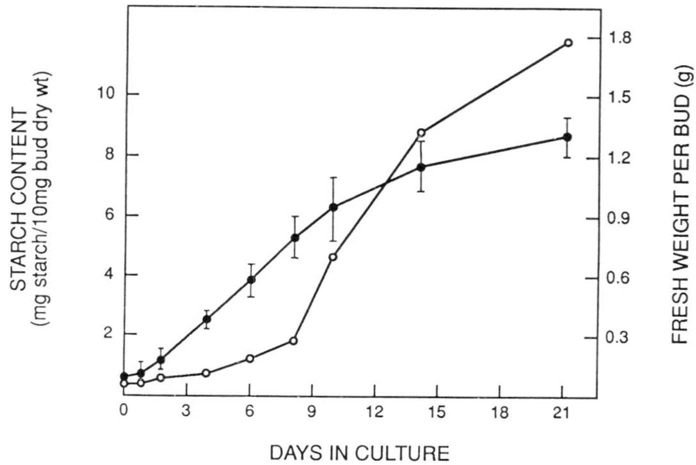

Figure 12.4. Fresh weights and starch contents of developing axillary bud tubers from the petiole-leaf cutting system harvested after 1-21 days of culture. Petiole-leaf cuttings with the axillary bud intact were taken from stock plants (cv. Superior) grown under a short day photoperiod (8 h light, 16 h dark) for 3 weeks. The cuttings were cultured in a Perlite/vermiculite mix under an 8 h photoperiod in a growth chamber. Zero day buds were excised directly from the stock plant; (o), fresh weight; (●), starch content.

2. Despite the evidence for enhanced metabolic activity in day 4 buds of this model tuberization system, transcript accumulation for three major tuber proteins (which make up 60-80% of the total protein in mature tubers) is relatively low up to day 6 (Fig. 12.5). By this time, the events of tuber initiation have already been activated. The low transcript levels for the abundant tuber storage proteins support the premise that this tuber system will facilitate the identification of early tuber-specific cDNAs by isolating very early stages in this developmental process.

To increase the chance of obtaining cDNA clones associated with tuber initiation, a cDNA library has been constructed from polyadenylated mRNA extracted from 4-day-old axillary bud tubers of petiole-leaf cuttings grown under an 8 h photoperiod using *Solanum tuberosum* cv 'Superior'. At 4 days, it is estimated that patatin protein makes up less than 2.0% of the total soluble protein present in the axillary tuber (Paiva *et al.*, 1982). The data of Fig. 12.5 indicate that only very low levels of tuber storage protein transcripts are present in day 4 buds. After cDNA synthesis and the addition of *Eco*RI linkers, the cDNA fragments were cloned into the λ ZAP II vector (Stratagene) and packaged into lambda phage. The library is being screened for early tuber cDNAs using a

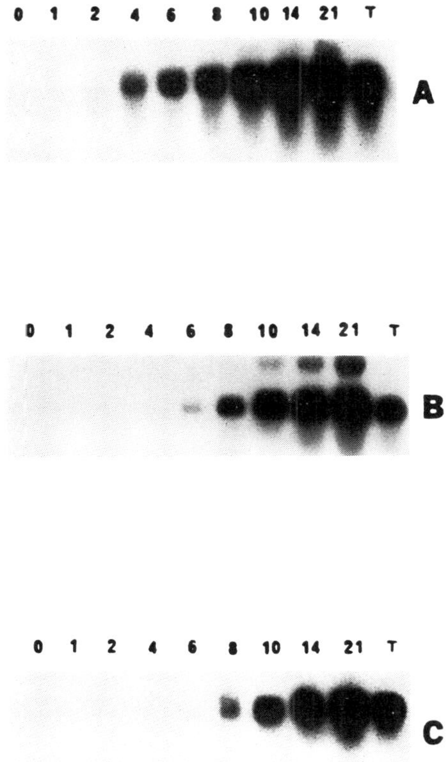

Figure 12.5. Northern blot hybridization of total RNA extracted from developing axillary tubers harvested after 1-21 days of culture. Petiole-leaf cuttings with the axillary bud intact were taken from stock plants (cv. 'Superior') grown under a short day photoperiod (8 h light) for 3 weeks. The cuttings were cultured in a perlite/vermiculite mix under an 8 h photoperiod in a growth chamber. Zero day buds (0) were excised directly from the stock plant. The filters were hybridized with ^{32}P-labeled DNA inserts for either patatin, pGM01 (A), the 22 kDa proteinase inhibitor, p34021 (B) or the cathepsin D inhibitor, p740 (C). Lane T represents total RNA extracted from a mature underground potato tuber. Ten micrograms of total RNA were loaded for each sample.

subtraction hybridization technique (Sive and St John, 1988) and conventional differential screening using labeled cDNAs from normal axillary bud mRNAs.

In addition, screening of the library is also being undertaken using plant homeotic and nodule-specific genes. Taylor *et al.* (1992a) have constructed a

Table 12.4. Accumulation of total RNA and protein in developing axillary bud tubers.

Days from start of incubation	mg total RNA $(gFW)^{-1}$	mg soluble protein $(gFW)^{-1}$
0	0.19 ± 0.01	8.2 ± 0.2
1	0.22 ± 0.02	7.8 ± 0.3
2	0.26 ± 0.08	8.0 ± 0.3
4	0.49 ± 0.06	12.7 ± 2.1
6	0.88 ± 0.16	14.6 ± 2.5
8	0.87 ± 0.42	11.2 ± 2.5
10	0.68 ± 0.04	9.8 ± 2.9
14	0.50 ± 0.06	9.1 ± 3.1
21	0.38 ± 0.12	9.1 ± 3.4

cDNA library from swollen stolons taken from whole plants during the early stages of tuberization. Using differential screening, they have identified several cDNA clones that are specifically expressed early in tuber development. Included in their group of early tuber cDNAs are an *S*-adenosyl-methionine decarboxylase, a ribosomal protein (Taylor *et al.*, 1992b) and a putative cell wall protein. They have also shown that changes in the pattern of α-tubulin isotype expression occurred during the early stages of tuber formation (Taylor *et al.*, 1991).

Conclusions

Clearly, the process of tuberization can be readily separated into morphological and biochemical events. The accumulation of patatin and the proteinase inhibitors is a reliable biochemical marker of tuber development. However, the synthesis of these storage proteins as well as the accumulation of starch are metabolic events associated with tuber enlargement and, in all likelihood, have nothing to do with controling tuber morphology. The morphological potential exhibited by the hybrid germplasm produced by crossing tuber-bearing and nontuber-bearing potato species indicates that the tuberization response is dose-dependent and genetically controled. Preliminary results from the screening of early tuber cDNA libraries demonstrates that there are a number of genes that are differentially expressed during the initial stages of tuber development. Candidates for these developmentally regulated genes could be cell division cycle or homeotic genes involved in regulating somatic cell differentiation.

References

Chapman, H.W. (1958) Tuberization in the potato plants. *Physiologica Plantarum* 11:215-224.

Duncan, D.A. and Ewing, E.E. (1984) Initial anatomical changes associated with tuber formation on single-node potato (*Solanum tuberosum* L.) cuttings. *Annals Botany* 53:607-610.

Ehlenfeldt, M.K. and Hanneman, R.E., Jr (1984) The use of endosperm balance number and 2n gametes to transfer exotic germplasm in potato. *Theoretical and Applied Genetics* 68:155-161.

Ewing, E.E. (1985) Cuttings as simplified models of the potato plant. In: Li, P. (ed.), *Potato Physiology*. Academic Press, Orlando, pp. 154-207.

Garcia-Torres, L. and Gomez-Campo, C. (1973) *In vitro* tuberization of potato sprouts as affected by Ethrel and gibberellic acid. *Potato Research* 16:73-79.

Goldberg, R.B. (1989) Plants: novel developmental processes. *Science* 240:1460-1467.

Gregory L.E. (1965) Physiology of tuberization in plants. In: Ruhland, W. (ed.) *Encyclopedia of Plant Physiology*, Vol. 15. Springer-Verlag, Berlin, pp. 1328-1354.

Hannapel, D.J. (1990) Differential expression of potato tuber protein genes. *Plant Physiology* 94:919-925.

Hannapel, D.J. (1991) Characterization of the early events of potato tuber development. *Physiologica Plantarum* 83:568-573.

Hannapel, D.J., Miller J.C. and Park, W.D. (1985) Regulation of potato tuber protein accumulation by gibberellic acid. *Plant Physiology* 78:700-703.

Hansen, J.D. and Hannapel, D.J. (1992) A wound-inducible potato proteinase inhibitor gene expressed in non-tuber-bearing species is not sucrose inducible. *Plant Physiology* 100:164-169.

Harmey, M.A., Crowley, M.P. and Clinch, P.E.M. (1966) The effect of growth regulators on tuberization of cultured stem pieces of *Solanum tuberosum*. *European Potato Journal* 9:146-151.

Hermsen, J.G.T. (1980) Recent progress and future plans for utilizing Mexican wild species for pest and disease resistance. In: *Utilization of the Genetic Resources of the Potato*,Vol. III, International Potato Center, Lima, Peru, pp. 129-139.

Höfgen, R. and Willmitzer, L. (1992) Transgenic potato plants depleted for the major tuber protein patatin via expression of antisense RNA. *Plant Science* 87:45-54.

Kumar, D. and Wareing, P.F. (1974) Studies on tuberization in *Solanum andigena*. II Growth hormones and tuberization. *New Phytologist* 73:833-840.

Lee, L., Hannapel, D.J., Mignery, G.A., Shumway, J., and Park, W.D. (1983) Control of tuber protein synthesis in potato. In: Goldberg, R.B. (ed), *Plant Molecular Biology, UCLA Symposium*, Alan R. Liss, Inc., New York. pp. 355-365.

Menzel, C.M. (1980) Tuberization in potato at high temperatures: responses to gibberellin and growth inhibitors. *Annals of Botany* 46:259-265.

Müller-Röber, B., Sonnewald, U. and Willmitzer, L. (1992) Inhibition of the ADP-glucose pyrophosphorylase in transgenic potatoes leads to sugar-storing tubers and influences tuber formation and expression of tuber storage protein genes. *EMBO Journal* 11:1229-1238.

Okazawa, Y. (1960) Studies on the relation between tuber formation of potato and its natural gibberellin content. *Proceedings of the Crop Science Society of Japan* 29:121-124.

Paiva, E., Lister, R.M. and Park, W.D. (1982) Comparison of the protein in axillary bud tubers and underground stolon tubers in potato. *American Potato Journal* 59:425-433.

Paiva, E., Lister, R.M. and Park, W.D. (1983) Induction and accumulation of major tuber proteins of potato in stem and petioles. *Plant Physiology* 71:161-168.

Park, W.D., Hannapel, D.J., Mignery, G.A. and Pikaard, C.S. (1985) Molecular approaches to the study of the major tuber proteins. In: Li, P. (ed.), *Potato Physiology*, Academic Press, Inc., Orlando, FL, pp. 261-278.

Racusen, D. (1984) Lipid acyl hydrolase of patatin. *Canadian Journal of Botany* 62:1640-1644.

Railton, I.D. and Wareing, P.F. (1973) Effects of daylength on endogenous gibberellins in leaves of *Solanum andigena* L. Changes in levels of free acidic gibberellin-like substances. *Physiologica Plantarum* 28:88-94.

Ryan, C.A. (1981) Plant proteinase inhibitors. In: Stumpf, P.K. and Conn, E.E. (eds), *The Biochemistry of Plants*, Academic Press, New York, Vol. 10, pp. 1-40.

Sive, H.L. and St John, T. (1988) A simple subtractive technique employing photoactivatable biotin and phenol extraction. *Nucleic Acids Research* 16:10,937.

Taylor, M.A., Davies, H.V. and Scobie, L.A. (1991) Molecular changes associated with tuberization in *Solanum tuberosum*. Differential expression of α-tubulin isotypes. *Physiologica Plantarum* 81:244-250.

Taylor, M.T., Arif, S.A.M., Kumar, A., Davies, H.V., Scobie, L.A., Pearce, S.R. and Flarell, A.J. (1992a) Expression and sequence analysis of cDNAs induced during the early stages of tuberization in different organs of the potato plant (*Solanum tuberosum* L.). *Plant Molecular Biology* 20:641-651.

Taylor, M.T., Arif, S.A.M., Pearce, S.R., Davies, H.V., Kumar, A. and George, L.A. (1992b) Differential expression and sequence analysis of ribosomal protein genes induced in stolon tips of potato (*Solanum tuberosum L.*) during early stages of tuberization. *Plant Physiology* 100:1171-1176..

Vreugdenhill, D. and Struik, P.C. (1989) An integrated view of the hormonal regulation of tuber formation in potato (*Solanum tuberosum*). *Physiologica Plantarum* 75:525-531.

Wheeler, R.M., Hannapel, D.J. and Tibbitts, T.W. (1988) Comparison of axillary bud growth and patatin accumulation in potato leaf cuttings as assays for tuber induction. *Annals of Botany* 62:25-30.

Chapter Thirteen:

Use of Ubiquitin Promoters for Transgene Expression in Potato

Joan E. Garbarino and William R. Belknap
USDA/ARS, Western Regional Research Center, Albany, CA, USA.

In the design and construction of a chimeric gene, the choice of promoter plays an important role in obtaining a desired phenotype. We have been investigating the potato ubiquitin gene family as a source of promoter elements for directing wound-activated transcription in the tuber. These promoters could be used in constructs designed to affect a number of tuber disorders, such as blackspot bruise and bacterial soft rot, which are linked to physical injury.

There are two categories of ubiquitin genes: ubiquitin-ribosomal protein fusion genes and polyubiquitin genes (Monia *et al.*, 1990; Ozkaynak *et al.*, 1987). Ubiquitin-ribosomal fusion genes contain a single ubiquitin coding unit fused to the coding region for either of two small proteins associated with ribosomes (Finley *et al.*, 1989; Redman and Rechsteiner, 1989). Polyubiquitin genes contain several ubiquitin direct repeats. Both types of ubiquitin genes are transcribed and translated as polyproteins (Ozkaynak *et al.*, 1987). The polyproteins are then processed by a specific hydrolase, resulting in free ubiquitin and ribosomal protein (for the ubiquitin-ribosomal protein fusion genes) or several free ubiquitins (for the polyubiquitin genes).

The potato ubiquitin gene family contains several members that are either transcriptionally activated or deactivated by wounding in the tuber (Rickey and Belknap, 1991; Garbarino *et al.*, 1992). We have characterized the expression of one ubiquitin-ribosomal protein (*ubi3*) and one polyubiquitin (*ubi7*) promoter in transgenic potatoes.

Ubiquitin-Ribosomal Protein Gene *ubi3*

Several plant ubiquitin-ribosomal protein fusion cDNAs or genomic clones have been isolated and characterized (Callis *et al.*, 1990; Chen and Rubenstein, 1991; Gausing and Jensen, 1990; Hoffman *et al.*, 1991; Garbarino *et al.*, 1992). They have previously been observed to be highly expressed in rapidly dividing tissue (Callis *et al.*, 1990; Chen and Rubenstein, 1991; Hoffman *et al.*, 1991; Garbarino *et al.*, 1992). In addition, a copy-specific probe from a potato

The Molecular and Cellular Biology of the Potato, Second Edition
Edited by W.R. Belknap, M.E. Vayda and W.D. Park

ubiquitin-ribosomal fusion protein cDNA hybridized to a transcript which accumulated in the tuber in response to both wounding and application of exogenous ethylene (Garbarino *et al.*, 1992). This potato cDNA encodes the larger of the two ribosomal proteins (80 amino acids) and is present in two copies in the potato genome (Garbarino *et al.*, 1992).

Construction of *ubi3*-GUS promoter fusions

A *ubi3* genomic clone was isolated from a λFIXII potato library by first prescreening aliquots with PCR, using primers specific for the N-terminus of the ubiquitin coding region and the C-terminus of the ribosomal protein coding region. Library aliquots producing the predicted PCR fragments were then screened with a probe containing the ribosomal protein coding region and 3' sequences (Garbarino and Belknap, 1994). A genomic *ubi3* clone was obtained using this probe, and the coding region and flanking regions were sequenced. The 3' sequence of the genomic clone differed at only two bases from the 170 bp 3' region of the *ubi3* cDNA (data not shown). Sequences from this genomic clone were fused to the reporter gene ß-glucuronidase (GUS) (Jefferson *et al.*, 1987) for evaluation of function in transgenic plants.

Two *ubi3*-GUS fusion transgenes were constructed using the genomic clone (Fig. 13.1). The first contained 920 bp of 5' sequence fused directly to GUS (Fig. 13.1B). The second contained this 5' sequence and the ubiquitin coding region fused in-frame to GUS (Fig. 13.1C). Previously published results (Butt *et al.*, 1989; Ecker *et al.*, 1989) suggested that inclusion of the GUS coding region would result in higher levels of transgene expression.

The two transgenes were cloned into the binary transformation vector pCGN1547 (McBride and Summerfelt, 1990) and mobilized into the potato cv. Russet Burbank via *Agrobacterium*-mediated transformation (Snyder and Belknap, 1993).

Expression of *ubi3*-GUS promoter fusions in transgenic potatoes

The level of GUS activity in the leaves of the transgenic clones was evaluated. Although there was a wide range in the relative activities among the individual clones, the plants containing the *ubi3*-GUS translational fusion (Fig. 13.1C) had GUS activities 5- to 10- fold higher than plants containing the *ubi3*-GUS promoter fusion (Fig. 13.1B).

Western blot analysis of the transgenic lines showed that the plants expressing the *ubi3*-GUS translational fusion produce appropriately processed GUS (Fig. 13.2). Both types of transgenic clones synthesized GUS protein which co-migrated with authentic protein. This indicates that the *ubi3*-GUS translational fusion is being hydrolyzed in the transgenic clones.

The developmental regulation of GUS activity in the transgenic clones was also examined (Table 13.2). As expected, GUS activity was observed to be

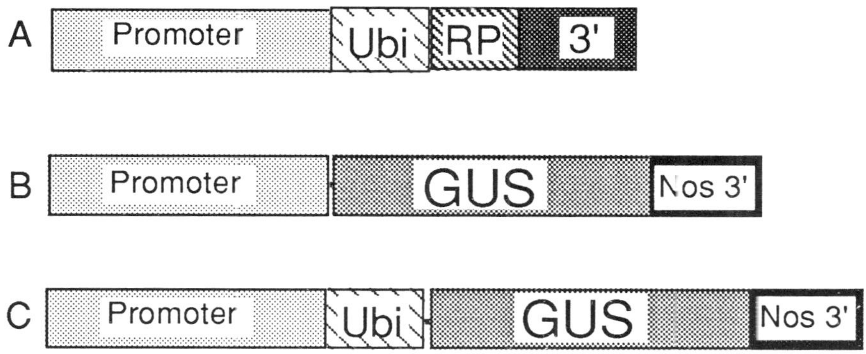

Figure 13.1. Maps of endogenous *ubi3* gene and GUS-fusion transgenes. (A) Endogenous *ubi3* gene containing promoter, ubiquitin (Ubi) and ribosomal protein (RP) coding region and 3' polyadenylation signal (3'). (B) *ubi3*-GUS promoter fusion with the nopaline synthase polyadenylation signal (Nos 3'). (C) *ubi3*-GUS translational fusion in which the GUS coding sequence is fused in-frame to the ubiquitin coding sequence.

considerably higher in meristematic tissue than in mature leaves. While this decrease in activity with leaf maturity is in agreement with previously published results, an unexpected increase in activity during leaf senescence was also observed (Table 13.2). This increase in activity was associated with increased levels of GUS mRNA (Garbarino and Belknap, 1994). Similarly, endogenous *ubi3* transcript levels were also observed to increase during senescence (Garbarino and Belknap, 1994).

While GUS expression was detected in transgenic plants containing both constructs, levels of expression for the clones containing ubiquitin-GUS translational fusion were higher than for the construct with the ubiquitin promoter only. The ubiquitin-GUS translational fusion was constructed with the expectation of higher levels of transgene expression, potentially mediated by increased translatability (Ecker *et al.*, 1989). However, transgenic lines expressing the translational fusion had higher levels of both transcript and translated product than those where GUS was fused directly to the promoter (Fig. 13.3). The data suggest that inclusion of the ubiquitin coding region resulted in an increase in either transcription or message stability. The translatability of mRNA in plants has previously been demonstrated to affect message stability (Vancanneyt *et al.*, 1990), indicating a potential mode for more efficient message accumulation in transgenic lines expressing the ubiquitin-GUS translational fusion. However, in addition to the presence of the ubiquitin coding region in the translational fusion, the two transgenes described here also differ in sequences surrounding the ATG initiation codon. These differences

Table 13.1. GUS activity in leaves of independent potato transgenic lines. Plants were transformed with GUS fused to the *ubi3* promoter (-ubi) or to the *ubi3* promoter plus the ubiquitin coding region (+ubi).

Construct type	GUS activity (nmol/min per mg protein)[1]				
-ubi	6.3 ± 1.0	2.0 ± 0.6	1.8 ± 0.2	1.4 ± 0.8	
+ubi	32.7 ± 8.7	11.5 ± 1.0	13.2 ± 3.8	9.8 ± 0.5	9.7 ± 1.0

[1]Leaf assayed was approximately fifth from the meristem (n = 3).
Fluorescent GUS assays were performed as described by Jefferson *et al.* (1987).

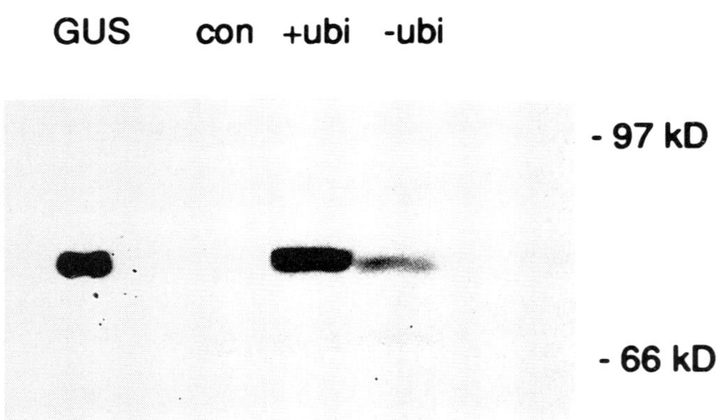

Figure 13.2. Immunodetection of GUS in transgenic potato plants. Total protein was extracted from leaf meristems from control (con) and transgenic clones containing the *ubi3* promoter ubiquitin translational fusion (+ubi, clone 3.5) or the *ubi3* promoter fusion (-ubi, clone 8.1). Total protein extracts (80 μg/lane) and authentic purified GUS (0.1 μg/lane) were fractionated by SDS-PAGE and probed with a GUS-specific antibody (Amersham).

have the potential to alter significantly the translatability of the mRNA (Kozak, 1986), and thus transcript stability.

In previous experiments we observed that *ubi3* expression increased in tubers after wounding (Garbarino *et al.*, 1992). In contrast, wounding of tubers from the transgenic plants did not result in an increase in GUS mRNA (Fig. 13.3) or activity (data not shown). When a duplicate blot was probed for endogenous *ubi3* expression, the expected increase in message was observed

Table 13.2. Comparison of GUS activities in potato plants transformed with the *ubi3* promoter without (-ubi) or with (+ubi) a ubiquitin coding unit.[1]

Transformant	Leaf meristem	Second leaf	Fifth Leaf	Senescent leaf	Tuber
3.2 -ubi	6.3 ± 0.7	2.5 ± 0.4	1.8 ± 0.2	5.4 ± 1.2	3.7 ± 0.3
8.1 -ubi	25.8 ± 2.1	10.0 ± 2.1	6.3 ± 1.0	19.2 ± 6.1	14.2 ± 1.3
3.5 +ubi	94.8 ± 12.6	60.3 ± 25.1	32.7 ± 8.7	50.1 ± 11.6	37.6 ± 10.4
9.8 +ubi	33.3 ± 0.5	18.9 ± 2.7	9.7 ± 1.0	22.7 ± 3.6	20.7 ± 3.5

[1]GUS activities determined as in Table 13.1.

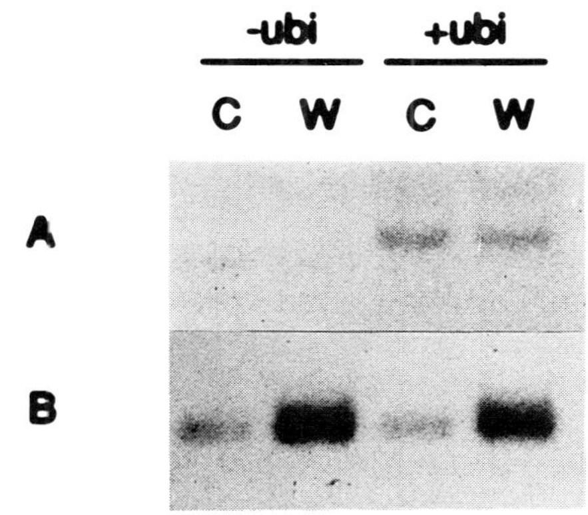

Figure 13.3. *Ubi3* transgenes are not induced by wounding. RNA was prepared from control (C) or wounded (W) (cut into 1-2 mm slices) tuber tissue from transgenic clones containing the *ubi3* promoter ubiquitin translational fusion (+ubi, clone 3.5) or the *ubi3* promoter fusion (-ubi, clone 3.2). RNA (10 µg/lane) was hybridized to probe for GUS (A) or a probe containing the 3' noncoding region of potato *ubi3* (B).

(Fig. 13.3). This suggests that the 920 bp of upstream sequence used in the chimeric constructs was insufficient to obtain wound dependent activation.

Potato Polyubiquitin Gene *ubi7*

Similar to the ubiquitin-ribosomal protein genes, several polyubiquitin cDNAs or genomic clones have been isolated from a variety of plant sources (Burke *et al.*, 1988; Watts and Moore, 1989; Binet *et al.*, 1991; Christensen *et al.*, 1992; Kawalleck *et al.*, 1993). We have isolated three different polyubiquitin cDNAs from potato (Garbarino *et al.*, 1992), each encoding an estimated six or

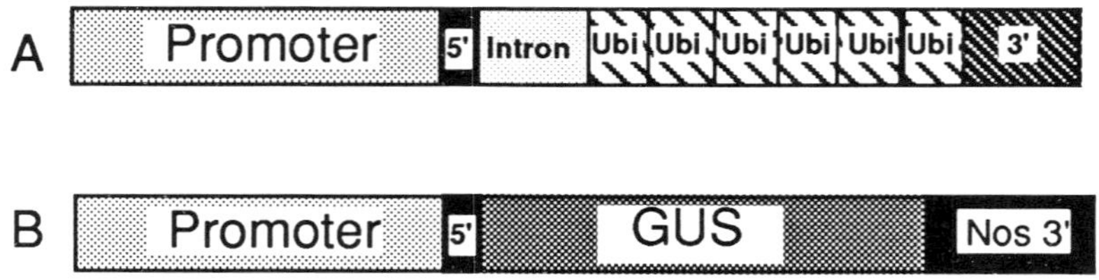

Figure 13.4. Maps of endogenous *ubi7* gene and GUS-fusion transgene. (A) Endogenous *ubi7* gene containing promoter, 5' untranslated region (5'), ubiquitin coding units (Ubi) and 3' polyadenylation signal (3'). (B) *ubi7*-GUS promoter fusion with the nopaline synthase polyadenylation signal (Nos3').

seven ubiquitin repeats. Tuber mRNA levels for two of these cDNAs were observed to increase with wounding. In experiments using the ubiquitin coding region as a probe, the increase in mRNA levels was shown to be localized to the wounded region of the tuber (Rickey and Belknap, 1991). For the third cDNA, mRNA levels decreased in the tuber after wounding. In contrast to the ubiquitin-ribosomal fusion gene which did not respond to heat shock, the wound-activated polyubiquitin genes were also activated by this stress. Each of these potato polyubiquitin genes appears to be present in one to three copies in the potato genome (Garbarino *et al.*, 1992).

Construction of *ubi7*-GUS promoter fusions

We have isolated a genomic clone for one of the wound-activated potato polyubiquitin cDNAs (*ubi7*). In common with other higher plant polyubiquitin genes, the *ubi7* gene contains an intron between the 5' untranslated end and the initiating methionine codon (Fig. 13.4). This intron has been reported to vary in size between approximately 300 bp in *Arabidopsis* (Norris *et al.*, 1993) to over 1600 bp in maize (Christensen *et al.*, 1992). The inclusion of this intron in polyubiquitin transgene constructs has been reported to increase transient expression 2.5-3-fold in *Arabidopsis* (Norris *et al.*, 1993) but had no effect on a sunflower polyubiquitin gene expressed in tobacco protoplasts (Binet *et al.*, 1991). The potato polyubiquitin intron is 575 bp in length.

To obtain the *ubi7* genomic clone, the λFIXII library was again prescreened by PCR, this time using primers from the 5' untranslated region and the amino terminal ubiquitin coding region (sequences derived from the *ubi7* cDNA). Library aliquots generating a positive PCR signal were screened with a ubiquitin probe, and positive clones further screened by PCR using the *ubi7* primers described above. The λFIXII clone encoding ubi7 exhibited an unusual phenotype (failure to lyse *Escherichia coli*) both on plates and in liquid culture, complicating the cloning process.

The ubiquitin coding region and adjacent sequences of the λFIXII clone were subcloned into pUC19, and the flanking regions were sequenced. A *ubi7*-GUS fusion transgene was constructed containing 1156 bp of upstream sequence, including the 5' untranslated end of the cDNA, fused to GUS (Fig. 13.4). The chimeric *ubi7*-GUS transgene was then mobilized into potato as described previously.

Expression of *ubi7*-GUS promoter fusion in transgenic potatoes

Previously, we reported that mRNA levels for the endogenous *ubi7* gene increased in tubers on wounding (Garbarino *et al.*, 1992). mRNA levels for the chimeric *ubi7*-GUS gene also increased on wounding of tubers from transgenic plants (Fig. 13.5), indicating that the intron is not required for wound activation.

A time course showing increased GUS activity on wounding in leaves and in tubers of an *ubi7*-GUS transformant is shown in Fig. 13.6. GUS activity increased approximately 2.5-4-fold within 24-48 h after wounding.

In common with a variety of other wound-induced plant genes (Sembdner and Parthier, 1993), expression from the *ubi7* promoter is activated by the application of exogenous methyl jasmonate (MeJas) (Fig. 13.7). In contrast, induction by abscisic acid (ABA) was not observed under these conditions. While many mRNAs which are MeJas-regulated also respond to exogenous ABA (Hildemann *et al.*, 1993; Sembdner and Parthier, 1993), wound-induced transcripts which respond to exogenous MeJas, but not ABA, have also been reported (Mason *et al.*, 1990; Bell and Mullet, 1991).

The presence of the wound and MeJas response with the *ubi7* promoter suggests that this promoter should contain elements previously associated with these responses in other promoter elements (Kim *et al.*, 1992; Loake *et al.*, 1992; Arias *et al.*, 1993; Kim *et al.*, 1993; Mason *et al.*, 1993). In particular 'G' box and 'H' box , as well as other motifs, would be expected (Staiger *et al.*, 1989; Kim *et al.*, 1992; Loake *et al.*, 1992; Arias *et al.*, 1993; Mason *et al.*, 1993). An analysis of the sequence of the *ubi7* promoter revealed the presence of both box types (Fig. 13.8), with an arrangement remarkably similar to that found in the 5' regions of the soybean vegetative storage proteins α and ß (Mason *et al.*, 1993). The *vspA* and *vspB* genes are regulated by wounding, MeJas and ABA in a manner similar to *ubi7* (Mason *et al.*, 1992). In addition, Mason *et al.* (1993) have demonstrated that the 50 bp sequence of *vspB* containing the 'G' box and Box 1 is sufficient to confer MeJas inducibility to a truncated cauliflower mosaic virus 35S promoter. While the location of the 'G' box of the potato gene is nearer to the transcription start site than the *vsp* genes, the relative location of this box is similar to the chalcone synthase gene from parsley (Schulze-Lefert *et al.*, 1989).

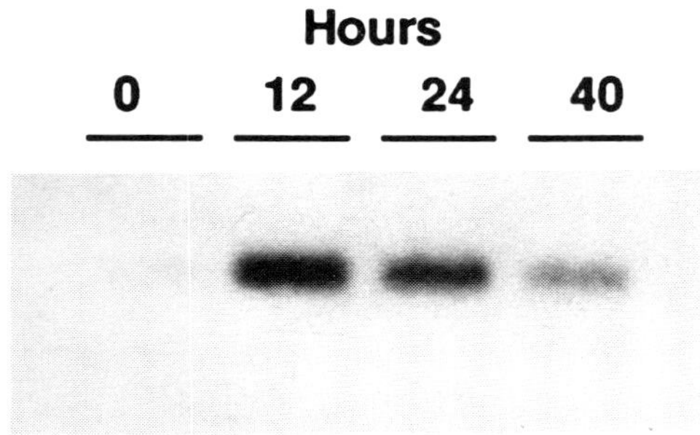

Figure 13.5. Wound-induction of the *ubi7* transgene in tubers. RNA was prepared from tubers at 0, 12, 24 and 40 h after wounding (cut into 1-2 mm slices). RNA (10 μg/lane) was hybridized to a probe for GUS.

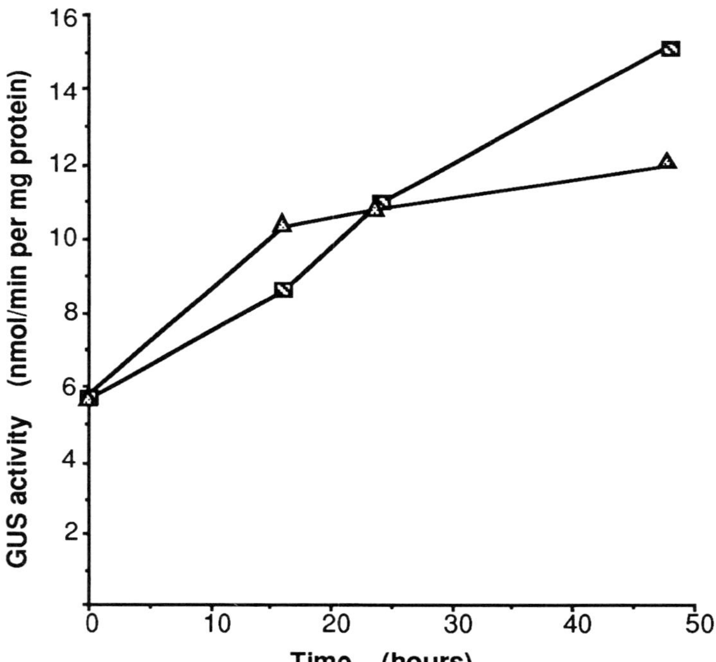

Figure 13.6. Wound induction of GUS activity in leaves (▲) and tubers (■) of plants transformed with the *ubi7*-GUS fusion. Leaves were wounded with multiple hemostat injuries, tubers were cut into 1-2 mm slices for the time period indicated prior to determination of GUS activity. GUS activities were determined as in Table 13.1.

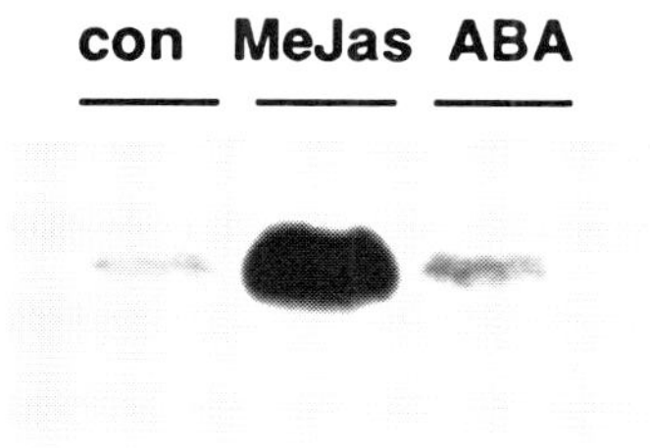

Figure 13.7. Regulation of the *ubi7* transgene by methyl jasmonate (MeJas). Total RNA (10 μg) was hybridized to a probe for GUS. RNA was isolated from leaves incubated in the dark for 16 h in water (control), 20 mM MeJas or 50 mM ABA.

	Box I		G-Box		Box III
vspB	CCCTAGAACC	- 28 bp -	CACGTG	- 39 bp -	AGAGTGGACCCC $^{-482}$
vspA	CCATACAACT	- 30 bp -	CACGTG	- 38 bp -	AGAGTGGACCCC $^{-1174}$
ubi7	CCCCACCACA	- 64 bp -	CACGTG	- 40 bp -	AGAGTCGGCCAT $^{-100}$

Figure 13.8. Comparison of the promoter structures of potato *ubi7* and soybean *vspA* and *vspB*. Numbers indicate distance to transcription start site. Promoter motifs for *vspB* and *vspA* from Mason *et al.* (1993).

Conclusions

The ubiquitin gene family contains several members that respond to wounding and developmental signals. We have used two different ubiquitin promoters to express GUS in potato plants. One of the promoters tested (*ubi3*) retained its developmental regulation, directing GUS expression at high levels in the meristems. However, the ability to direct increased expression in response to wounding was lost. A construct in which a ubiquitin coding unit was included on the *ubi3* promoter and translationally fused to GUS was also tested. The GUS protein appeared to be properly processed and expression was higher than with the *ubi3* promoter alone. The second ubiquitin promoter tested (*ubi7*) was able to direct increased expression of GUS in response to wounding in both leaves and tubers, despite the absence of an intron present in the native gene.

Expression of the *ubi7*-GUS gene also increased in response to methyl jasmonate. Future experiments are designed to test the activity of the *ubi7* promoter by inclusion of its intron and/or a ubiquitin.

Acknowledgements

The authors wish to express their gratitude to D.R. Rockhold for expert technical assistance and to Calgene Inc. for providing pCGN1547. References to a company and/or product by the USDA is only for purposes of information and does not imply approval or recommendation of the product to the exclusion of others that may also be suitable.

References

Arias, J.A., Dixon, R.A. and Lamb, C.J. (1993) Dissection of the functional architecture of a plant defense gene promoter using a homologous *in vitro* transcription initiation system. *Plant Cell* 5:485-496.

Bell, E. and Mullet, J.E. (1991) Lipoxygenase gene expression is modulated in plants by water deficit, wounding and methyl jasmonate. *Molecular and General Genetics* 230:456-462.

Binet, M.-N., Weil, J.-H. and Tessier, L.-H. (1991) Structure and expression of sunflower ubiquitin genes. *Plant Molecular Biology* 17:395-407.

Burke, T.J., Callis, J. and Vierstra, R.D. (1988) Characterization of a polyubiquitin gene from *Arabidopsis thaliana. Molecular and General Genetics* 213:435-443.

Butt, T.R., Sobhanaditya, J., Monia, B.P., Sternberg, E.J., Marsh, J.A., Stadel, J.M., Ecker, D.J. and Crooke, S.T. (1989) Ubiquitin fusion augments the yield of cloned gene products in *Escherichia coli. Proceedings of the National Academy of Sciences USA* 86:2540-2544.

Callis, J., Raasch, J.A. and Vierstra, R.D. (1990) Ubiquitin extension proteins of *Arabidopsis thaliana. Journal of Biological Chemistry* 265:12486-12493.

Chen, K. and Rubenstein, I. (1991) Characterization of the structure and transcription of an ubiquitin fusion gene from maize. *Gene* 107:205-212.

Christensen, A.H., Sharrock, R.A. and Quail, P.H. (1992) Maize polyubiquitin genes: structure, thermal perturbation of expression and transcript splicing, and promoter activity following transfer to protoplasts by electroporation. *Plant Molecular Biology* 18:675-689.

Ecker, D.J., Stadel, J.M., Butt, T.R., Marsh, J.A., Monia, B.P., Powers, D.A., Gorman, J.A., Clark, P.E., Warren, F., Shatzman, A. and Crooke, S.T. (1989) Increasing gene expression in yeast by fusion to ubiquitin. *Journal of Biological Chemistry* 264:7715-7719.

Finley, D., Bartel, B. and Varshavsky, A. (1989) The tails of ubiquitin precursors are ribosomal proteins whose fusion to ubiquitin facilitates ribosome biogenesis. *Nature* 338:394-401.

Garbarino, J.E. and Belknap, W.R. (1994) Isolation of a ubiquitin-ribosomal protein gene (*ubi3*) from potato and expression of its promoter in transgenic plants. *Plant Molecular Biology*. 24:119-127.

Garbarino, J.E., Rockhold, D.R. and Belknap, W.R. (1992) Expression of stress-responsive ubiquitin genes in potato tubers. *Plant Molecular Biology* 20:235-244.

Gausing, K. and Jensen, C.B. (1990) Two ubiquitin-long-tail fusion genes arranged as closely spaced direct repeats in barley. *Gene* 94:165-171.

Hildemann, T., Ebneth, M., Pena-Cortes, H., Sanchez-Serrano, J.J., Willmitzer, L. and Prat, S. (1993) General roles of abscisic and jasmonic acids in gene activation as a result of mechanical wounding. *The Plant Cell* 4:1157-1170.

Hoffman, N.E., Ko, K., Milkowski, D. and Pichersky, E. (1991) Isolation and characterization of tomato cDNA and genomic clones encoding the ubiquitin gene *ubi3*. *Plant Molecular Biology* 17:1189-1201.

Jefferson, R.A., Kavanaugh, T.A. and Bevan, M.W. (1987) GUS fusions: ß-glucuronidase as a sensitive and versatile gene fusion marker in higher plants. *EMBO Journal* 6:3901-3907.

Kawalleck, P., Sommssich, I.E., Feldbrugge, M., Hahlbrock, K. and Weisshaar, B. (1993) Polyubiquitin gene expression and structural properties of the *ubi*4-2 gene in *Petroselinum crispum*. *Plant Molecular Biology* 21:673-684.

Kim, S.-R., Choi, J.-L., Costa, M. and An, G. (1992) Identification of G-Box sequence as an essential element for methyl jasmonate response of potato proteins inhibitor II promoter. *Plant Physiology* 99:627-631.

Kim, S.-R., Kim, Y. and An, G. (1993) Identification of methyl jasmonate and salicylic acid response elements form the nopaline synthase (*nos*) promoter. *Plant Physiology* 103:97-103.

Kozak, M. (1986) Point mutations define a sequence flanking the AUG initiator codon that modulates translation by eukaryotic ribosomes. *Cell* 44:283-292.

Loake, G.J., Faktor, O., Lamb, C.J., Dixon, R.A. (1992) Combination of H-box [CCTACC(N)$_7$CT] and G-box (CACGTG) cis elements is necessary for feed-forward stimulation of a chalcone synthase promoter by the phenylpropanoid-pathway intermediate *p*-coumaric acid. *Proceedings of the National Academy of Sciences USA* 89:9230-9234.

McBride, K.E. and Summerfelt, K. (1990) Improved binary vectors for *Agrobacterium*-mediated plant transformation. *Plant Molecular Biology* 14:269-276.

Mason, H.S., DeWald, D.B., Creelman, R.A. and Mullet, J.E. (1990) Expression of two soybean vegetative storage protein genes during development and in response to water deficit, wounding and jasmonic acid. *The Plant Cell* 2:569-579.

Mason, H.S., DeWald, D.B. and Mullet, J.E. (1993) Identification of a methyl jasmonate-responsive domain in the soybean *vspB* promoter. *The Plant Cell* 5:241-251.

Monia, B.P., Ecker, D.J. and Crooke, S.T. (1990) New perspectives on the structure and function of ubiquitin. *Bio/technology* 8:209-215.

Norris, S.R., Meyer, S.E. and Callis, J. (1993) The intron of *Arabidopsis thaliana* polyubiquitin genes is conserved in location and is a quantitative determinant of chimeric gene expression. *Plant Molecular Biology* 21:895-906.

Ozkaynak, E., Finley, D., Solomon, M.J. and Varshavsky, A. (1987) The yeast ubiquitin genes: a family of natural gene fusions. *EMBO Journal* 6:1429-1439.

Redman, K.L. and Rechsteiner, M. (1989) Identification of the long ubiquitin extension as ribosomal protein S27a. *Nature* 338:438-440.

Rickey, T.M. and Belknap, W.R. (1991) Comparison of the expression of several stress responsive genes in potato tubers. *Plant Molecular Biology* 16:1009-1018.

Schulze-Lefert, P., Dangl, J.L., Becker-Andre, M., Hahlbrock, K. and Schulz, W. (1989) Inducible *in-vivo* DNA footprints define sequences necessary for UV light activation of the parsley chalcone synthase gene. *EMBO Journal* 8:651-656.

Sembdner, G. and Parthier, B. (1993) The biochemistry and the physiological and molecular actions of jasmonates. *Annual Review of Plant Physiology and Plant Molecular Biology* 44:569-589.

Snyder, G.W. and Belknap, W.R. (1993) A modified method for routine *Agrobacterium*-mediated transformation of *in vitro* grown potato microtubers. *Plant Cell Reports* 12:324-327.

Staiger, D., Kaulen, H. and Schell, J. (1989) A CACGTG motif of the *Antirrhinum majus* chalcone synthase promoter is recognized by an evolutionarily conserved nuclear protein. *Proceedings of the National Academy of Sciences USA* 86:6930-6934.

Vancanneyt, G.E., Rosahl, S. and Willmitzer, L. (1990) Translatability of a plant-mRNA strongly influences its accumulation in transgenic plants. *Nucleic Acids Research* 18:2917-2921.

Watts, F.Z. and Moore A.L. (1989) Nucleotide sequence of a full length cDNA clone encoding a polyubiquitin gene from *Pisum sativum*. *Nucleic Acids Research* 17:10100.

Chapter Fourteen:

Regulation of Translation in Potato Tubers in Response to Environmental Stress

Michael E. Vayda and James K. Morelli
Department of Biochemistry, University of Maine, Orono, ME, USA.

Translational control is an important aspect of gene regulation. For example, we have shown that translational activity of potato tubers varies with stress conditions and impacts upon the susceptibility of the crop to disease (Crosby and Vayda, 1991; Vayda *et al.*, 1992). Others have shown that the translational activity of plants varies with organ type, age, stress, temperature change, light cycle, growth conditions and other environmental cues (Okimoto *et al.*, 1980; Scharf and Nover, 1982; Kirk and Kirk, 1985; Berry *et al.*, 1988, 1990; Mason *et al.*, 1988; Ursin *et al.*, 1991; Wang *et al.*, 1992). Such environment-dependent fluctuations in translational competence are likely to affect the expression of desirable crop traits and the performance of transgenes under field conditions. Thus, an understanding of the mechanisms by which translational processes are regulated is necessary for optimizing gene expression in plants.

The potato tuber is a good model system for the study of translational control. Bulk quantities of tuber material that respond in a uniform manner are available for analysis, and distinct effects upon translational competency have been observed in the mature tuber in response to mechanical wounding and hypoxia (Vayda and Schaeffer, 1988; Butler *et al.*, 1990; Crosby and Vayda, 1991).

Translational competence has been shown to affect the ability of tuber tissue to withstand infection by bacterial pathogens (Zucker and Hanklin, 1970; Vayda *et al.*, 1992). Mechanical wounding triggers a response that decreases the susceptibility of tubers to bacterial rot. One prominent aspect of this response is a 4-fold increase in translational activity within 1 h of wounding and selective translation of newly synthesized wound response mRNAs. The blockage of translation by the addition of cycloheximide at the time of wounding severely compromises the ability of tubers to resist bacterial infection and tissue maceration (Zucker and Hanklin, 1970; Vayda *et al.*, 1992). Translation is arrested when tubers are incubated in poorly oxygenated environments, that is,

The Molecular and Cellular Biology of the Potato, Second Edition
Edited by W.R. Belknap, M.E. Vayda and W.D. Park

atmospheres with less than 2% O_2 (Vayda and Schaeffer, 1988; Butler *et al.*, 1990; Crosby and Vayda, 1991). Such hypoxic stress prevents expression of the wound response and consequently renders tubers susceptible to attack by opportunistic bacterial pathogens. In this chapter, we describe events that occur in response to wounding and hypoxia, and suggest a mechanism by which translational processes may be regulated in potato tubers.

The Wound Response of Tubers

Mechanical wounding is the primary route by which bacterial pathogens gain access to the tuber parenchyma. However, two observations make it clear that the wounding event triggers a response in the tuber that reduces growth of and maceration by the invading pathogen. Tubers wounded with a sterile pipette tip several hours prior to inoculation with bacteria are significantly less susceptible to tissue maceration than tubers wounded and inoculated concurrently. Further, tubers prevented from expressing the wound response by addition of the protein synthesis inhibitor cycloheximide exhibit an even greater degree of bacterial growth and tissue maceration than control tubers wounded and inoculated concurrently (Zucker and Hanklin, 1970; Vayda *et al.*, 1992).

The specific gene products involved in limiting bacterial growth and maceration are not known. Wounding initiates a complex response in potato leading to the synthesis of a diversity of products involved in several physiological processes. Numerous transcripts accumulate soon after wounding. Some of these mRNAs encode proteins whose physiological function is not known, such as *wun1* (Logemann *et al.*, 1988, 1989) and *win2* (Stanford *et al.*, 1989, 1990; Weiss and Bevan, 1991), others encode proteins whose functions are known. These include phenylalanine ammonia lyase (PAL) (Vayda and Schaeffer, 1988; Bevan *et al.*, 1989; Butler *et al.*, 1990; Crosby and Vayda, 1991; Ishizuka *et al.*, 1991; Vayda *et al.*, 1992), HMG-CoA reductase (HMGR) (Oba *et al.*, 1985; Stermer and Bostock, 1987; Yang *et al.*, 1991; Choi *et al.*, 1992), DHAP synthetase (Dyer *et al.*, 1989), polyphenol oxidase (PPO) (Roberts *et al.*, 1988), lipoxygenase (Lojkowska, 1988), ubiquitin and the 70 kDa heat shock protein chaperonin (Belknap and Rickey, 1990; Rickey and Belknap, 1991; Garbarino *et al.*, 1992), extensin (HRGP) (Butler *et al.*, 1990; Rumeau *et al.*, 1990; Crosby and Vayda, 1991), the glycine-rich cell wall protein (GRP) (Butler *et al.*, 1990; Crosby and Vayda, 1991) and histones (Butler *et al.*, 1990). There is no definitive evidence that any of the identified gene products are involved directly in disease resistance. However, the diversity of genes induced by wounding suggests that the response is a generalized reaction to physical injury. Indeed, the response is essentially the same whether or not pathogens are introduced to the wound (Butler *et al.*, 1990; Rumeau *et al.*, 1990; Yang *et al.*, 1991).

Many of the gene products induced by wounding are involved in the processes of suberization, cell wall strengthening and wound periderm development. PAL and HMGR are involved in the production of phenolic compounds and lipids, respectively, that partake in the free radical polymerization initiated by PPO to form suberin (Kolattukudy, 1981; Roberts *et al.*, 1988). HRGP and GRP are integral cell wall proteins, and histones are required for cell division. DNA replication has been observed in tuber tissue following wounding (Butler *et al.*, 1990), consistent with an earlier report of cell division at the wound site (Barkhausen, 1978). Taken together, these observations suggest that a significant role of the wound response is formation of a suberized layer and a wound periderm, as physical barriers to pathogens (Vayda *et al.*, 1990; Yang *et al.*, 1991).

The wound response of tubers is similar to, although distinct from, that of leaves and stems: several genes induced by the wounding of plant leaves are not induced by wounding potato tubers. For example, chalcone synthase (CHS) and chalcone isomerase (CHI) involved in production of flavanoid compounds, and the proteinase inhibitors (PI I and PI II) which act as insect antifeedants, are expressed in leaves in response to wounding (Keil *et al.*, 1989), UV irradiation (Ryder *et al.*, 1984; Bowell *et al.*, 1985; Edwards *et al.*, 1985) or stress hormones (Peña-Cortes *et al.*, 1989, Sanchez-Serrano *et al.*, 1990). These genes are not expressed in tubers in response to wounding (Butler *et al.*, 1990). Expression of PI I and PI II genes is induced by high sucrose concentrations (Keil *et al.*, 1989; Johnson and Ryan, 1990; Kim *et al.*, 1991), accordingly, these proteins and their mRNAs accumulate to high levels in tuber, a sink tissue. Transcription of PI I, PI II and patatin (PAT) genes is suspended upon wounding (Logemann *et al.*, 1988) and their abundant mRNAs are displaced from tuber polysomes and degraded completely within 6 h after wounding (Butler *et al.*, 1990; Crosby and Vayda, 1991). Further, the wound response of tubers is limited to cells within a few millimeters of the wound (W. Butler, J.S. Crosby and M.E. Vayda, unpublished observations) whereas the wound response of leaves is systemic (Sanchez-Serrano *et al.*, 1990). These observations indicate that there are qualitative differences in the response of potato organs to wounding.

Translational Activity of Tubers is Stimulated by Wounding

We have established that the overall translational activity of wounded tubers is significantly greater than that of nonstressed tubers (Vayda and Schaeffer, 1988; Butler *et al.*, 1990; Crosby and Vayda, 1991). Wounded tuber tissue *in vivo* incorporates up to 20-fold more [35S]methionine into labeled, TCA-precipitable polypeptides than do nonstressed tubers (Vayda and Schaeffer, 1988). By 1 h after wounding, tuber tissue exhibits a 2 to 4-fold increase in the incorporation of [35S]methionine/g fresh weight (FW) (Fig. 14.1). A modest but reproducible decrease in translational activity is observed 4-6 h after wounding, with a secondary increase in activity beginning at 9 h and rising through 24 h after wounding (Fig. 14.1). This biphasic induction is consistent with the observed

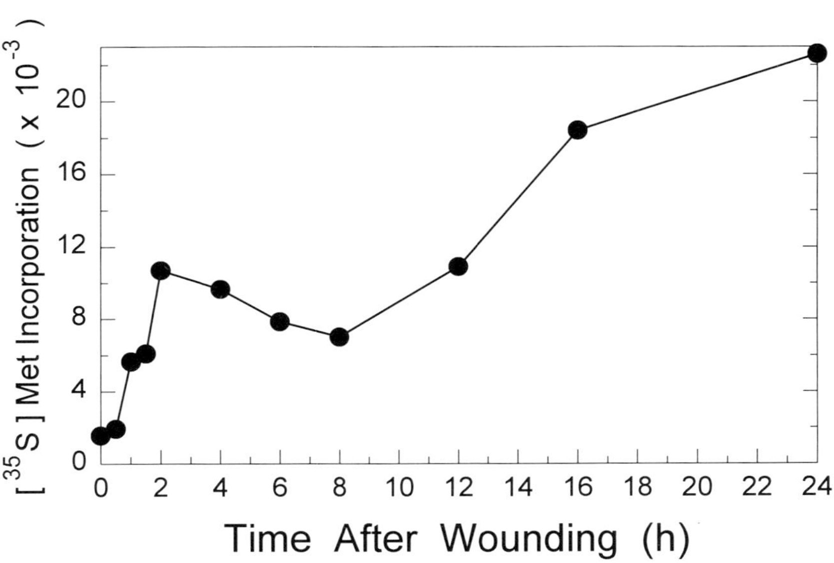

Figure 14.1. Translational activity of potato tubers in response to wounding. 0.5 mCi of [35S]methionine was presented to a tuber wound, at the time of wounding or at the indicated times after wounding. After a 15 min incubation at 22°C, the 2 g core of tuber tissue surrounding the wound was extracted as described previously (Vayda and Schaeffer, 1988). Incorporation was measured as TCA-precipitable counts per g FW of tuber tissue. Results shown are the average of three trials.

induction profiles of wound response mRNAs: mRNAs encoding PAL and HMGR accumulate and are expressed within the first hour after wounding, but mRNAs encoding HRGP, GRP and histones do not accumulate until 9-20 h after wounding (Butler *et al.*, 1990; Crosby and Vayda, 1991). The increase in translational activity may be explained partially by the synthesis of new ribosomal components and translation factors: rRNA and mRNA encoding ribosomal proteins (Rickey and Belknap, 1990; Garbarino *et al.*, 1992) and elongation factors (J. Morelli and M.E. Vayda, unpublished observation) have been observed upon wounding.

The stimulation of the translational machinery upon wounding is also evident *in vitro*. Crude extracts of wounded tubers are 4-8-fold more active in the translation of exogenously added mRNA than are extracts prepared from nonstressed mature tubers (S.-Y. Zheng and M.E. Vayda, unpublished observations). Further, polysomes isolated from wounded tubers exhibit a greater translational competence than polysomes from nonstressed tubers, as is evident by run-off translation in a rabbit reticulocyte lysate system (Crosby and Vayda, 1991) that is capable of elongation but not reinitiation (Berry *et al.*, 1988,

1990). Polysomes isolated from wounded tubers consistently produce more translation products than an equal amount of polysomes isolated from nonstressed tubers (Crosby and Vayda, 1991) indicating that in wounded tuber cells there are more polysome complexes actively engaged in translation than in nonstressed tuber cells. The basis for the difference in specific activity between wounded and nonstressed tuber extracts is not known.

The products produced by run-off translation of polysomes from wounded tubers are distinct from those produced by polysomes from nonstressed tubers. The only prominent products from nonstressed tubers are the 40 kDa patatin, the 22 kDa and 10 kDa proteinase inhibitor proteins, and an unidentified 100 kDa polypeptide, whereas the complement of polypeptides produced by run-off translation of wounded tuber polysomes (Crosby and Vayda, 1991) is essentially the same as those synthesized *in vivo* by wounded tubers (Vayda and Schaeffer, 1988). Indeed, the mRNA populations associated with polysomes from nonstressed and wounded tubers are distinct (Crosby and Vayda, 1991). The tuberization-associated mRNAs such as PI II and PAT are displaced from polysomes into a slowly sedimenting 'unbound' RNA fraction within 1 h after wounding (Crosby and Vayda, 1991). The wound response mRNAs that accumulate during the same time period are largely associated with the high S value polysome complexes (Crosby and Vayda, 1991). These observations indicate that there is a general activation of translational competency upon wounding as well as a discrimination of selected messages to be translated.

A 32 kDa Polypeptide is Associated with Polysomes of Wounded Tubers

The polypeptide composition of polysomes isolated from wounded tubers is essentially identical to that of nonstressed tubers with the exception of a prominent 32 kDa protein that is not evident in the polysome fraction of nonstressed tubers (Crosby and Vayda, 1991; Fig. 14.2, panel A). This protein becomes associated with polysomes within 30 min of wounding, and remains associated for at least 48 h after wounding, paralleling the presence of wound response mRNAs (Crosby and Vayda, 1991). The 32 kDa protein is in 5-10-fold molar excess relative to ribosomal proteins (Fig. 14.2, panel A), and is stripped from the polysome fraction by a 0.1% Triton X-100 wash that does not cause polysomes to dissociate into free subunits (Crosby and Vayda, 1991). These observations lead us to suggest that the 32 kDa protein is not an integral ribosomal protein, but becomes peripherally associated upon wounding.

The 32 kDa polypeptide has been isolated and used to generate antiserum in rabbits. The anti-32 kDa antibody was purified from the rabbit serum by affinity chromatography (Harlow and Lane, 1988) using gel purified 32 kDa protein covalently conjugated to Sephadex (ImmunoPure Antibody-Antigen Immobilization Kit, Pierce). The affinity-purified antibody strongly recognized a single band in 1-D SDS-PAGE analysis of polysome-associated proteins from

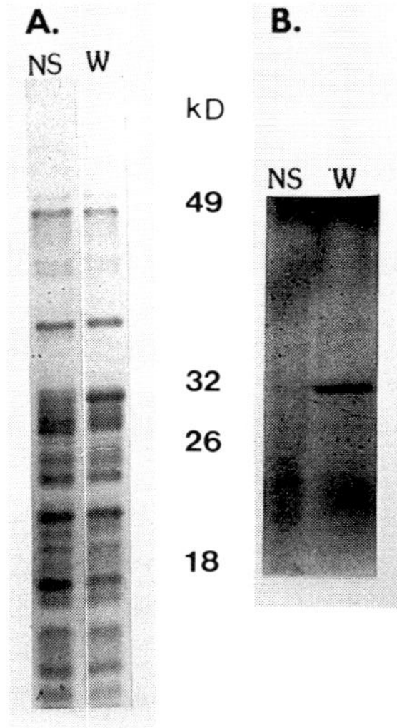

Figure 14.2. Accumulation of the 32 kDa protein in the polysome fraction of wounded tubers. Polysomes were isolated from non-stressed (NS) and 6 h wounded (W) tubers by sedimentation through sucrose gradients as described previously (Crosby and Vayda, 1991). Proteins were resolved by 15% SDS-PAGE. (A) Polypeptides detected by staining with Coomassie Blue. (B) Immunoblot of polysome protein resolved by SDS-PAGE and transferred to PVDF paper. Blot was incubated with affinity-purified anti-32 kDa protein antibody. Binding of the antibody was detected by the ECL (Enhanced Chemi-Luminescence) System (Amersham).

wounded tubers (Fig. 14.2, panel B). This antibody established that the 32 kDa protein is present in low amounts in nonstressed tuber polysome fractions and that its abundance is greatly stimulated upon wounding (Fig. 14.2, panel B). The antibody did not detect the 32 kDa protein in the supernatant fractions of either wounded or nonstressed tubers, suggesting that the 32 kDa protein is synthesized in response to wounding. We have also observed that the 32 kDa polypeptide is associated with polysomes isolated from sprouting tubers and developing tubers, suggesting that this protein is not 'wound specific' but rather associates with ribosomes under conditions when translational competence is increased.

The results shown in Fig. 14.3 demonstrate that the 32 kDa protein has an acidic isoelectric point (pI ~6.0-6.5) in 2-D electrophoretic analysis. Only one prominent species increases greatly in abundance upon wounding (Fig. 14.3,

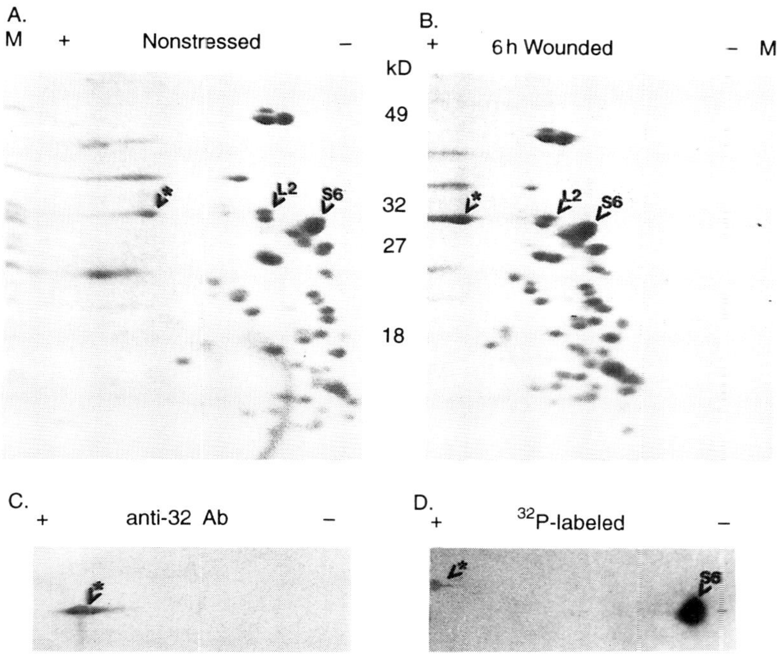

Figure 14.3. Identification of the 32 kDa protein by 2-D NEPHGE electrophoresis. (A) Polysome-associated proteins isolated from a nonstressed potato tuber, resolved by 2-D NEPHGE gel analysis (Sanders *et al.*, 1980) and stained with Coomassie Blue. (B) Coomassie blue-stained NEPHGE 2-D gel of polysome associated proteins isolated from a 6 h wounded tuber. S6 and L2 indicate the positions of known ribosomal component polypeptides. * indicates the position of the 32 kDa wound-induced protein. (C) Immunoblot of a parallel gel to that shown in panel B, electrophoretically transferred to PVDF and incubated with affinity-purified anti-32 kDa protein antibody. Bound antibody was detected by a goat-antirabbit IgG antibody conjugated to horseradish peroxidase and the Western Blue colorimetric reagent (Promega, stabilized BCIP/NBT). (D) Autoradiogram of 2-D NEPHGE gel of 2 h wounded tuber polysome-associated proteins labeled by incubation with [^{32}P] orthophosphate.

compare panels A and B) and reacts with the affinity purified anti-32 kDa antiserum (Fig. 14.3, panel C). Ribosomal proteins S6 or L2, which exhibit similar mobilities in 1-D SDS-PAGE (Scharf and Nover, 1982; Bailey-Serres and

Freeling, 1990; Spedding, 1990) do not vary greatly in abundance upon wounding and are not reactive with the antiserum in immunoblots (Fig. 14.3). The 32 kDa protein becomes phosphorylated within 30 min after wounding (Crosby and Vayda, 1991); however, it is clear from these analyses that the major phosphoprotein associated with wounded tuber polysomes is S6 (Fig. 14.3, panel D). We have isolated the 32 kDa polypeptide by 2-D isoelectric focusing/SDS-PAGE and electroblotting to PVDF paper. However, our initial attempts to obtain the amino acid sequence of this polypeptide were unsuccessful because the N - terminus is blocked, presumably by acetylation. Thus, the identity of the 32 kDa protein is not yet known.

The role that the 32 kDa polypeptide plays in translational activation is also not known, however, there are two possibilities based upon the known characteristics of the protein. Two classes of acidic ribosomal proteins from yeast and animals have been described: 13-18 kDa species (Zinker, 1980; Vidales *et al.*, 1981; Tsurugi and Ogata, 1985; Mitsui *et al.*, 1988; Saenz-Robles *et al.*, 1990) and 30-38 kDa species (Mitsui *et al.*, 1988). These polypeptides bind to ribosomes in non-stoichiometric amounts and become more abundant when growth is stimulated, and the low molecular weight species is phosphorylated in yeast (Zinker, 1980; Vidales *et al.*, 1981). The potato 32 kDa protein may be a plant analog of these proteins and mediate translational activity by direct association with ribosomal proteins. Conversely, the 32 kDa protein may be an mRNA binding protein. The presence of the 32 kDa protein parallels that of the wound-induced mRNAs on polysomes and the 32 kDa protein is displaced from ribosomes by puromycin which causes premature termination and release of mRNA (Crosby and Vayda, 1991). However, a role for the 32 kDa protein in translational activation is yet to be demonstrated.

The Abundance of Elongation Factor EF-1α Increases Upon Wounding

Elongation factor EF-1α is known to increase in abundance in several plant and animal systems when translation is stimulated; conversely, decreased expression of this factor correlates with loss of translational competency (Webster, 1985; Cavallius *et al.*, 1986; Brady, 1988; Pokalsky *et al.*, 1989; Ursin *et al.*, 1991). For example, EF-1α accumulates in regenerating liver following hepatectomy, and decreases as cell cultures or whole organisms become senescent (Webster, 1985; Cavallius *et al.*, 1986; Brady, 1988). Indeed, immortal cell cultures maintain high levels of EF-1α and, in *Drosphila*, overexpression of EF-1α correlates with longer life span and delayed onset of senescence (Shepard *et al.*, 1989). In plants, the gene encoding EF-1α is expressed at the highest levels in tissues with high translational activity (meristems, growing leaf tissue, root tips) and lowest in senescing organs (Pokalsky *et al.*, 1989; Ursin *et al.*, 1991). Similarly, we have determined that the expression and accumulation of EF-1α increase upon wounding of potato tubers.

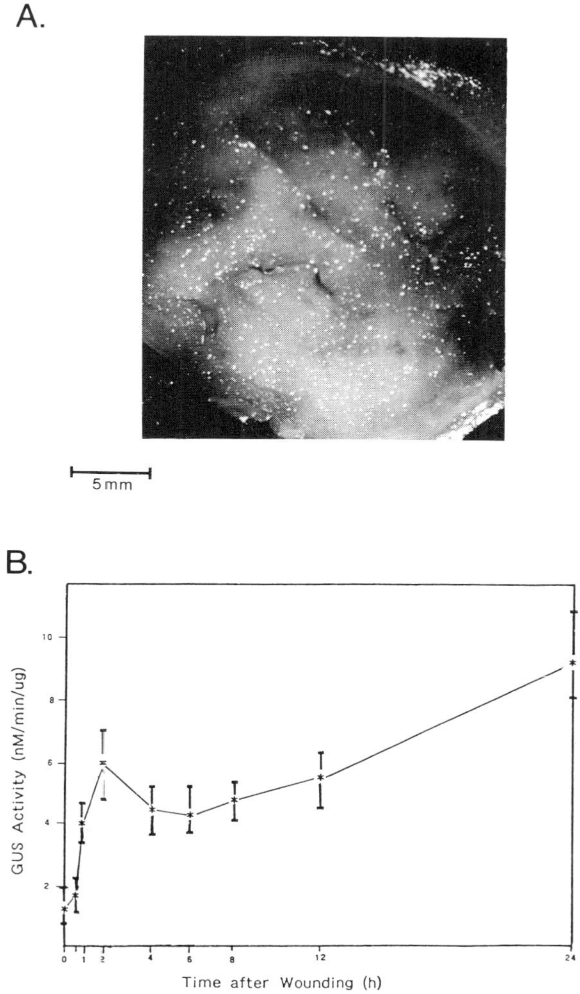

Figure 14.4. Expression of a chimeric tomato EF-1α promoter/GUS gene construct in wounded potato tubers. Tubers were obtained from potato clones transformed with the EF-1α/GUS construct described by Ursin *et al.* (1991). (A) Photograph of a 1 h wounded tuber histochemically stained with X-Gluc to demonstrate GUS activity (Jefferson *et al.*, 1987). (B) Fluorometric assay of GUS activity in wounded tubers. 0.05 g of nonstressed tuber tissue or 0.05 g of the same tuber removed at the times indicated after wounding were ground in extraction buffer (Jefferson *et al.*, 1987). Specific activity of GUS was determined by the rate of hydrolysis of 1 mM 4-methylumbelliferyl-β-D-galactoside, with spectrofluorometric determination of fluorescence relative to 4-methylumbelliferone standards.

Wounding of transgenic potato tubers rapidly elicits the expression of a chimeric construct containing 6.5 kb of tomato EF-1α upstream region, the β-glucuronidase (GUS) coding region, and 0.6 kb of EF-1α 3' noncoding region (Ursin *et al.*, 1991; generously provided by C. Shewmaker and Calgene, Inc.). Histochemical staining of transgenic tubers for GUS activity shows that the construct is expressed within 1 h after wounding, but in a highly localized fashion: GUS induction is evident only within 3 mm of the wound site (Fig. 14.4, panel A). Fluorometric assay of GUS activity also indicates rapid induction of this construct for the first 2 h after wounding, followed by a slight but reproducible decrease in activity by 6 h after wounding, and a secondary induction 9 h after wounding (Fig. 14.4, panel B). This pattern of induction correlates with the biphasic stimulation of translational activity evident in Fig. 14.1. The transient decrease in GUS activity is surprising. However, both RNA gel blot hybridization and protein immunoblots confirm this biphasic fluctuation (Fig. 14.5). Endogenous EF-1α mRNA and protein fluctuate in a similar fashion (Fig. 14.5). We conclude that EF-1α is induced when the level of translation is increased, and that the observed turnover of GUS 4-8 h after wounding must be due to a wound-associated protease activity.

Hypoxic Stress Arrests Translation

We have established that translational processes are inhibited in tubers within 30 min of transfer to hypoxic conditions, regardless of whether these tubers are wounded prior to onset of hypoxia (Vayda and Schaeffer, 1988; Butler *et al.*, 1990; Crosby and Vayda, 1991). Tubers resume a low level of translational activity when subjected to hypoxic conditions for prolonged periods of 14-24 h (Vayda and Schaeffer, 1988; Butler *et al.*, 1990). However, the proteins synthesized after prolonged hypoxia do not include those encoded by the wound-response mRNAs. Rather, a distinct set of proteins is synthesized that includes alcohol dehydrogenase (ADH) and aldolase (ALD) (Vayda and Schaeffer, 1988; Butler *et al.*, 1990; Crosby and Vayda, 1991) whose role in fermentative metabolism may be necessary for survival in low oxygen conditions. In this regard, the hypoxic response is similar to that reported for other crop plants (Sachs *et al.*, 1980; Hake *et al.*, 1984, 1985; Ricard *et al.*, 1986; Kennedy *et al.*, 1992). However, translation of the wound-response mRNAs is not detected although these species remain present in the cytoplasm at constant steady-state levels for at least 12 h of hypoxia (Butler *et al.*, 1990). This failure to translate wound-response mRNAs is the apparent reason that potato tubers become susceptible to opportunistic bacterial pathogens during hypoxic stress. Although the high degree of maceration observed under moist, poorly oxygenated conditions can be attributed in part to the cell density-dependent, and low O_2-augmented activation of bacterial genes encoding pectate lyase and other degradative enzymes, it is clear that the inability of tuber tissue to mount a wound response is a major factor contributing to the susceptibility of tuber

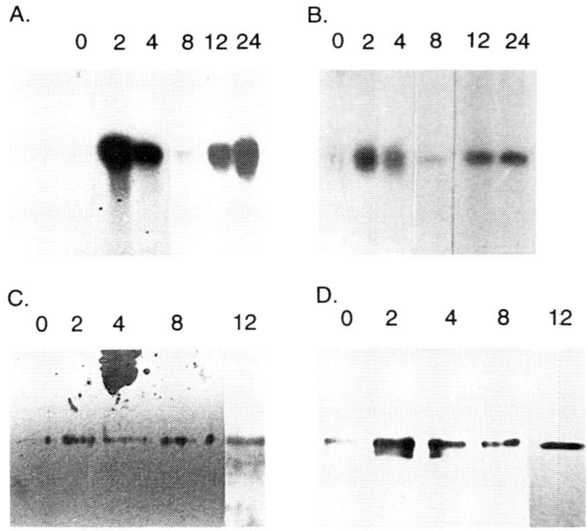

Figure 14.5. Accumulation of GUS and EF-1a mRNA and protein in wounded potato tubers. Total RNA (Butler *et al.*, 1990) and soluble protein (Vayda and Schaffer, 1988) were isolated from tubers of the transformed potato clones described in Fig. 14.4, prior to wounding and at the times indicated after wounding. (A) RNA gel blot hybridized to ^{32}P-labeled GUS cDNA. (B) RNA gel blot hybridized to ^{32}P-labeled EF-1a cDNA (Pokalsky *et al.*, 1989). (C) Immunoblot probed with commercial anti-GUS antibody (Clonetech) and visualized by the ECL system. (D) Immunoblot probed using anti-wheat germ EF-1a antiserum and visualized by the ECL system (Amersham).

tissue to bacterial rot (Zucker and Hanklin, 1970; Vayda *et al.*, 1992). Treatments that block translation under aerobic conditions, such as addition of cycloheximide or brief hypoxic shock prior to aerobic inoculation, allow pathogens to cause as much maceration as if the tubers had been incubated hypoxically for the entire period.

The mechanism of translational inhibition must be an arrest of elongation or termination because wound-response mRNAs remain bound to polysomes for at least 24 h after transfer to hypoxic conditions although they are not translated (Crosby and Vayda, 1991). The inhibition of translation *in vivo* is complete within 30 min of hypoxic incubation (Vayda and Schaeffer, 1988) without detectable change in the abundance of wound response mRNAs (Butler *et al.*, 1990). The wound response mRNAs are translatable *in vitro* using a rabbit reticulocyte lysate (Butler *et al.*, 1990), indicating that the inhibition of translation is not due to degradation of mRNAs species. It is not known whether

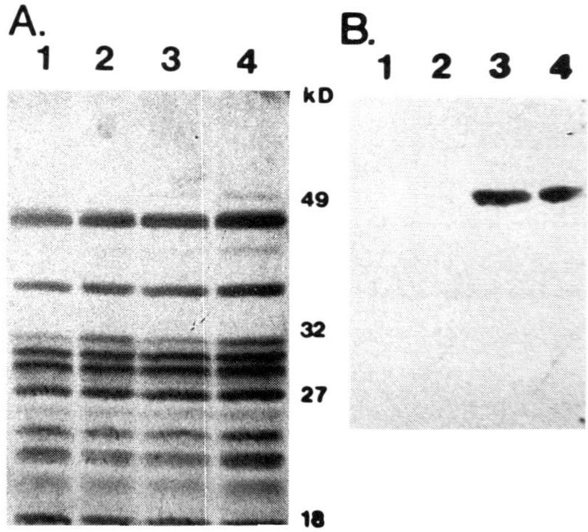

Figure 14.6. EF-1α is associated with the polysome fraction of hypoxic tubers. Polysomes were obtained by sedimentation through sucrose as described previously (Crosby and Vayda, 1991), and polysome-associated proteins were resolved by SDS-PAGE. Lanes 1, nonstressed tubers ; lanes 2, 2 h wounded tubers; lanes 3, 14 h hypoxic tubers; lanes 4, tubers wounded for 2 h then incubated in a hypoxic argon atmosphere for 14 h. (A) Silver stained SDS-polyacrylamide gel of polysome samples. (B) Immunoblot of an identical gel transferred to PVDF paper, incubated with anti-wheat germ EF-1α antiserum and visualized by the ECL system (Amersham).

translation initiation is inhibited as well. Polysomes isolated from these wounded, hypoxic tubers are very inefficient at run-off translation in the rabbit reticulocyte lysate (Crosby and Vayda, 1991), and extracts from these tubers are incompetent at supporting translation *in vitro* (S.-Y. Zheng and M.E. Vayda, unpublished results). These results indicate the presence of an inhibitory factor, the absence of a factor that is not complemented by the heterologous system, or the irreversible inactivation of one or more of the translational components. Similar arrest of translation by blockage of the elongation cycle has been observed in amaranth seedlings shifted to darkness (Berry *et al.*, 1988, 1990; Wang *et al.*, 1992) and in maize roots when subjected to hypoxic conditions (Webster *et al.*, 1991).

Table 14.1. pH and lactate content of nonstressed, wounded and hypoxic tuber tissues.

Sample	Average pH	Lactate (µmol/g FW)
Nonstressed tubers	6.85	0.28
Tubers hypoxic		
30 min	6.40	0.73
1 h	6.40	0.68
2 h	6.25	0.83
4 h	6.35	0.86
7 h	6.40	1.18
10 h	6.30	1.21
14 h	6.35	1.44
16 h	6.40	1.63
24 h	6.35	1.92
48 h	6.45	2.88
Tubers wounded		
30 min	6.80	0.45
1 h	6.75	0.41
2 h	6.85	0.46
4 h	6.80	0.38
20 h	6.80	0.40

Tubers were wounded in air or incubated in a hypoxic argon atmosphere as described previously (Vayda *et al.*, 1992). 2 g of tubers were ground in 1.0 ml of distilled water and the pH was measured immediately. For lactate determination, extracts were clarified of protein by incubation with chilled 10% TCA. Lactate in the supernatants was determined using a Lactate Determination Kit (Sigma). Results shown are the average of four trials.

EF-1α Is Stably Associated with Polysomes During Hypoxic Arrest of Translation

The protein complement of hypoxic tuber polysomes exhibits no obvious lack of components that might explain the loss of translational competency. Specifically, the 32 kDa protein remains bound to polysomes when wounded tubers are transferred to hypoxic atmospheres (Crosby and Vayda 1991; Fig. 14.6, panel A), and thus, arrest is not due to dissociation of this putative activation factor from polysomes. However, two polypeptides do accumulate in the polysome fraction of hypoxic tubers with apparent molecular weights of 48 kDa and 100 kDa (Crosby and Vayda, 1991; Fig. 14.6, panel A). The 48 kDa species is identified in immunoblots as EF-1α (Fig. 14.6, panel B): this

Table 14.2. Changes in tuber tissue ATP level in response to hypoxia or wounding. ATP levels were determined by luciferase assay of extract fractions.

Tuber sample	ATP (nmol/g)	Variance
Nonstressed	13.2	±1.2
Hypoxic 1 h	13.7	±1.5
Hypoxic 2 h	8.2	±2.2
Hypoxic 4 h	6.3	±2.7
Hypoxic 6 h	7.5	±2.5
Hypoxic 8 h	7.9	±1.8
Hypoxic 12 h	4.0	±1.3
Hypoxic 24 h	3.8	±1.2
Hypoxic 48 h	2.0	±1.3
Hypoxic 48 h	2.0	±1.3
Nonstressed	12.7	±2.0
Wounded 1 h	11.5	± 1.6
Wounded 2 h	6.7	± 2.1
Wounded 4 h	8.4	± 3.1
Wounded 6 h	10.4	± 2.5
Wounded 8 h	8.3	± 2.2
Wounded 12 h	3.7	± 3.6
Wounded 24 h	10.5	± 2.6
Wounded 48 h	11.8	± 1.6

polypeptide reacts with two different antisera directed against heterologous EF-1α polypeptides (generously provided by W. Moller and K. Browning).Immunoblots indicate that the total amount of cellular EF-1α in the 40-70% ammonium sulfate extracts of aerobic and hypoxic wounded tubers is the same (J. Morelli and M.E. Vayda, unpublished observations). These data indicate that EF-1α accumulates in the polysome fraction of hypoxic tubers and is depleted from the soluble pool. This accumulation of EF-1α is an aberrant event. EF-1α is a soluble translation factor that presents amino acyl-tRNAs to the A site of ribosomes and, thus, functions by only transient association with ribosomes. The stable association of this factor with ribosomes suggests that some event during hypoxia prevents dissociation of EF-1α from ribosomes. Such an association would arrest ribosomes at one stage in the elongation cycle and block ribosome progress on mRNA, consistent with the translational arrest observed in hypoxic tubers.

Table 14.3. Amino acid content of nonstressed, hypoxic and wounded tuber tissue.[1] (moles amino acid per g of lyophilized tuber tissue).

Amino acid	Moles in Tuber[2] (x 10⁴)	Relative content (%)			
		Hours hypoxic			
		2	5	24	48
Total	5.500	54.44	53.34	58.72	61.16
Gln	1.034	71.11	43.91	77.22	77.46
Asn	0.968	83.94	90.28	93.55	93.90
Asp	0.541	34.80	32.38	26.53	30.86
Arg	0.494	33.73	41.99	33.75	40.29
GABA	0.453	27.91	82.97	39.64	63.06
Glu	0.431	43.20	102.48	45.17	36.57
Val	0.415	42.87	40.32	46.05	46.88
Tyr	0.155	58.36	45.91	54.88	61.74
Thr	0.147	44.30	39.95	48.99	57.27
Ser	0.147	60.52	58.67	50.35	47.92
His	0.136	35.38	46.13	39.87	43.76
Ile	0.127	43.97	43.47	49.68	45.00
Phe	0.123	55.58	40.81	57.20	52.81
Gly	0.101	24.43	32.43	22.44	23.94
Leu	0.095	43.39	38.01	52.45	50.09
Ala	0.0601	46.31	278.85	218.05	212.53
Met	0.049	32.54	8.55	36.68	42.19
Trp	0.024	51.56	51.46	44.28	83.04
		Hours wounded			
		2	5	24	48
Total	5.500	101.44	64.42	52.64	46.93
Gln	1.034	98.16	74.91	69.53	69.31
Asn	0.968	102.12	104.35	84.34	73.59
Asp	0.541	118.81	61.82	37.37	36.83
Arg	0.494	85.63	43.08	34.44	26.35
GABA	0.453	59.04	41.78	36.90	45.91
Glu	0.431	89.23	39.16	24.69	34.57
Val	0.415	120.79	44.27	37.48	35.86
Tyr	0.155	116.94	46.44	55.35	29.54
Thr	0.147	128.40	58.19	38.98	35.70
Ser	0.147	108.86	80.58	112.81	38.09
His	0.136	112.98	48.08	35.15	36.10
Ile	0.127	123.67	40.04	34.82	19.98
Phe	0.123	123.82	40.73	30.15	12.73
Gly	0.101	56.66	23.55	26.38	17.49
Leu	0.095	139.55	54.80	34.765	25.40
Ala	0.060	157.58	227.47	62.23	31.57
Met	0.049	102.43	15.06	27.285	15.82
Trp	0.024	116.92	31.86	83.54	15.63

[1] 100 mg of lyophilized tuber tissue was extracted with acidified ethanol (80% ethanol, 0.25M HCl) at 40°C for 1 h. Extracts were filtered (0.2 μm) and analyzed by HPLC. Amino acids were derivatized using *o*-phthaldialdehyde mercaptopropionic acid. Amino acids were detected by fluorescence spectroscopy, quantified relative to an internal norvaline standard.

[2] Nonstressed.

Metabolic Acidosis May Cause Translational Arrest During Hypoxia

The arrest of translation during hypoxic stress may be caused by a pH-mediated inactivation of elongation factors. We have noted that non-buffered hypoxic tuber extracts consistently show a pH that is 0.4-0.6 units more acidic than nonstressed tubers (Table 14.1). Hypoxic tubers exhibit an average pH of 6.4-6.2 whereas the pH of the aerobic wounded tubers remains at pH 6.8. The pH of tuber tissue drops 0.2-0.4 units within the first 30 min after transfer to hypoxic conditions. Lactate accumulates in parallel with this change in pH (Table 14.1). A 3-fold increase in lactate is evident within 30 min of hypoxic incubation, followed by a steady, slow increase in lactate to 48h. Sieber and Brandle (1991) reported a similar increase in lactate during the first 6 h of potato tuber hypoxia. Although the change in pH and lactate accumulation measured resides largely in the vacuole, it is likely that significant acidification of the cytoplasm occurs as well. The cytoplasmic pH of maize roots drops by 0.6 units within 25 min of hypoxia as demonstrated by ^{31}P- and ^{13}C-NMR spectroscopy (Roberts *et al.*, 1984a 1984b, 1992). This acidification correlates with the loss of translational competency. Webster *et al.* (1991) have shown that decreasing the pH of an aerobic corn root extract *in vitro* reduces translational competence to a low level comparable to that of hypoxic root extracts. This treatment decreased both the competence of the extract fraction containing the elongation factors (the 40-70% ammonium sulfate precipitate) and the ability of isolated polysomes to direct run-off translation *in vitro* (Webster *et al.*, 1991). The similarity of the maize and potato tuber systems under hypoxia suggests that metabolic acidosis may be responsible for the observed inactivation of translation.

The translational arrest in hypoxic tubers cannot be due solely to fluctuations in ATP levels or amino acid content. Tuber ATP and energy charge do drop during hypoxic incubation (Sieber and Brandle, 1991). As shown in Table 14.2, tuber ATP levels drop 2-fold within the first hour and 3-fold within 4 h of hypoxia. However, a similar decrease in tuber ATP level is observed in response to wounding when translation is known to be stimulated (Table 14.2).

Prolonged hypoxia of 14-24 h decreases tuber ATP levels to 10-20% that of nonstressed tubers, yet during this period tubers resume translation to produce ADH, ALD and other hypoxia-induced species (Butler *et al.*, 1990; Crosby and Vayda, 1991). Similarly, the free amino acid content of tubers fluctuates in response to hypoxia but not in a manner that correlates with translational arrest (Table 14.3). Alanine content increases presumably because of an accumulation of pyruvate and other fermentative products. The levels of other amino acids are decreased 2-4-fold relative to nonstressed tubers, but similar decreases are observed 24 h after wounding when tubers are very active in protein synthesis (Table 14.4). Thus, we conclude that a decrease in pH, but not the minor fluctuations in amino acid and ATP content we have detected, mediates translational inhibition of hypoxic tubers.

Conclusions

The translational activity of potato tubers is stimulated upon wounding and arrested by hypoxia. The activation of translation in response to wounding is correlated with the association of a 32 kDa protein with polysomes and an increase in the expression and cellular abundance of EF-1α. The abundance of these proteins is not affected by hypoxic stress. Rather, translational arrest is apparently caused by the non-cyclic association of EF-1α with ribosomes, simultaneously inhibiting polysome capability for run-off translation and depleting extracts of components necessary for translation *in vitro*. It is not clear whether the failure to cycle is due to an impairment of EF-1α structure or the ribosome-associated GTPase involved in EF-1α release. However, it seems likely that translational arrest is due to the accumulation of lactate and the pH decrease that occurs immediately upon the onset of hypoxia.

The identity and role of the 32 kDa protein is not known. Further, it is not understood how wound response mRNAs are selected for translation whereas pre-existing tuberization-associated mRNAs are displaced from polysomes. Nor is it clear how ADH and ALD mRNAs are translated during prolonged hypoxia while the wound response mRNAs remain arrested in translation. Resolution of these questions is necessary for a full understanding of how environmental stress affects plant gene expression.

Acknowledgements

This work was supported by grants from the USDA (#91-371006622) and the Maine Agricultural Experiment Station (#38402). The author wishes sincerely to thank Dr W. Moller and Dr K. Browing for generously providing anti-EF-1α antisera, Dr W Belknap for generating antisera to the 32 kDa protein, Dr H. Davies for the amino acid analyses, and Dr C Shewmaker and Calgene Inc for providing transgenic potato clones expressing the EF-1α/GUS chimeric construct. This is MAES publication #1810.

References

Bailey-Serres, J. and Freeling, M. (1990) Hypoxic stress-induced changes in ribosomes of maize seedling roots. *Plant Physiology* 94:1237-1243.

Barkhausen, R. (1978) Ultrastructural changes in wounded plant storage tissue. In: Kahl, G. (ed.), *Biochemistry of Wounded Plant Tissues*. DeGruyter, Berlin, pp. 1-42.

Belknap, W. and Rickey, T. (1990) Physiological and stress-induced changes in potato tuber gene expression. In: Vayda, M.E. and Park, W.D. (eds), In. *Molecular and Cellular Biology of the Potato*. CAB International, Wallingford, UK, pp. 89-95.

Berry, J.O., Carr, J.P. and Klessig, D.F. (1988) mRNAs encoding ribulose-1,5-bisphosphate carboxylase remain bound to polysomes but are not translated in amaranth seedlings transferred to darkness. *Proceedings of the National Academy of Sciences USA* 85:4190-4194.

Berry, J.O., Breiding, D.E. and Klessig, D.F. (1990) Light-mediated control of translational initiation of ribulose-1,5-bisphosphate carboxylase in amaranth cotyledons. *Plant Cell* 2:795-803.

Bevan, M., Shufflebottom, D., Edwards, K., Jefferson, R. and Schuch, W. (1989) Tissue- and cell-specific activity of a phenylalanine ammonia-lyase promoter in transgenic plants. *EMBO Journal* 8:1899-1906.

Bowell, G.P., Bell, J.N., Cramer, C.L., Schuch, W., Lamb, C.J. and Dixon, R.A. (1985) L-Phenylananine ammonia-lyase from *Phaseolus vulgaris*: characterisation and differential induction of multiple forms from elicitor-treated cell suspension cultures. *European Journal of Biochemistry* 149:311-419.

Brady, C.J. (1988) Nucleic acid and protein synthesis. In: Nooden, L.D. and Leopold, A.C. (eds), *Senescence and Aging in Plants*. Academic Press, New York, pp. 147-179.

Butler, W., Cook, L. and Vayda, M.E. (1990) Hypoxic stress inhibits multiple aspects of the potato tuber wound response. *Plant Physiology* 93:264-270.

Cavailius, J., Rattan, S.I.S. and Clark, B.F.K. (1986) Changes in activity and amount of active elongation factor 1a in aging and immortal human fibroblast cultures. *Experimental Gerontology* 21:149-157.

Choi, D., Ward, B.L. and Bostock, R.M. (1992) Differential induction and suppression of potato 3-hydroxy-3-methylglutaryl coenzyme A reductase genes in response to *Phytophthora infestans* and to its elicitor arachidonic acid. *Plant Cell* 4:1333-1344.

Crosby, J.S. and Vayda, M.E. (1991) Stress-induced translational control in potato tubers may be mediated by polysome-associated proteins. *Plant Cell* 3:1013-1023.

Dyer, W.E., Henstrand, J.M., Handa, A.K. and Herrmann, K.M. (1989) Wounding induces the first enzyme of the shikimate pathway in Solanaceae. *Proceedings of the National Academy of Sciences USA* 86:7370-7373.

Edwards, K., Cramer, C.L., Bolwell, G.P., Dixon, R.A., Schuch, W. and Lamb, C.J. (1985) Rapid transient induction of phenylalanine ammonia-lyase mRNA in elicitor-treated bean cells. *Proceedings of the National Academy of Sciences USA* 82:6731-6735.

Garbarino, J., Rockhold, D. and Belknap, W.R. (1992) Expression of stress responsive ubiquitin genes in potato tubers. *Plant Molecular Biology* 20:235-244.

Hake, S, Taylor, W.C. and Freeling, M. (1984) Molecular analyses of genetically stable mutants of the maize *Adh1* gene. *Molecular and General Genetics* 194:42-48.

Hake, S., Kelley, P.M., Taylor, W.C. and Freeling, M. (1985) Coordinate induction of alcohol dehydrogenase I, aldolase and other anaerobic RNAs in maize. *Journal of Biological Chemistry* 260:5050-5054.

Harlow, E. and Lane, D. (1988) Immunoblotting and immunoaffinity purification. In: *Antibodies, A Laboratory Manual*. Cold Spring Harbor Laboratory Press, Cold Spring Harbor, NY, pp. 471-552.

Ishizuka, M., Yamada, F., Tanaka, Y., Takeuchi, Y. and Imaseki, H. (1991) Sequential induction of mRNAs for phelylalanine ammonia-lyase in slices of potato tuber. *Plant Cell Physiology* 32:5764.

Jefferson, R.A., Kavanagh, T.A. and Bevan, M.W. (1987) GUS fusions: ß-glucuronidase as a sensitive and versatile gene fusion marker in higher plants. *EMBO Journal* 13:3901-3907.

Johnson, R. and Ryan, C.A. (1990) Wound-inducible potato inhibitor II genes: enhancement of expression by sucrose. *Plant Molecular Biology* 14:527-536.

Keil, M., Sanchez-Serrano, J.J. and Willmitzer, L. (1989) Both wound-inducible and tuber-specific expression are mediated by the promoter of a single member of the potato proteinase inhibitor II gene family. *EMBO Journal* 8:1323-1330.

Kennedy, R.A., Rumpho, M.E. and Fox, T.C. (1992) Anaerobic metabolism in plants. *Plant Physiology* 100:1-6.

Kim, S.-R., Costa, M.A. and An, G. (1991) Sugar response element enhances wound response of potato proteinase inhibitor II promoter in transgenic tobacco. *Plant Molecular Biology* 17:973-983.

Kirk, M.M. and Kirk, D.L. (1985) Translational regulation of protein synthesis in response to light, at a critical stage of *Volvox* development. *Cell* 41:419-428.

Kolattukudy, P.E. (1981) Structure, biosynthesis and biodegradation of cutin and suberin. *Annual Review of Plant Physiology* 83:539567.

Logemann, J., Mayer, J.E., Schell, J. and Willmitzer, L. (1988) Differential expression of genes in potato tubers after wounding. *Proceedings of the National Academy of Sciences USA* 85:1136-1140.

Logemann, J., Lipphardt, S., Lorz, H., Hauser, I., Willmitzer, L. and Schell. (1989) 5' upstream sequences from the *wun1* gene are responsible for gene activation by wounding in transgenic plants. *Plant Cell* 1:151-158.

Lojkowska, E. (1988) Lipid composition and post-wounding degradation in potato slices from cultivars differing in susceptibility to autolysis. *Potato Research* 31:541-549.

Mason, H.S., Mullet, J.E. and Boyer, J.S. (1988) Water deficit-induced changes in abscisic acid, growth, polysomes and translatable RNA in soybean hypocotyls. *Plant Physiology* 88:289-294.

Mitsui, K., Nakagawa, T. and Tsurugi, K. (1988) On the size and the role of a free cytosolic pool of acidic ribosomal proteins in yeast *Saccharomyces cerevisiae*. *Journal of Biochemistry* 104:908-911.

Oba, K., Kondo, K., Doke, N. and Uritani, I. (1985) Induction of 3-hydroxy-3-methylglutaryl CoA reductase in potato tubers after slicing, fungal infection or chemical treatment, and some properties of the enzyme. *Plant Cell Physiology* 26:873-880.

Okimoto, R., Sachs, M.M., Porter, E.K. and Freeling, M. (1980) Patterns of polypeptide synthesis in various maize organs under anaerobiosis. *Planta* 150:89-94.

Peña-Cortes, H., Sanchez-Serrano, J.J., Mertens, R., Willmitzer, L. and Prat, S. (1989) Abscisic acid is involved in the wound-induced expression of the proteinase inhibitor II gene in potato and tomato. *Proceedings of the National Academy of Sciences USA* 86:9851-9855.

Pokalsky, A.R., Hiatt, W.R., Ridge, N., Rasmussen, R., Houck, C.M. and Shewmaker, C.K. (1989) Structure and expression of elongation factor 1a in tomato. *Nucleic Acids Research* 17:4661-4673.

Ricard, B., Mocquot, B., Fournier, A., Delseny, M. and Pradet, A. (1986) Expression of alcohol dehydrogenase in rice embyros under anoxia. *Plant Molecular Biology* 7:321-329.

Rickey, T.M. and Belknap, W.R. (1991) Comparison of the expression of several stress-responsive genes in potato tubers. *Plant Molecular Biology* 16:1009-1018.

Roberts, E., Kutchan, T. and Kolattukudy, P.E. (1988). Cloning and sequencing of cDNA for a highly anionic peroxidase from potato and the induction of its mRNA in suberizing potato tubers and tomato fruits. *Plant Molecular Biology* 11:15-26.

Roberts, J.K.M., Callis, J., Wemmer, D., Walbot, V. and Jardetzky, O. (1984a) Mechanism of cytoplasmic pH regulation in hypoxic maize root tips and its role in survival under hypoxia. *Proceedings of the National Academy Sciences USA* 81:3379-3383.

Roberts, J.K.M., Callis, J., Jardetzy, O., Walbot, V. and Freeling, M. (1984b) Cytoplasmic acidosis as a determinant of flooding intolerance in plants. *Proceedings of the National Academy of Sciences USA* 81:6029-6033.

Roberts, J.K.M., Hooks, M.A., Miaullis, A.P., Edwards, S. and Webster, C. (1992) Contribution of malate and amino acid metabolism to cytoplasmic pH regulation in hypoxic maize root tips studied using nuclear magnetic resonance spectroscopy. *Plant Physiology* 98:480-487.

Rumeau, D., Maher, E.A., Kelman, A. and Showalter, A.M. (1990) Extensin and phenylalanine ammonia-lyase gene expression altered in potato tubers in response to wounding, hypoxia, and *Erwinia carotovora* infection. *Plant Physiology* 93:1134-1139.

Ryder, T.B., Cramer, C.L., Bell, J.N., Robbins, M.P., Dixon, R.A. and Lamb, C.J. (1984) Elicitor rapidly induces chalcone synthase mRNA in *Phaseolus vulgaris* cells at the onset of phytoalexin defense response. *Proceedings of the National Academy of Sciences USA* 81:5724-5728.

Sachs, M.M., Freeling, M. and Okimoto, R. (1980) The anaerobic proteins of maize. *Cell* 20:761-767.

Saenz-Robles, M.T., Remacha, M., Vilella, M.D., Zinker, S. and Ballesta, J.P.G. (1990) The acidic ribosomal proteins as regulators of the eukaryotic ribosomal activity. *Biochimica et Biophysica Acta* 1050:51-55.

Sanchez-Serrano, J.J., Amati, S., Keil, M., Peña-Cortes, H., Prat, S., Recknagel, C. and Willmitzer, L. (1990) Promoter elements and hormonal regulation of proteinase inhibitor II gene expression in potato. In: Vayda, M.E. and Park, W.D. (eds), *The Molecular and Cellular Biology of the Potato*. CAB International, Wallingford, pp. 57-69.

Sanders, M.M., Groppi, V.E. and Browning, E.T. (1980) Resolution of basic cellular proteins including histone variants by two-dimensional gel electrophoresis: evaluation of lysine to arginine ratios and phosphorylation. *Analytical Biochemistry* 103:157-165.

Scharf, K.D. and Nover, L. (1982) Heat-shock-induced alterations of ribosomal protein phosphorylation in plant cell cultures. *Cell* 30:427-437.

Shepard, J.C.W., Walldorf, U.P.H. and Gehring, W.J. (1989) Fruit flies with additional expression of the elongation factor EF-1α live longer. *Proceedings of the National Academy of Sciences USA* 86:7520-7521.

Sieber, M. and Brandle, R. (1991) Energy metabolism in rhizomes of *Acorus calamus* (L.) and tubers of *Solanum tuberosum* (L.) with regard to their anoxia tolerance. *Botanica Acta* 104:279-282.

Spedding, G. (1990) *Ribosomes and Protein Synthesis: A Practical Approach* IRL Press, Oxford, UK.

Stanford, A., Bevan, M. and Northcote, D. (1989) Differential expression within a family of novel wound-induced genes in potato. *Molecular and General Genetics* 215:200-208.

Stanford, A.C., Northcote, D.H. and Bevan, M.W. (1990) Spatial and temporal patterns of transcription of a wound-induced gene in potato. *EMBO Journal* 9:593-603.

Stermer, B.A. and Bostock, R.M. (1987) Involvement of 3-hydroxy-3-methylglutaryl coenzyme A reductase in the regulation of sesquiterpenoid phytoalexin synthesis in potato. *Plant Physiology* 84:404-408.

Tsurugi, K. and Ogata, K. (1985) Evidence for the exchangeability of acidic ribosomal proteins on cytoplasmic ribosomes in regenerating rat liver. *Journal of Biochemistry* 98:1427-1431.

Ursin, V.M., Irvine, J.M., Hiatt, W.R. and Shewmaker, C.K. (1991) Developmental analysis of elongation factor EF-1a expression in transgenic tobacco. *Plant Cell* 3:583-591.

Vayda, M.E. and Schaeffer, H.J. (1988) Hypoxic stress inhibits the appearance of wound-response proteins in potato tubers. *Plant Physiology* 88:805-809.

Vayda, M.E., Antonov, L.S., Yang, Z., Butler, W.O. and Lacy, G.H. (1992) Hypoxic stress inhibits aerobic wound-induced resistance and activates hypoxic resistance to bacterial soft rot. *American Potato Journal* 1: 239-253.

Vidales, F.J., Sanchez-Madrid, F. and Ballesta, J.P.G. (1981) The acidic proteins of eukaryotic ribosomes: a comparative study. *Biochimica et Biophysica Acta* 656:28-35.

Wang, J.-L., Klessig, D.F. and Berry, J.O. (1992) Regulation of C4 gene expression in developing amaranth leaves. *Plant Cell* 4:173-184.

Webster, C., Kim, C.-Y. and Roberts, J.K.M. (1991) Elongation and termination reactions of protein synthesis on maize root tip polyribosomes studied in a homologous cell-free system. *Plant Physiology* 96:418-425.

Webster, G.C. (1985) Protein synthesis in aging organisms. In: Sohal, R.S. (ed.), *Molecular Biology of Aging: Gene Stability and Gene Expression.* Raven Press, New York, pp. 263-289.

Weiss, C. and Bevan, M. (1991) Ethylene and a wound signal modulate local and systemic transcription of *win2* genes in transgenic potato plants. *Plant Physiology* 96:943-951.

Yang, Z., Park, H., Lacy, G.H. and Cramer, C.L. (1991) Differential activation of potato 3-hydroxy-3-methylglutaryl coenzyme A reductase genes by wounding and pathogen challenge. *Plant Cell* 3:397-405.

Zinker, S. (1980) P5/P5', the acidic ribosomal phosphoproteins from *Saccharomyces cerevisiae*. *Biochimica et Biophysica Acta* 606:76-82.

Zucker, M. and Hanklin, L. (1970) Physiological basis for a cycloheximide induced soft rot of potatoes by *Pseudomonas fluorescens*. *Annals of Botany* 34:1047-1062.

Chapter Fifteen:

Expression of the Brazil Nut Methionine-Rich Protein in Transgenic Potato Plants

H.M. Tu, L.W. Godfrey and S.M. Sun
Department of Plant Molecular Physiology, University of Hawaii, Honolulu, HI, USA.

Plant proteins are an important source of dietary protein for humans as well as livestock. The major sources of dietary plant protein are from cereals, legumes and root/tuber crops. However, most are deficient in certain essential amino acids with respect to human nutrition. Cereals are usually deficient in lysine and tryptophan, and legumes and root/tuber crops in the sulphur amino acids, methionine and cysteine (FAO, 1970; Yamaguchi, 1980). The potato, a tuber crop, is the fourth major food crop in the world. In 1991, 251 million metric tons of potato were produced (FAO, 1991). However, although it is a good source of carbohydrate and contributes a significant amount of protein, the potato is deficient in the essential sulfur amino acids methionine and cysteine.

Methods to improve the nutritional quality of crops include standard plant breeding methods, mutagenesis and, more recently, new molecular genetic approaches. In all cases, the existing genotype of the plant is altered either by the introduction of novel genes from related or phylogenetically distant species, by altering the expression of existing genes to create new patterns of gene expression, or by the expression of variants of the existing proteins (Sun and Larkins, 1993). Each method has demonstrated merits and faults which can only be dealt with case by case. In this chapter, we explore the molecular genetic approach as a means of increasing the sulfur amino acid content in potatoes.

For our purpose, we utilized the Brazil nut 2S protein (BN2S) gene from *Bertholletia excelsa* H.B.K. which expresses a protein with a methionine content of 19% and cysteine content of 8%, and the two potato cultivars, Russet Burbank and Atlantic. BN2S is one of a family of sulfur-rich seed storage albumins that are composed of two subunits, a large (9 kDa) and a small (3 kDa) polypeptide. These proteins are synthesized as a precursor molecule of 18 kDa which is sequentially post-translationally processed from 18 kDa to 15 kDa to 12 kDa and finally to the 9 kDa and 3 kDa polypeptide (Figure 15.1A) (Altenbach *et al.*, 1986; de Castro *et al.*, 1987). At least with respect to the Brazil nut, these modifications are accompanied with the localization of the final

The Molecular and Cellular Biology of the Potato, Second Edition
Edited by W.R. Belknap, M.E. Vayda and W.D. Park

protein into the plant's protein bodies or protein storage vacuoles (Sun *et al.*, 1988). From protein sequence analysis, the BN2S gene family encodes a group of proteins that have a methionine content of 19-20% and a cysteine content of 8% (Ampe *et al.*, 1986; Altenbach *et al.*, 1987; Gayler *et al.*, 1990; Lilley *et al.*, 1990). Based on protein sequence analysis of the purified 2S proteins and comparison of these with the protein sequences translated from cDNA sequences, the processing also results in the removal of a linker region of five amino acids present between the small subunit and the large subunit and four amino acids at the carboxy-terminus (Ampe *et al.*, 1986; Altenbach *et al.*, 1987; de Castro *et al.*, 1987). For our studies, we utilized a BN2S cDNA clone encoding a protein with 19% methionine (Altenbach *et al.*, 1987).

Transgenic expression of chimeric BN2S genes has previously been demonstrated in several heterologous systems including rapeseed (*Brassica napus*) (De Clercq *et al.*, 1990; Guerche *et al.*, 1990; Altenbach *et al.*, 1992), *Arabidopsis* (De Clercq *et al.*, 1990), and tobacco (*Nicotiana tabacum*) (Altenbach *et al.*, 1989; De Clercq *et al.*, 1990). In *Arabidopsis*, the chimera consisted of the AT2S1 (*Arabidopsis* 2S-1 genes) 5' flanking and signal peptide encoding regions with the BN2S cDNA deleted for its signal peptide or the AT2S1 promoter and a portion of the AT2S1 small subunit encoding region with the small subunit of the BN2S cDNA. In rapeseed, in addition to the *Arabidopsis* chimera, either the soybean lectin gene's 5' flanking region and the signal peptide were used with the BN2S propeptide encoding region or the French bean phaseolin regulatory regions were used with the BN2S cDNA. In tobacco, the chimeras tested in *Arabidopsis* were utilized in addition to the French bean phaseolin construct. In each case, the BN2S chimera was expressed and correctly processed. These findings suggest that the processing mechanism for the BN2S protein is not unique but rather is present in a variety of other dicots.

Together the results obtained for the expression of the BN2S cDNA and the processing and stability of the 2S protein support the use of this small storage protein as a candidate to potentially enhance the nutritional quality of plants. For our studies, we engineered a chimera composed of the cauliflower mosaic virus 35S promoter (35S-p), BN2S cDNA, and the nopaline synthase polyadenylation and termination (NOS terminator) sequences. This chimera was then placed in a binary vector and introduced into two potato cultivars via *Agrobacterium tumefaciens* transformation.

Construction of the Chimeric BN2S Gene

The cDNA for the BN2S protein was chosen as our target gene for transformation due to its high methionine and cysteine content. The BN2S cDNA described by Altenbach *et al.* (1987) was subcloned first into pTZ18U (Mead *et al.*, 1986) to form pHT804 (Figure 15.1B); this resulted in the convenient flanking of the BN2S cDNA with *Xba*I sites. From pHT804, the

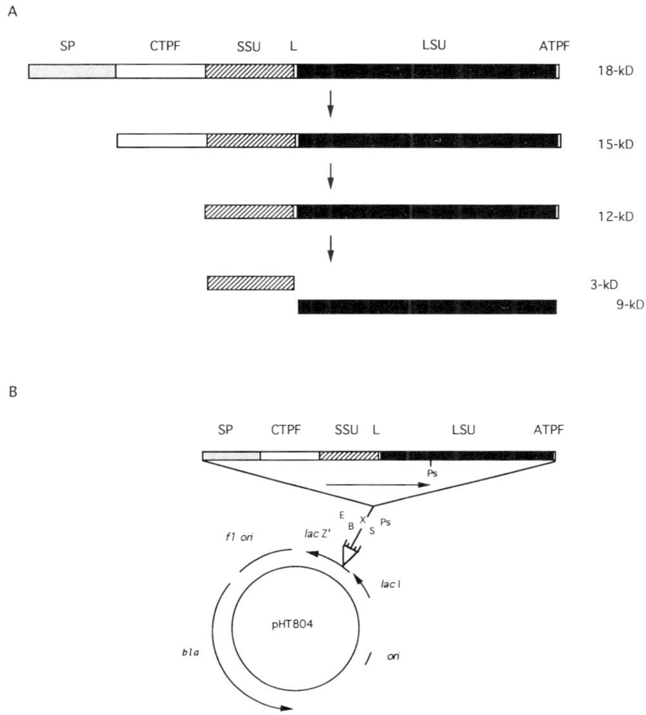

Figure 15.1. The processing of the BN2S protein and subcloning of the BN2S cDNA. (A) As shown schematically, the BN2S protein is processed sequentially from a precursor molecule of 18 kDa to 15 kDa to 12 kDa to its two mature subunits of 9 kDa and 3 kDa. During this processing, the linker and four amino acids of the C-terminus are removed. ATPF, amino-terminal processed fragment; CTPF, carboxy-terminal processed fragment; L, linker; LSU, large (9 kDa) subunit; SP, signal peptide; SSU, small (3 kDa) subunit. (B) The BN2S cDNA was subcloned from pHS-3 (Altenbach *et al.*, 1987) as a *Hin*fI fragment. The *Hin*fI sites flank the ATG start codon and the TAG stop codon. The fragment was made blunt by Klenow treatment and then ligated with the *Xba*I and Klenow-treated pTZ18U (Mead *et al.*, 1986) to form the plasmid pHT804. These sets of fill-ins and ligations resulted in the recovery of both the *Hin*fI and *Xba*I sites, such that the *Xba*I sites flank the two interior *Hin*fI sites. The BN2S cDNA can be readily recovered as a 453 bp *Xba*I or 447 bp *Hin*fI fragment. B, *Bam*HI; E, *Eco*RI; Ps, *Pst*I; S, *Sal*I; X, *Xba*I.

BN2S cDNA was placed into the vector pHT825, a derivative of pBI121 lacking the ß-glucuronidase (GUS) sequence (Jefferson *et al.*, 1987), to form pHT843 (Figure 15.2). The plasmid pHT843 was transferred via *Agrobacterium tumefaciens* into the *Solanum tuberosum* L. cultivars Russet Burbank and Atlantic. The vector pHT843 contains unique *Xba*I and *Bam*HI sites present between the 35S promoter and NOS terminator. The vector confers kanamycin

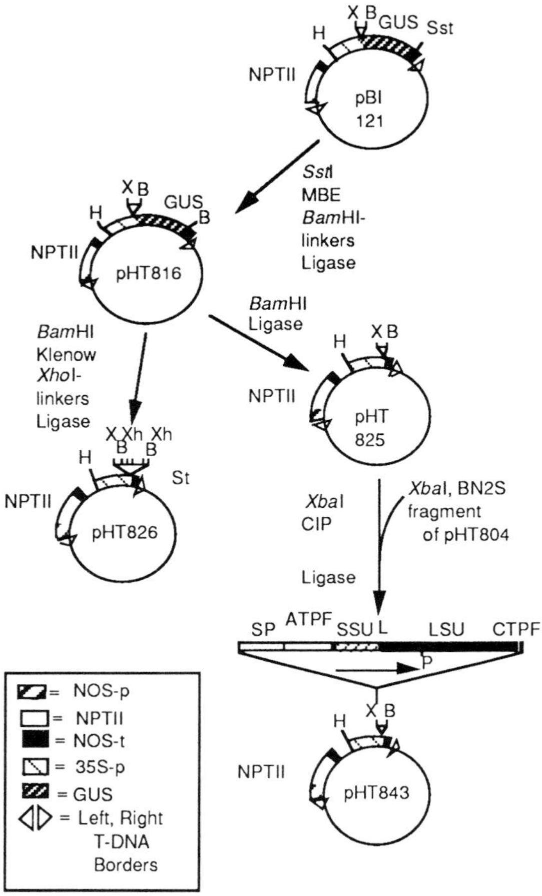

Figure 15.2. The vector pHT825 and insertion of the BN2S cDNA. pBI121 was cleaved with *Sst*I and the cleaved ends were treated with mung bean exonuclease (MBE) and then with T4 DNA ligase in the presence of *Bam*HI linkers (pCGGATCCG) to create pHT816. pHT825 was constructed by treating the plasmid pHT816 with *Bam*HI to drop out the ß-D-glucuronidase (GUS) coding sequence and then religating the ends under dilute conditions. Cloning into the *Bam*HI or *Xba*I sites of pHT825 will result in the flanking of the insert with the full cauliflower mosaic virus 35S promoter (35S-p) and the nopaline synthase terminator and polyadenylation sequence (NOS-t). As an alternative to pHT825, pHT826 was constructed whereby *Xho*I linkers (pCCTCGAGG) were inserted between the Klenow-treated *Bam*HI cleaved ends resulting in the addition of the cloning sites of *Xho*I and *Stu*I (i.e. the result of multiple *Xho*I linkers) flanked by *Bam*HI sites. The above manipulations were done according to the standard methods stated in Maniatis *et al.* (1982), Miller (1972) and Cohen *et al.* (1972). B, *Bam*HI; CIP, calf intestinal alkaline phosphatase; H, *Hin*dIII; Ligase, T4 DNA ligase; MBE, mung bean exonuclease; NPTII, neomycin phosphotransferase II; Sst, *Sst*I; St, *Stu*I; X, *Xba*I; Xh, *Xho*I.

resistance to its host bacterium and plant but does not allow for the histochemical/colorimetric identification associated with the GUS gene. As such, transformants were determined based on their growth pattern on high concentrations of kanamycin (100-200 mg/l). Non-transformed plants were found to produce aerial roots rather than true roots on media containing 50 mg/l kanamycin, whereas transformed plants produce true roots on media containing up to at least 350 mg/l kanamycin. This phenotypic screening of plantlets was initially verified by ELISA and Western determination of the test plants and later used as the sole method of screening. However, because this type of screening is time-consuming (as the plants must be allowed to grow to at least 5 cm)), a GUS version of the vector pHT825 (Figure 15.3) is suggested for future use as it allows for very young transformants to be screened histochemically with the substrate X-glucuronide (Jefferson, 1987; Gould and Smith, 1989; Martin *et al.*, 1992).

As a control for the transformation, pBI121, a GUS-containing vector was transferred into both cultivars in parallel with the BN2S-containing construct. For these plants, transformants could be identified by their production of the GUS gene product. In the presence of X-glucuronide, a blue precipitate is formed in the tissue. Leaves, stems, roots and tubers were all tested for the production of the GUS gene product. The intensity of the precipitate varied from pale to dark in both cultivars and showed a variety of staining patterns from continuous to spotty. GUS activity was also used to corroborate the validity of the phenotypic screening used above.

BN2S Protein Analysis

Expression of the BN2S construct in leaves and petioles was determined by Western analysis with a monoclonal antibody specific for the 9 kDa subunit of the BN2S protein (Sun *et al.*, 1987). Independent isolates arising from the *Agrobacterium*-treated tuber disc(s) were tested for their ability to correctly express and process the BN2S protein. Eighteen transformants from cv. Atlantic, representing isolates taken from eight discs, and forty transformants from cv. Russet Burbank, representing nine discs, were assayed. For each of these transformants, the 9 kDa product of the BN2S gene was detected indicating that the BN2S protein had been correctly processed in the leaves (and petioles) of the two cultivars. The same product was not observed in the control plants, pBI121-transformed cv. Atlantic or non-transformed cv. Russet Burbank. With respect to the expression levels, for both cultivars, they varied between 0.2% and 0.01%. Lower levels may have been present but were not detected under the conditions used. For the purpose of illustration, a high expressor of each cultivar was chosen and shown in Figure 15.4A and C.

Expression of the BN2S construct was determined for *in vitro* grown microtubers from a subset of the transformants from each of the two cultivars. Microtubers were initiated and maintained according to Ishida *et al.* (1989) with

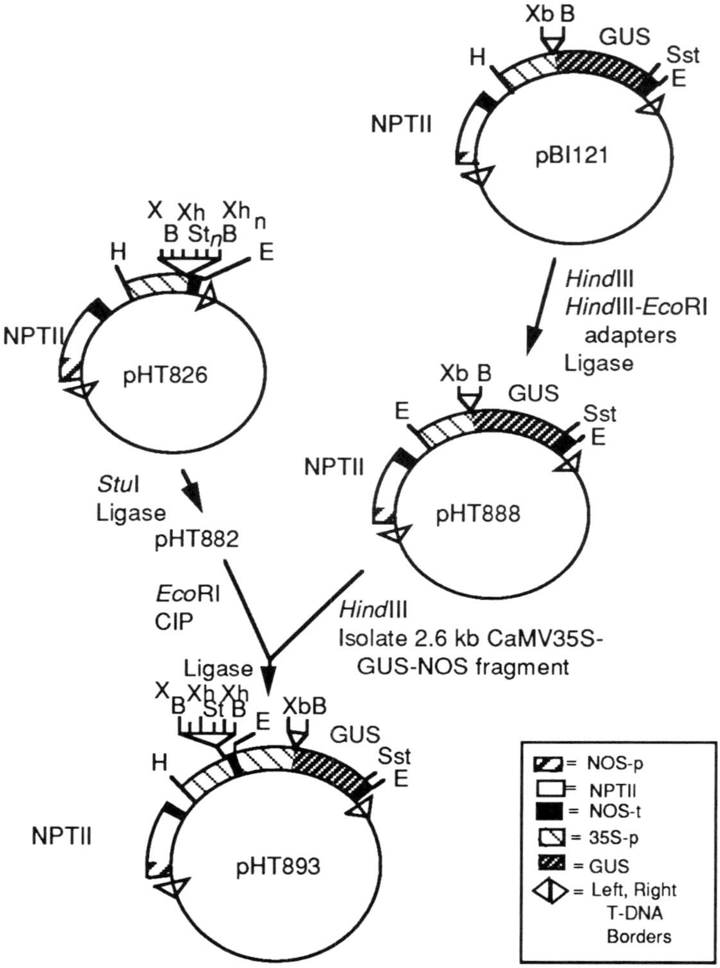

Figure 15.3. The GUS vector pHT893. pBI121 was cleaved with *Hind*III and the cleaved ends were ligated in the presence of the *Hind*III-*Eco*RI adaptors (pAGCTCGAATTCG) to create pHT888. The plasmid pHT882 was obtained from pHT826 (Fig. 15.2) by sequentially cleaving with excess *Stu*I and then ligating under dilute conditions to ensure the presence of a single *Stu*I site. The 2.6 kb *Eco*RI fragment containing the marker gene 35S-p-GUS-NOS-t was isolated from pHT888 and ligated to the *Eco*RI-cleaved and calf alkaline phosphatase-treated plasmid pHT882 to form pHT893. B, *Bam*HI; CIP, calf alkaline phosphatase; E, *Eco*RI; H, *Hind*III; Ligase, T4 DNA ligase; *n*, number ≥1; Sst, *Sst*I; St, *Stu*I; X, *Xho*I; Xb, *Xba*I.

the addition of 1.86% filter-sterilized A-rest (Dow-Elanco). After a period of 5-6 weeks, the tubers were harvested for analysis. Six transformants from the cv. Atlantic set representing five discs and five transformants from the cv. Russet Burbank set, representing four discs were tested. For each of these transformants, expression of the 9 kDa subunit of the final protein was detected This is illustrated in Figure 15.4B and D. While absolute amounts of protein were not determined, in all cases the expression level of the BN2S protein in tubers was found to be up to 8-fold lower than that present in the leaves and petioles.

Southern Analysis

Six cv. Atlantic and 16 cv. Russet Burbank transformants were selected for Southern analysis to determine how many copies of the BN2S chimera were present and whether any rearrangements had occurred. This represented isolates from five discs for each cultivar. Genomic DNA from each plant was isolated and digested to release the full BN2S chimera inclusive of its promoter and terminator. The copy number varied from 1 to 40 for the cv. Atlantic set (Figure 15.5A), as compared with from 1 to only 5 for the cv. Russet Burbank set (Figure 15.5B). A correlation between copy number and expression level of the BN2S protein was not observed. Also, no rearrangements were noted for any of these transformants.

Conclusion

In this study, the BN2S gene was utilized as a target gene for the introduction into potato of a protein rich in methionine. The BN2S protein was also analyzed for its processing and stability in the leaves (and petioles) and tubers of potato. Based on immunoblot analysis, the BN2S protein was correctly processed in both the leaves (and petioles) and the tubers of both cultivars with expression levels in the tubers up to 8-fold lower than those in the leaves and petioles. The observed protein expression levels both in the leaves (and petioles) and in the tubers were consistent with that expected for a gene driven by the 35S promoter. Position effects (Jones *et al.*, 1985) may explain the lack of correlation between the expression levels and copy number of the integrated BN2S chimera and the differences in the expression levels. Northern analysis (data not shown) of total RNA isolated from both the leaves (and petioles) and tubers was also consistent with a lower level of steady-state transcripts in the tubers than in the leaves and petioles; tubers contained about 3-6-fold less transcript than the leaves and petioles.

At the levels of expression that we observed, no adverse affect was seen with respect to the morphology and development of the plant either in tissue culture or under growth chamber conditions. The expression levels that we observed, though consistent with those expected for a gene driven by the 35S promoter, are insufficient to alter the methionine content in the tuber. Therefore, it will be

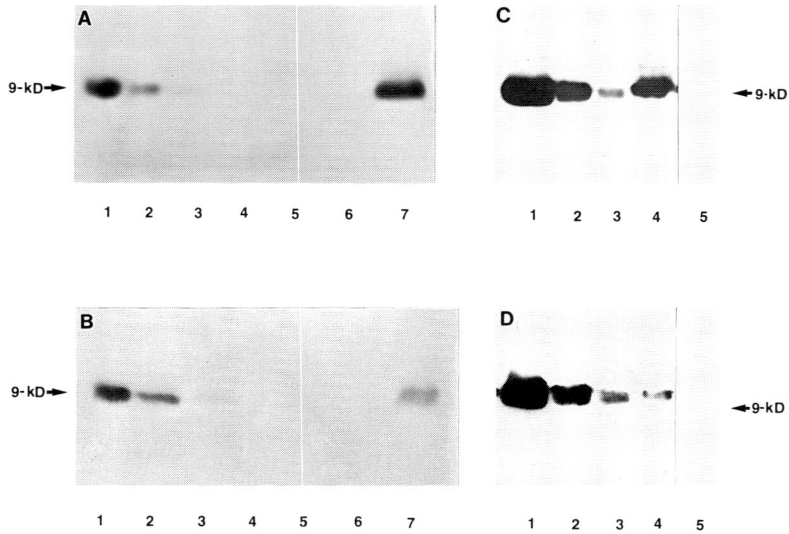

Figure 15.4. Western immunoblot analysis of the cv. Atlantic and cv. Russet Burbank transformants. *Agrobacterium tumefaciens* LBA4404 harboring the plasmid pHT843, a derivative of the binary vector pHT825 (Figure 15.2) containing the BN2S construct, or pBI121 was used to transform potato cells following the procedure of Ishida *et al.* (1989) with several modifications (W. Belknap, personal communication). Transformants were screened either initially by ELISA (Amersham) for the NPTII protein or as shown here by Western analysis for the BN2S 9 kDa subunit. Western analysis of the BN2S protein was done utilizing the ECL kit (Amersham) with a monoclonal mouse antibody specific for the 9 kDa subunit of the BN2S protein following electrophoretic transfer of 100-200 μg of protein from a 20% polyacrylamide-SDS gel to a nitrocellulose paper (pore size of 0.1 μm, Shleicher and Schuell). Protein extracts for leaves (and petioles) were prepared in 0.25 M Tris (pH 7.8)/1 mM phenylmethylsulfonyl fluoride (PMSF), and for *in vitro* grown tubers, in 25 mM $NaPO_4$ buffer (pH 7.0)/ 2 mM $NaHSO_3$/1.25 mM PMSF and 0.01 g polyvinylpyrrolidone per g of tissue. A and B: 100 μg of total proteins from either leaves (and petioles) (A) or *in vitro*-grown tubers (B) of pBI121- or pHT843-transformed Atlantic was analyzed by Western blotting. Lanes 1-3, 100, 50 and 20 ng of purified BN2S protein as standards, respectively; 4, low molecular weight marker (Sigma); 5, pBI121 transformant #79; 6, pBI-121-transformant #94; and 7, pHT843 transformant #32. C and D: 200 μg of total protein from either leaves (and petioles) (C) or *in vitro* grown tubers (D) of pHT843-transformed or non-transformed Russet Burbank was analyzed as in A. Lanes 1-3, 500, 100 and 50 ng of purified BN2S protein as standards, respectively; 4, pHT843 transformant #171; and 5, non-transformed cv. Russet Burbank.

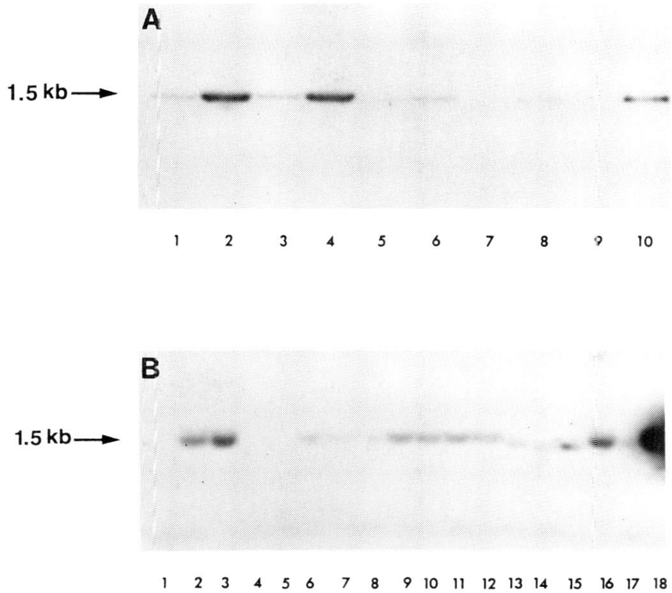

Figure 15.5. Southern blot analysis of the cv. Atlantic and cv. Russet Burbank transformants. Genomic DNA was prepared according to Doyle and Doyle (1989) using the leaf and stem tissues. Five or ten micrograms of genomic DNA (quantified by fluorimetry) was cleaved with the restriction endonucleases *Eco*RI and *Hin*dIII and separated in 1.0% agarose. The separated DNA was transferred onto Hybond-N$^+$ nylon membrane (Amersham) following alkaline treatment of the gel according to Church and Gilbert (1984). Hybridization was done according to the manufacturer's recommendation. Copy number was determined based on the genome size of *Solanum tuberosum* L. of 2 x 10^9 basepair/haploid genome (Bennett and Smith, 1976). The DNA probe was prepared by random priming of the gel-purified *Xba*I fragment containing the BN2S cDNA sequence (Fig. 15.1). (A) 10 μg of genomic DNA from Atlantic transformed with pHT843 was digested with *Eco*RI and *Hin*dIII, separated on a 1.0% agarose gel and transferred to a nylon membrane. The membrane was probed with ^{32}P-labeled BN2S cDNA. The pBI121-transformed isolate #79 is not shown. Lanes 1-6, pHT843 transformants #32, 83, 89, 92, 115, and 121, respectively; lanes 7 and 9, blanks; lanes 8 and 10 represent two and 20 copies of the chimeric gene. (B) 5 μg of genomic DNA from non-transformed and pHT843-transformed Russet Burbank was digested and treated as in A. Lane 1, molecular weight marker; 2, non-transformed Russet Burbank; 3-17, pHT843 transformants #60, 86, 87, 93, 136, 137, 162, 163, 164, 165, 167, 169, 170, 171, 177; and 18, the equivalent of ten copies of the chimeric gene.

necessary to increase the production of the BN2S protein by at least 10-fold. This may be accomplished by utilizing a stronger general promoter or tuber-specific promoter. In either case, the promoter may be a naturally occurring one or a chimeric one. The patatin promoter represents a class of tuber-specific promoters that expresses 5-10-fold higher than the 35S promoter in the tuber (Wenzler *et al.*, 1989; Jefferson *et al.*, 1990; Liu *et al.*, 1990). Alternatively, the 35S promoter or another promoter may be altered to confer tuber-specificity with increased promotion. In addition to increasing the promoter's strength, sequences that increase the half-life of the RNA or protein or both may be useful. However, since the effect of increasing the production and concentration of the BN2S is unknown with respect to the potato, for each of these changes, the plant's growth and development will need to be determined.

In conjunction with strategies that increase the concentration of the protein in the tuber, changes to the coding region of the BN2S protein may also be attempted. To this end, we have made mutations within the BN2S coding region that increase the methionine content of the final protein from 19% to 27%. Preliminary results indicate that changes that increase the methionine content to 25% are tolerated and are correctly processed in the potato.

Although further increases in the production of the BN2S protein will necessary to alter significantly the methionine content in potato, the demonstration of a predicted level of stable protein and its correct processing supports our approach of utilizing the BN2S cDNA and mutants to enhance the sulfur amino acid content in potato. In the future, it will be interesting to investigate by immuno-localization the subcellular location of these seed proteins in the tubers, leaves and petioles of the transgenic potato.

References

Altenbach, S.B., Pearson, K.W., Leung, F.W. and Sun, S.S.M. (1986) The stepwise processing of a sulfur-rich seed protein from Brazil nut (*Bertholletia excelsa*). In: Shannon, L.M. and Chrispeels, M.J. (eds), *Molecular Biology of Seed Storage Proteins and Lectins.* Waverly Press, Baltimore, Maryland. pp.137-146.

Altenbach, S.B., Pearson, K.W., Leung, F.W. and Sun, S.S.M. (1987) Cloning and sequence analysis of a cDNA encoding a Brazil nut protein exceptionally rich in methionine. *Plant Molecular Biology* 8:239-250.

Altenbach, S.B., Pearson, K.W., Meeker, G., Staraci, L.C. and Sun, S.S.M. (1989) Enhancement of the methionine content of seed proteins by the expression of a chimeric gene encoding a methionine-rich protein in transgenic plants. *Plant Molecular Biology* 13:513-522.

Altenbach, S.B., Kuo, C.-C., Staraci, L.C., Pearson, K.W., Wainwright, C., Georgescu, A. and Townsend, J. (1992) Accumulation of a Brazil nut albumin in seeds of transgenic canola results in enhanced levels of seed protein methionine. *Plant Molecular Biology* 18:235-245.

Ampe, C., Van Damme, J., de Castro, L.A.B., Sampaio, M.J.A.M., Van Montagu, M. and Vandekerckhove, J. (1986) The amino-acid sequence of the 2S sulphur-rich proteins from seeds of Brazil nut (*Bertholletia excelsa* H.B.K.). *Europern Journal of Biochemistry* 159:597-604.

Bennett, M.D. and Smith, J.B. (1976) Nuclear DNA amounts in angiosperms. *Philosophical Transactions of the Royal Society*, Series B 274:227-274.

Church, G.M. and Gilbert, W. (1984) Genomic sequencing. *Proceedings of the National Academy of Sciences USA* 81:1991-1995.

Cohen, S.N., Chang, A. and Hsu, L. (1972) Nonchromosomal antibiotic resistance in bacteria: genetic transformation of *E. coli*. *Proceedings of the National Academy of Sciences USA* 69:2110-2114.

de Castro, L.A.B., Lacerda, Z., Aramayo, R.A., Sampaio M.J.A.M. and Gander, E.S. (1987) Evidence for a precursor molecule of Brazil nut 2S seed proteins from biosynthesis and cDNA analysis. *Molecular and General Genetics* 206:338-343.

De Clercq, A., Vandewiele, M., Van Damme, J., Van Montagu, M., Vandekerckhove, J. and Krebbers, E. (1990) Stable accumulation of modified 2S albumin seed storage proteins with higher methionine contents in transgenic plants. *Plant Physiology* 94:970-979.

Doyle, J.J. and Doyle, J.L. (1989) Isolation of plant DNA from fresh tissue. *BRL Focus 12*:13-15.

FAO (1970) *Nutritional Studies: Amino Acid Content of Foods*. Food and Agriculture Organization of the United Nations, Rome, Vol. 24, 285 pp.

FAO (1991) *Yearbook: Production*. Food and Agriculture Organization of the United Nations, Rome, Vol. 45, p. 90.

Gayler, K.R., Kolivas, S., Macfarlane, A.J., Lilley, G.G., Baldi, M., Blagrove, R.J. and Hohnson, E.D. (1990) Biosynthesis, cDNA and amino acid sequences of a precursor of conglutin delta, a sulphur-rich protein from *Lupinus angustifolius*. *Plant Molecular Biology* 15:879-893.

Gould, J.H. and Smith, R.H. (1989) A non-destructive assay for GUS in the media of plant tissue cultures. *Plant Molecular Biology Repoerter* 7:209-216.

Guerche, P., De Almeida, E.R.P., Schwarztein, M.A., Gander, E., Krebbbers, E. and Pelletier, G. (1990) Expression of the 2S albumin from *Bertholletia excelsa* in *Brassica napus*. *Molecular and General Genetics* 221:305-314.

Ishida, B.K., Snyder, G.W. and Belknap, W.R. (1989) The use of *in vitro*-grown microtuber discs in *Agrobacterium*-mediated transformation of Russet Burbank and Lemhi potatoes. *Plant Cell Reports* 8:325-328.

Jefferson, R.A. (1987) Assaying chimeric genes in plants: the GUS gene fusion system. *Plant Molecular Biology Reporter* 5:387-405.

Jefferson, R.A., Kavanagh, T.A. and Bevan, M.W. (1987) GUS fusions: ß-glucuronidase as a sensitive and versatile gene fusion marker in higher plants. *EMBO Journal* 6:3901-3907.

Jefferson, R., Goldsbrough, A. and Bevan, M.W. (1990) Transcriptional regulation of a patatin-1 gene in potato. *Plant Molecular Biology* 14:995-1006.

Jones, J.D.G., Dunsmuir, P. and Bedbrook, J. (1985) High level expression of introduced chimaeric genes in regenerated transformed plants. *EMBO Journal* 4:2411-2418.

Lilley, G.G., Caldwell, J.B., Kortt, A.A., Higgins, T.J. and Spencer, D. (1990) Isolation and primary structure for a novel methionine-rich protein from sunflower seeds (*Helianthus annuus.* L). In: T.H. Applewhite (ed.), *Proceedings of the World Congress on Vegetable Protein Utilization in Human Foods and Animal Feedstuffs.* American Oil Chemists' Society, Champaign, IL, pp. 497-502.

Liu, X.J., Prat, S., Willmitzer, L. and Frommer, W.B. (1990) *Cis*-regulatory elements directing tuber-specific and sucrose-inducible expression of a chimeric class I patatin promoter/GUS-gene fusion. *Molecular and General Genetics* 223:401-406.

Maniatis, T., Fritsch, E.F. and Sambrook, J. (1982) *Molecular Cloning: A Laboratory Manual.* Cold Spring Harbor Laboratory Press, Cold Spring Harbor, NY.

Martin, T., Schmidt, R., Altmann, T. and Frommer, W.B. (1992) Non-destructive assay systems for detection of ß-glucuronidase activity in higher plants. *Plant Molecular Biology Reporter* 10:37-46.

Mead, D.A., Szczesna-Skorupa, E. and Kemper, B. (1986) Single-stranded DNA 'blue' T7 promoter plasmids: a versatile tandem promoter system for cloning and protein engineering. *Protein Engineering* 1:67-74.

Miller, J. (1972) *Experiments in Molecular Genetics.* Cold Spring Harbor Laboratory Press, Cold Spring Harbor, NY.

Sun, S.S.M. and Larkins, B.A. (1993) Transgenic plants for improving seed storage proteins. In: Kung, S.D. and Wu, R. (eds), *Transgenic Plants.* Academic Press, San Diego, CA, pp. 339-372.

Sun, S.S.M., Altenbach, S.B. and Leung, F.W. (1987) Properties, synthesis and processing of a sulfur-rich protein in Brazil nut (*Bertholletia excelsa* H.B.K.). *European Journal of Biochemistry* 162:477-483.

Sun, S.S.M., Altenbach, S.B., Pearson, K.W., Leung, F.W., Meeker, G. and Staraci, S. (1988) Quality improvement of crop proteins through genetic engineering. In: Chang S.T. *et al.* (eds), *Recent Advances in Biotechnology and Applied Biology.* The Chinese University Press, Hong Kong, pp. 621-628.

Wenzler, H.C., Mignery, G.A., Fisher, L.M. and Park, W.D. (1989) Analysis of a chimeric class-I patatin-GUS gene in transgenic potato plants: high level expression in tubers and sucrose-inducible expression in cultured leaf and stem explants. *Plant Molecular Biology* 12:41-50.

Yamaguchi, M. (1980) *World Vegetable.* AVI Publishing Co., Westport, CT. p. 395.

Chapter Sixteen:

Strategies Towards Introducing Resistance to Bacterial Pathogens in Transgenic Potatoes

Klaus Düring
University of Hamburg, Center for Applied Plant Molecular Biology, Hamburg, Germany. Present Address: Federal Center for Breeding of Cultivated Plants, Institute for Breeding Methods in Vegetables, Quedlinburg, Germany.

Bacterial Diseases of the Potato

Phytopathogenic bacteria mostly belong to the genera *Erwinia, Pseudomonas, Xanthomonas* and *Clavibacter*. Several of these bacteria are able to infect potato and cause severe disease symptoms. Given the worldwide agricultural importance of the potato crop and its high susceptibility to bacterial infection, bacterial infections of potato represent a significant economic factor. These are especially significant in regions of warm climate.

Erwinia carotovora spp. and *Erwinia chrysanthemi* are the causal agents of soft rot and black leg, *Pseudomonas solanacearum* of bacterial wilt (in subtropical climates) and *Clavibacter michiganensis sepedonicus* of bacterial ring rot. Soft rot and black leg are economically very important diseases in the warm climates of many third world countries, as conditions for growth and storage of potatoes are not as sophisticated as in industrialized countries. Environmental conditions are very favorable to bacterial growth in subtropical and tropical areas because of high humidity and high temperatures.

While the plant-pathogen interaction between *Solanum tuberosum* and *Erwinia carotovora* spp. has been well studied, we are still far from understanding the molecular mechanisms of pathogenicity and/or resistance.

Molecular Pathology of *Erwinia carotovora*

Erwinia carotovora (*E. c.*)is detectable in all potato tubers. This organism can survive in soil for extended time periods only in association with plant tissue (Pérombelon, 1973; Burr and Schroth, 1977). The potato seed tuber, therefore, represents the primary source of infection. *Erwinia* bacteria survive in the lenticels in numbers ranging from a few up to a hundred cells per site. The

The Molecular and Cellular Biology of the Potato, Second Edition
Edited by W.R. Belknap, M.E. Vayda and W.D. Park

determining factors for rapid multiplication of the bacterial cells include environmental conditions (wounding, high temperature, high humidity, low oxygen, etc.), but not the density of the original inoculum. However, only when the bacteria have been able to multiply to a density of ~10^6 cells per gram of tissue do disease symptoms becomes detectable (Pérombelon, 1982).

Maceration of the infected tissue is due to the production of secreted pectolytic enzymes by the pathogenic bacteria (Collmer and Keen, 1986). This is true not only for the genus *Erwinia*, but also for most other plant-bacteria interactions. The complex cascade of enzyme gene activation and biochemical enzyme action, as well as mechanisms of signaling cascades, are currently under investigation. As yet, no specific signal molecule for the interaction between potato and *Erwinia carotovora* has been described.

Plant Protection Approaches

Potato cultivars used for agriculture today generally do not have any resistance to *Erwinia*. They are all more or less susceptible to soft rot and black leg. Classical breeding for bacterial resistance has not been possible because of the lack of resistance traits in breeding material. In addition, chemical means of plant protection are not available. The application of antibiotics or of copper salts as chemical agents is known to result in reduced disease development, but is environmentally unfavorable. Only in a few wild *Solanum* species has resistance towards *Erwinia* been described. However, these species are sexually incompatible with today's potato cultivars and are difficult to exploit using classical breeding methods.

Plant Lysozymes

Plant antibacterial resistance mechanisms have yet to be well characterized at a molecular level. The lysozymes represent putative candidates for interaction with invading bacteria. These enzymes have been detected in several plant species, and are most probably ubiquitous. Generally, plant lysozymes are bifunctional enzymes with a predominant chitinase activity (Jollès and Jollès, 1984) and are classified as pathogenesis related (PR) proteins. The plant chitinases can be subdivided into two classes: one contains a set of acidic enzymes which are located in the extracellular space, while the other contains the basic counterparts which are located in the vacuole (Cutt and Klessig, 1992). Only the basic chitinases have been reported to possess lysozyme activity (Trudel *et al.*, 1989; Majeau *et al.*, 1990). This implies that plant lysozymes are generally located in the vacuole. Only a few exceptions have been described to date (Audy *et al.*, 1988; Stintzi *et al.*, 1993). Most of these bifunctional enzymes possess a rather weak lysozyme activity. The extracellular lysozymes isolated from tobacco represent the first reported to be localized in the intercellular spaces and to have

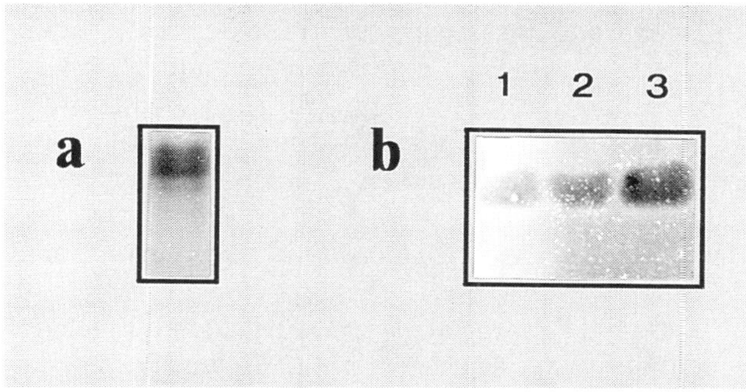

Figure 16.1. Detection of endogenous bacteriolytic activities (lysozymes) in potato tissues using a gel overlay technique. The substrate bacteria were polymerized within a polyacrylamide gel (overlay gel). The protein was run on a non-denaturing polyacrylamide gel and blotted through the overlay gel at 37°C and pH 7.2. (a) 15 μg of total protein extract from potato leaves has been loaded and *Micrococcus lysodeikticus* used as a substrate. Strong signals are detectable at low relative mobility. (b) The same with extracts from potato tuber tissue in healthy (lane 1) and wounded state (lane 2: 24 h and lane 3: 48 h after wounding) and *Pseudomonas solanacearum* as substrate. The signals were much weaker than in (a) and were enhanced by photography. The observed signals are of the same relative mobility as in (a).

high enzymatic activity (Stintzi *et al.*, 1993).

The ability of endogenous plant lysozymes to participate actively in defense reactions has yet to be demonstrated. However, because of their enzymatic specificity an antibacterial role appears reasonable. On the other hand, as most of these enzymes are located in the vacuolar compartment, the interaction with the invading bacteria can occur only after the invading bacteria have multiplied to high densities. At this point, high levels of lysozyme activity would probably be required to effect the resistance reaction (Düring, 1993).

Enzymatic analysis of potato bacteriolytic enzymes using the gel overlay technique demonstrated a varying activity towards different substrates. Total protein extract was separated on a discontinuous polyacrylamide gel for basic proteins (Reisfeld *et al.*, 1962) and subsequently blotted through a 7.5% polyacrylamide overlay gel (0.5 mm thick) containing the respective bacteria. The bacterial substrates employed in these experiments included: *Micrococcus lysodeikticus*, the standard substrate for lysozyme assays (Sigma M-3770): 2 mg lyophilized cells per ml; and bacterial suspensions (70-80 μl of A_{550} ~2/ml in H_2O) of *E. c. atroseptica* (National Collection of Plant Pathogenic Bacteria,

Harpenden, UK, strain 549) and *Pseudomonas solanacearum* (strain 325). When using *M. lysodeikticus* potato leaf tissue displayed multiple strong bacteriolytic activities at low relative mobility (Fig. 16.1a). At least three bands were repeatedly detected. In contrast, in both uninfected mechanically wounded tubers and wounded tubers inoculated with *E. c. atroseptica*, only weak lysozyme activities were detected.

In addition, the detectable bacteriolytic activities were significantly dependent on the substrate. Only weak signals were found with *E. c. atroseptica* and no signal at all with *Pseudomonas solanacearum* in leaf samples (data not shown). Low activities have also been detected in tuber samples with all these bacteria as substrates. Nevertheless, a slight induction was observed following wounding of tuber slices in all cases (Fig. 16.1b). In summary, in leaves, bacteriolytic enzymes are present which are significantly active *in vitro* against at least one phytopathogenic bacterial species (*E. c. atroseptica*). But, in tubers where the invasion of important bacterial phytopathogens predominantly occurs, only a very weak antibacterial capacity is present. In addition, this activity is not induced to significantly higher levels following infection by *E. c. atroseptica.* Indeed, preliminary data (K. Düring, unpublished) suggest that the bacterial slime growing on potato discs may repress the slight enhancement observed after wounding.

Thus, no strong antibacterial enzymes seem to be present at the site of infection in the case of bacterial diseases of potato. It is well known that artificial infections of potatoes with *E. c.* sspp. via leaves are very difficult to obtain. This might correlate with the pattern of lysozyme distribution. The low levels of endogenous bacteriolytic activity in potato tubers might be a reason for the susceptibility of the tubers towards soil-borne phytopathogenic bacteria. The expression patterns of endogenous lysozymes, as well as their subcellular localization and their substrate range may be important factors in determination of the plant defense capacity.

Biotechnological Approaches Using Lysozyme to Achieve Antibacterial Resistance

Biotechnology enables the specific modification of components of the plant. The introduction of a foreign lysozyme into the intercellular spaces of a transgenic potato plant, especially in tubers, might allow an early interaction of the enzyme with the invading bacteria. The subcellular localization of the foreign protein is thought to be a major determinant of any antibacterial resistance approach. As described above, the initial inoculum was present only at a low density. At this time point low levels of an efficient antibacterial agent located in the intercellular spaces of transgenic plants might lead to a successful inhibition of the multiplication of phytopathogenic bacteria.

The localization of a suitable foreign protein in the intercellular spaces of transgenic plants can generally be achieved by fusing the DNA sequence coding

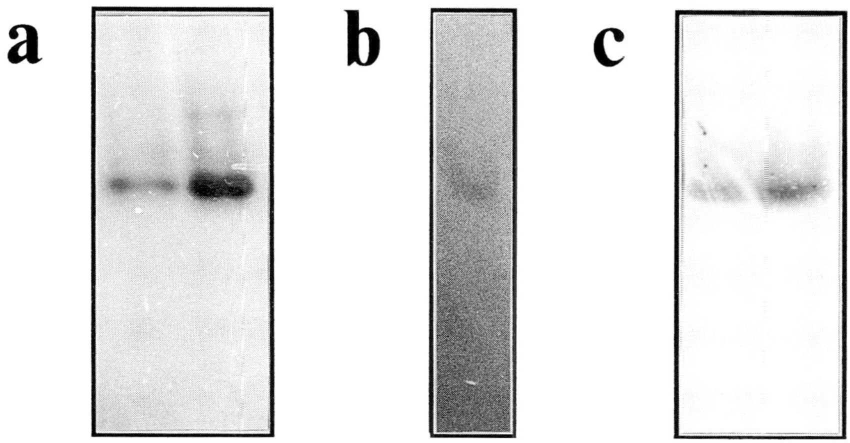

Figure 16.2. Bacteriolytic activity of T4 lysozyme as detected by the overlay gel assay. Substrates: (a) *Micrococcus lysodeikticus*, 200 ng (left lane) and 2 µg (right lane) T4 lysozyme; (b) *E. c. atroseptica*, 2 µg T4 lysozyme; (c) *Pseudomonas solanacearum*, 200 ng (left lane) and 2 µg (right lane) T4 lysozyme. T4 lysozyme tends to form a dimer which is visible in the right hand lane of panel (a).

for a plant signal peptide on to the 5' end of the respective gene (Düring, 1988; Hippe *et al.*, 1989). The fusion of the signal peptide of barley aleurone α-amylase to the N-terminus of the bacteriophage T4 lysozyme led to an efficient secretion of the foreign protein in transgenic tobacco plants, as demonstrated by electron microscopic immunogold labeling (Düring, 1988). T4 lysozyme represents a good candidate for an antibacterial resistance approach in transgenic potato plants for several reasons. Its activity *in vitro* against *M. lysodeikticus*, *Escherichia coli* and some other bacteria is known (Tsugita, 1971). Furthermore, it could be demonstrated by the gel overlay assay technique that T4 lysozyme possesses bacteriolytic activity against both *E. c. atroseptica* and *P.solanacearum*. The activity against *P. solanacearum* is comparable in strength to that against *Micrococcus lysodeikticus* whereas that against *E. c. atroseptica* is much weaker (Fig. 16.2). In comparison, hen egg white lysozyme does not show any activity against *P.solanacearum* but only against *E. c. atroseptica*. The latter activity is similar to that of T4 lysozyme (data not shown).

The application of the biotechnological lysozyme secretion system to potato plants has been performed by using a construct with the chimeric barley α-amylase signal peptide - T4 lysozyme gene under the control of the CaMV 35S promoter. A tetraploid genotype (Z2) has been used for *Agrobacterium tumefaciens*-mediated gene transfer (Düring *et al.*, 1993). Molecular analysis of

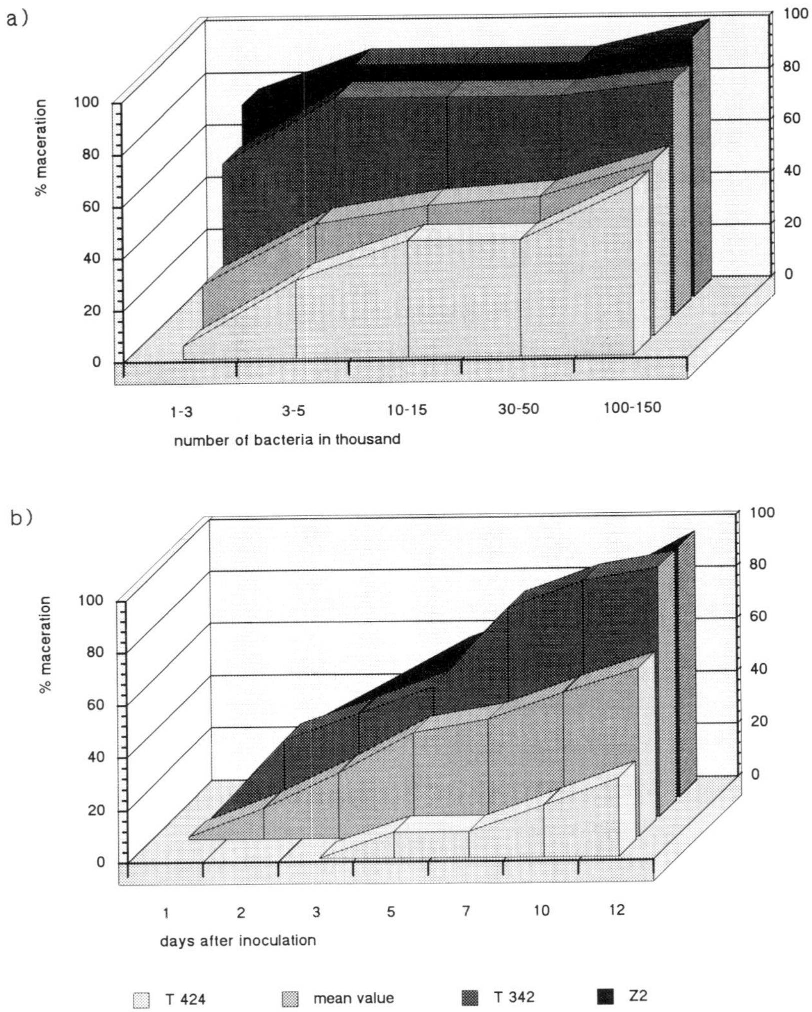

Figure 16.3. Tuber disc inoculation with *Erwinia carotovora atroseptica*. (a) The infected tuber discs were incubated for 3 days at ~20°C under high humidity. Rating was done by checking the amount of rotten tissue. (b) Inoculation with 25,000 bacteria was assayed for 12 days under the same conditions. Z2: non-transgenic control; T 342: transgenic control; mean value: mean value of all transgenic plants used in the experiment including the low expressing ones; T 424: example for the highest observed effect.

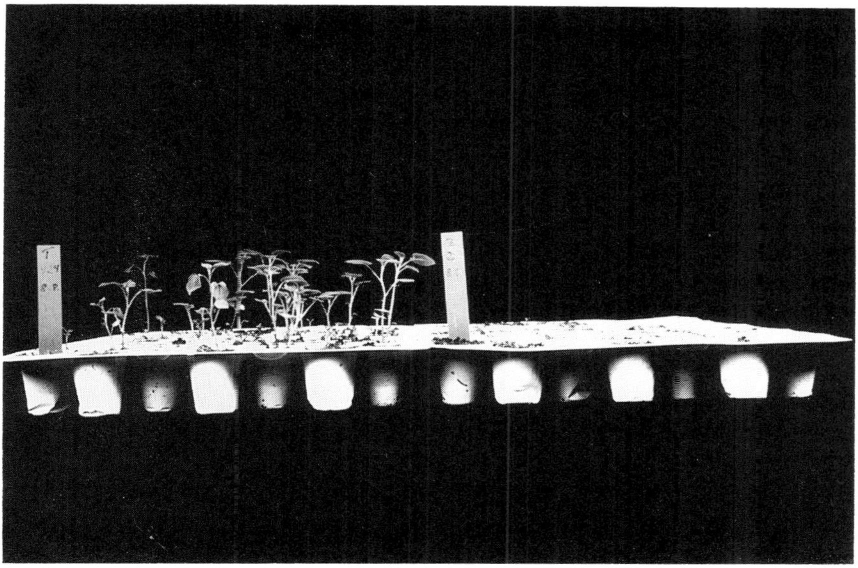

Figure 16.4. Sprouting bioassay with T 424 transgenic potato plant and Z2 non-transgenic control plant. Tuber discs containing one eye were prepared and inoculated with 1.25 x 10^6 *E. c. atroseptica* cells/ml for 2 h. The inoculated discs were transferred to soil and cultivated in the greenhouse under humid conditions. After 7 weeks half of the 78 explants of the transgenic plant had produced strong healthy plants whereas the control explants did not produce sprouts.

the transgenic plants revealed three groups of regenerants, those with a nearly non-detectable transgene expression, those with intermediate expression, and those with an easily detectable expression level when assayed for mRNA and protein. Nevertheless, the overall expression levels are very low (~0.001% of total soluble protein). Localization of the foreign lysozyme in the intercellular spaces has been confirmed by analysis of intercellular washing fluids and electron microscopic immunogold labeling.

The phytopathological analysis of the transgenic plants was performed *in vitro* and in the greenhouse. It is well known that quantitative analysis of resistance to *E. c.* similar to other bacterial infections is difficult to assess under these conditions. The reproducibility of the established tests in quantitative terms is rather low. This is due to the environmental conditions required for bacterial invasion and disease development. Therefore, two types of tests were used for assaying the biological effects in transgenic plants. The inoculation of potato tuber discs with defined amounts of bacteria and incubation at ~20°C under high humidity allows a quantitative comparison within only one experiment. The results obtained throughout several series of experiments display a qualitatively reproducible enhancement of resistance in accordance with the classes of regenerants, as deduced from molecular analysis. Maceration of tuber discs from transgenic plants is significantly reduced (Fig. 16.3). The results from the molecular and phytopathological experiments provide a consistent line of evidence for an antibacterial activity of the foreign lysozyme in the transgenic potato plants.

Another type of experiment was performed in the greenhouse demonstrating the sprouting capability of infected tuber pieces from transgenic plants in contrast to the controls (Fig. 16.4). In an experiment with 78 explants of the T 424 transformant and the Z2 non-transgenic control plants it was demonstrated that more than half of the transgenic explants were able to sprout under these conditions, whereas the controls did not produce any viable shoots (Düring *et al.*, 1993). The observation of these shoots over the entire period of growth did not reveal any disease symptoms. In addition, the control explants which had not sprouted obviously were macerated.

Set in relation to the data available from the literature for natural infection conditions, the results obtained from these experiments suggest that in the field it might be possible to obtain enhanced resistance to phytopathogenic bacteria by this biotechnological approach. The analysis of other potato-bacteria interactions will be used to evaluate the possibility of preventing the development of other bacterial diseases. Bacterial wilt in potato, caused by *P. solanacearum*, might be efficiently inhibited because of the high bacteriolytic activity of T4 lysozyme against this bacterium. It is interesting to observe that the low levels of foreign T4 lysozyme produced in the transgenic plants are sufficient to cause a significant reduction in susceptibility towards a phytopathogenic bacterium. The expression level of the foreign enzyme might be enhanced by molecular biological strategies using different promoters or by adapting the coding sequence to the optimal set of codons used in dicots. Another goal of current research is to find regulated expression systems which allow restriction of the T4 lysozyme biosynthesis *in planta* to only infected tissue.

A similar biotechnological approach using hen egg white lysozyme has been started by the group of Asselin (Trudel *et al.*, 1992). Until now, expression of hen egg white lysozyme with its native signal peptide in transgenic tobacco plants has been described. However, only a small portion of the produced foreign lysozyme has been located in the intercellular spaces. Obviously, in this case the signal peptide is not fully compatible with the plant secretory apparatus. No phytopathological data have been reported yet from transgenic tobacco or potato plants with this system.

Is It Possible to Construct Incompatible Interactions Using Biotechnology?

The results obtained with the T4 lysozyme-expressing transgenic potato plants demonstrate that the biological effect is dependent on the level of foreign protein expression. The extent of resistance decreases with increasing bacterial inoculum. The observed effect has to be classified, therefore, as relative resistance. It will be of interest to see if future research employing modified constructs will be able to provide higher levels of T4 lysozyme in transgenic potatoes and if those will lead to a plant-pathogen system which can be classified as completely resistant under high infection pressures. This would imply that high levels of T4 lysozyme are able to destroy every bacterium invading the host plant and that T4 lysozyme could function as a single antibacterial determinant in plants. The resulting interaction should then be termed an incompatible interaction.

Other approaches evaluating specific components of the molecular processes involved in the pathogenicity of the invading bacterium are under development. They might allow inhibition of the growth of the invading bacteria by specific molecular interactions instead of killing them. These strategies could be classified as incompatible interactions in the sense of gene-for-gene interactions.

Other Biotechnological Approaches

The antibacterial capacity of cecropin and other lytic peptides from different origins has been reported (Nordeen *et al.*, 1992; Jaynes *et al.*, 1993). But, as described by Nordeen *et al.* (1992), the lytic peptide cecropin in general displays a toxic effect not only to bacterial cells but also to plant protoplasts. Therefore, only a limited 'window' of cecropin levels has to be achieved in the transgenic plants which allows the growth of the bacteria to be inhibited without affecting the development of the plant itself.

Jaynes *et al.* (1993) demonstrated that in transgenic tobacco plants a designed cecropin B analog could effect a reduction in the susceptibility to *P. solanacearum*. Delayed wilt symptom formation, reduced disease severity and reduced mortality after infection were observed in the F1 progeny of two transgenic plants. The phytopathological tests employed were the 'wounded root bacterial challenge assay' and the 'stem inoculation assay'. It is not clear from the publication if transgenic plants carrying the gene construct with the CaMV 35S promoter or the proteinase inhibitor II promoter were used for the assay. The authors mention that the cecropin analog gene detected by PCR was observed to be repeatedly truncated. In all tested seedlings the codons for amino acids numbers 37 and 38 were missing although those directly following the termination codon and the polyadenylation sequence were intact. This instability of the foreign gene, suggesting that there exists a highly specific and reproducible deletion mechanism, is difficult to explain. The consequences for the use of this construct in terms of reproducibility and stability of the foreign trait were not discussed. It was only mentioned that the cecropin analog remained lytically active when these last two amino acids of the polypeptide are missing. No effect was observed with transgenic plants expressing the native cecropin, SB-37. A significant increase in lytic activity of cecropin has been reported for the combined action *in vitro* together with hen egg white lysozyme (Jaynes *et al*, 1993). This might be due to the mechanical weakening of the bacterial cell wall by the action of lysozyme. Cecropin might then interact more effficiently with the membrane.

Another approach using the barley α-thionin has been reported to confer enhanced resistance to bacterial pathogens (Carmona *et al.*, 1993). Thionins are small cysteine-rich polypeptides which are synthesized from larger precursor molecules. Constitutive expression from the CaMV 35S promoter in transgenic tobacco plants led to a significant decrease in bacterial growth of *Pseudomonas syringae syringae* and a reduced percentage of necrotic lesions with *Pseudomonas syringae tabaci.* The bactericidal activity of extracts from transgenic plants could be demonstrated *in vitro* using *Clavibacter michiganensis sepedonicus.* But, in comparison with wheat α-thionin, no resistance effect was detectable(Carmona *et al.*, 1993).

In neither report were data presented for the specific subcellular localization of the foreign gene products. Therefore, no conclusions can be drawn in terms of efficacy. These two biotechnological resistance strategies based on cecropin and α-thionin could also be applied to other more important plant-bacteria interactions. Phytopathological tests will reveal their importance for agricultural application.

Future Perspectives

The described biotechnological approaches might be used in combination to generate more efficient multigenic resistance. This should minimize the risk of pathogens overcoming the resistance. In addition, as already described, a combination of two antibacterial factors can enhance the biological effect produced in transgenic plants by interaction at the molecular level.

More specific approaches may include expressing compounds which specifically interact with pathogenicity factors of the bacterium or complement specific components of the plant's metabolism which are sensitive to bacterial toxins (e.g. de la Fuente-Martinez *et al.*, 1992). In these cases a thorough knowledge of the molecular aspects of the plant-bacterium interaction is necessary. The approaches described above aim at a general, non-specific antibacterial resistance reaction and are therefore more widely applicable. On the other hand, an inherent ecological risk of damaging beneficial members of the soil microflora has to be ruled out before this approach merits a broad application in agriculture.

Acknowledgement

The excellent technical assistance of Mrs Hilke Döpke is gratefully acknowledged.

References

Audy, P., Benhamou. N., Trudel. J. and Asselin, A. (1988) Immunocytochemical localization of a wheat germ lysozyme in wheat embryo and coleoptile cells and cytochemical study of its interaction with the cell wall. *Plant Physiology* 88:1317-1322.

Burr, T.J. and Schroth, M.N. (1977) Occurence of soft-rot *Erwinia* spp. in soil and plant material. *Phytopathology* 67:1382-1387.

Carmona, M.J., Molina, A., Fernández, J.A., López-Fando, J.J. and Garcia-Olmedo, F. (1993) Expression of the α-thionin gene from barley and tovacco confers enhanced resistance to bacterial pathogens. *The Plant Journal* 3:457-462.

Collmer, A. and Keen, N.T. (1986) The role of pectic enzymes in plant pathogenesis. *Annual Review of Phytopathology* 24:383-409.

Cutt, J.R. and Klessig, D. (1992) Pathogenesis-related proteins. In: Boller, T. and Meins, F. (eds), *Plant Gene Research, Genes Involved in Plant Defense.* Springer-Verlag, Wien, pp. 209-243.

de la Fuente-Martinez, J.M., Mosqueda-Cano, G., Alvarez-Morales, A. and Herrera-Estrella, L. (1992) Expression of a bacterial phaseolotoxin-resistant ornithyl transcarbamylase in transgenic tobacco confers resistance to *Pseudomonas syringae* pv. *phaseolicola*. *Bio/Technology* 10:905-909.

Düring, K. (1988) Wundinduzierbare Expression und Sekretion von T4 Lysozym und monoklonalen Antikörpern in *Nicotiana tabacum*; Ph.D. thesis, Universität Köln.

Düring, K. (1993) Can lysozymes mediate antibacterial resistance in plants? *Plant Molecular Biology* 23:209-214.

Düring, K., Porsch, P., Fladung, M. and Lörz, H. (1993) Transgenic potato plants resistant to the phytopathogenic bacterium *Erwinia carotovora*. *The Plant Journal* 3:587-598.

Hippe, S., Düring, K. and Kreuzaler, F. (1989) *In situ* localization of a foreign protein in transgenic plants by immunoelectron microscopy following high pressure freezing, freeze substitution and low temperature embedding. *European Journal of Cell Biology* 50:230-234.

Jaynes, J.J., Nagpala, P., Destefano-Beltran, L., Hong Huang, J., Kim, J., Denny, T. and Cetiner, S. (1993) Expression of a cecropin B lytic peptide analog in transgenic tobacco confers enhanced resistance to bacterial wilt caused by *Pseudomonas solanacearum*. *Plant Science* 89:43-53.

Jollès, P. and Jollès, J. (1984) What's new in lysozyme research? *Molecular and Cellular Biochemistry* 63:165-189.

Majeau, N., Trudel, J. and Asselin, A. (1990) Diversity of cucumber chitinase isoforms and characterization of one seed basic chitinase with lysozyme activity. *Plant Science* 68:9-16.

Nordeen, R.D., Sinden, S.L., Jaynes, J.M. and Owens, L.D. (1992) Activity of cecropin SB37 against protoplasts from several plant species and their bacterial pathogens. *Plant Science* 82:101-107.
Pérombelon, M.C.M. (1973) Sites of contamination and numbers of *Erwinia carotovora* present in stored seed potato stocks in Scotland. *Annals of Applied Biology* 74:59-65.
Pérombelon, M.C.M. (1982) The impaired host and soft rot bacteria. In: Mount, M.S. and Lacy, G.H. (eds), *Phytopathogenic Prokaryotes*. Academic Press, New York, pp. 55-69.
Reisfeld, R.A., Lewis, K.J. and Williams, W.E. (1962) Disk electrophoresis of basic proteins and peptides on polyacrylamide gels. *Nature* 195:281-283.
Stintzi, A., Geoffroy, D., Bersuder, D., Fritig, B. and Legrand, M. (1993) cDNA cloning and expression studies of tobacco class III chitinases-lysozymes. In: Fritig, B. and Legrand, M. (eds), *Developments in Plant Pathology, Vol. 2, Mechanisms of Plant Defense Responses* (Proceedings of the 2nd EFPP Conference, Strasbourg, France, August 1992. Kluwer Academic Publishers, Dordrecht, pp. 312-315.
Trudel, J., Audy, P. and Asselin, A. (1989) Electrophoretic forms of chitinase activity in Xanthi-nc tobacco healthy and infected with tobacco mosaic virus. *Molecular Plant-Microbe Interactions* 2:315-324.
Trudel, J., Potvin, C. and Asselin, A. (1992) Expression of hen egg white lysozyme in transgenic tobacco. *Plant Science* 87:55-67.
Tsugita, A. (1971) Phage lysozyme and other lytic enzymes. In: Boyer, P.D. (ed.), *The Enzymes*, Vol. 5. Academic Press, New York, pp. 344-411.

Chapter Seventeen:

Field Performance of Transgenic Potatoes

William R. Belknap, Dennis Corsini†, Joseph J. Pavek†, Gordon W. Snyder and David R. Rockhold

USDA/ARS, Western Regional Research Center, Albany, CA, USA.
†University of Idaho Research and Extension Center, USDA/ARS, Aberdeen, ID, USA.

The practical application of molecular genetics to potato improvement can be effected by two methods. First, transgenes can be introduced into suitable breeding lines, and new varieties with improved performance generated by standard crossing and selection procedures. Second, transgenes can be introduced directly into existing cultivars to confer specific phenotypic changes. The second approach has the advantage of producing transgenic lines within established genetic backgrounds, potentially expediting the introduction of this technology into the marketplace.

In order to produce transgenic lines with commercial potential for a given potato cultivar, two basic requirements must be met. The first prerequisite is that the transgene introduced confers improved properties to the parent cultivar. A number of transgenes designed to improve both agronomic and quality properties have already been introduced into the potato, and evaluated both in the laboratory and in the field (Vayda and Belknap, 1992). The second requirement for commercial application is the regeneration of transgenic lines which, in addition to expressing the transgene-dependent phenotype, maintain key characteristics of the original cultivars.

Agrobacterium tumefaciens-mediated transformation is the conventional method for introducing transgenes into potato. This technique is dependent upon the regeneration of transgenic plantlets from single somatic cells in the infected explant tissue (Draper *et al.*, 1988). Within a population of regenerated plants, variability resulting in altered performance could arise from three sources: somaclonal variation (Rietveld *et al.*, 1991), T-DNA insertional mutagenesis (Feldman, 1991) and effects dependent upon the expression of the introduced genes (Dale and McPartlan, 1992).

In the absence of transformation, a high frequency of somaclonal variation has been observed in potato plants regenerated *in vitro*. Somaclonal variation is observed in potato plants regenerated either from protoplasts (Secor and

The Molecular and Cellular Biology of the Potato, Second Edition
Edited by W.R. Belknap, M.E. Vayda and W.D. Park

Shephard, 1981), or from a variety of explant materials (Karp *et al.*, 1989; Rietveld *et al.*, 1991). Similarly, somaclonal variations were observed in field studies of populations of transgenic potato regenerants (Dale and McPartlan, 1992).

In contrast, T-DNA-mediated insertional mutagenesis would be expected to be of limited importance in commercial potato cultivars. These types of mutations would be largely recessive (Feldman, 1991), and would not be likely to be expressed in tetraploid varieties. Finally, as demonstrated by Dale and McPartlan (1992), the expression of individual transgenes has the potential to negatively affect the performance of transgenic lines.

In a population of plants expressing a given transgene, considerable variability is expected in both the level of transgene expression (Odell *et al.*, 1987) and relative phenotypic response among regenerants (Lawson *et al.*, 1990). For this reason it is desirable to have a number of transgenic plants which maintain a normal (typical) phenotype from which to select commercially viable lines. A recent study has evaluated the field performance of transgenic potato populations relative to non-transformed regenerants and plants derived from shoot cuttings (Dale and McPartlan, 1992). The purpose of the experiments presented here was to determine the relative frequency of typical phenotypes within transgenic potato populations.

Transformation of Potato Cultivars

While a variety of methods has been used to introduce transgenes into higher plants, *Agrobacterium*-mediated transformation remains the predominant protocol for the potato. This procedure offers the advantage of mobilizing defined regions of DNA into the nuclear genomes of the plant. To generate the transgenic lines in this study two cultivars, Russet Burbank and Lemhi Russet (Pavek *et al.*, 1981), were transformed using microtuber disks as explant material (Snyder and Belknap, 1993). The three transgenes shown in Fig. 17.1 were introduced into potato using binary vectors and a disarmed *Agrobacterium tumefaciens* strain harboring the disarmed Ti plasmid pAL4404 (Hoekema *et al.*, 1983). The marker gene ß-glucuronidase (GUS), transcribed from the cauliflower mosaic virus (CaMV) 35S promoter, was introduced using the vectors pCGN1547 (McBride and Summerfelt, 1990) and pBI121 (Jefferson *et al.*, 1987). Two additional transgenes designed to increase resistance to blackspot bruise (pCLaSP) and bacterial infection (pCA38) were introduced using the pCGN1547 vector. The CLaSP transgene (Snyder and Belknap, 1993), contains cDNA sequences encoding a tyrosine-rich arylphorin (Tefler *et al.*, 1983), a larval serum protein (LSP) gene from the wax moth *Galleria mellonella* (Ray *et al.*, 1987). In the chimeric construct, the LSP cDNA sequences are translationally fused to the signal peptide from bean (*Phaseolus vulgaris* L.) phytohemagglutinin (Dorel *et al.*, 1989). CLaSP transcription is directed from a Class I patatin promoter

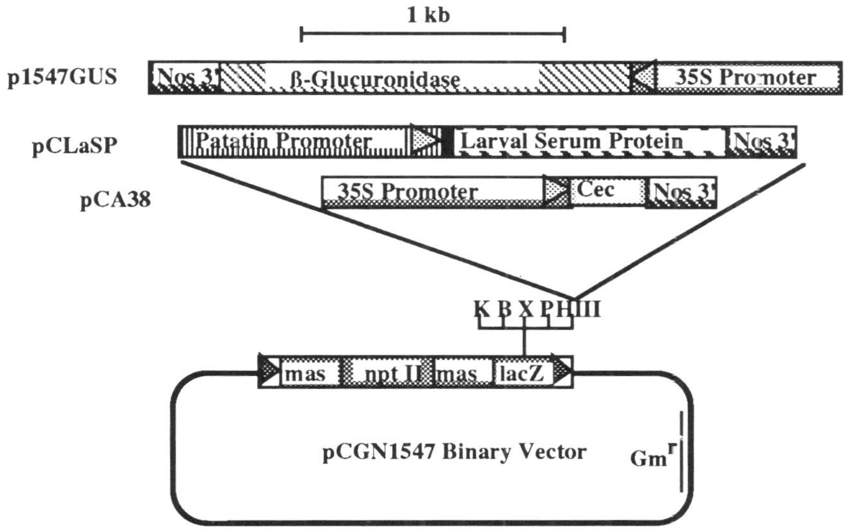

Figure 17.1. Maps of transgenes introduced into potato. All transgenes were introduced into the *Hin*dIII site of the binary transformation vector pCGN1547. The bar indicates the relative size of the introduced transgenes only, the vector (14.7 kb) is represented on a smaller scale. Unique restriction sites in pCGN1547 are indicated as K (*Kpn*I), B (*Bam*HI), X (*Xba*I), P (*Pst*I) and HIII (*Hind*III). Cec, cecropin; lacZ, lac α peptide; mas, manopine synthase 5' and 3' regions, Nos3', nopaline synthase 3' region; nptII, neomycin phosphotransferase II.

synthesized by PCR (Perkin-Elmer Cetus, Norwalk, CN) using cv. Lemhi Russet genomic DNA as a template and oligonucleotide primers based on published sequence (Bevan *et al.*, 1986). The pCA38 transgene encodes a lytic peptide (cecropin B) transcribed from a duplicate 35S promoter. The transgene was constructed as described in Destefano-Beltran *et al.* (1990).

Shoots were removed from the microtuber discs during the transformation and transgenic plants were identified by their ability to produce vigorous roots in the presence of 200 mg/l kanamycin (Snyder and Belknap, 1993). Stable integration and expression of the transgenes in these clones was verified by both Northern and Southern blotting. The transgenic clones contained from one to five copies of T-DNA per tetraploid genome (data not shown).

Two types of control plants were employed. In order to evaluate the effects of transgene insertion on phenotype, 'escapes' from the transformation procedure were used as controls. These plants were grown from shoots removed from the microtuber discs which failed to root in the presence of 200 mg/l kanamycin. The controls for the field evaluations were the wild-type parent clones not exposed to the transformation protocol.

Table 17.1. Viability of transgenic and control clones under tissue culture, greenhouse and field conditions.

Cultivar/transgene	No. of clones in culture[a]	Minituber-producing clones[b]	Field seed producing clones[c]	Normal (typical) clones[d]
Russet Burbank GUS	10	8	3	0
Lemhi Russet GUS	15	11	7	1
Russet Burbank CLaSP	5	5	5	0
Lemhi Russet CLaSP	21	18	17	3
Russet Burbank cecropin	6	6	6	0
Russet Burbank control	4	4	4	2
Lemhi Russet control	16	16	12	10

[a]Total clones generated.
[b]Clones capable of producing seed tubers under greenhouse conditions.
[c]Clones capable of producing seed tubers under field conditions.
[d]Clones with performance characteristics similar to the non-transformed cultivar.

Performance of Transgenic Clones in the Greenhouse

Transgenic and control plantlets were propagated in selective and non-selective shoot media respectively. Following root formation, plantlets were transferred to closed Magenta vessels containing vermiculite hydrated with Murashige and Skoog salts for approximately one week. The plantlets were then hardened off for a period of one week and transferred to soil in the greenhouse. Cuttings taken in the greenhouse were transplanted and used to produce minitubers for field planting. Several clones, especially those containing the GUS gene, failed to produce significant tubers under greenhouse conditions (Table 17.1). The relatively small number of cecropin-expressing transgenic lines obtained (Table 17.1) may reflect toxicity of the transgene product (Nordeen *et al.*, 1992).

Performance of Transgenic Clones in the Field

The transgenic minitubers, and minitubers from control plants, were introduced into the field site at Aberdeen, Idaho during the normal planting period (May). Minitubers containing the GUS transgene were introduced into the field in the 1990 growing season. Minitubers containing the CLaSP and pCA38 transgenes were introduced into the field in the 1991 growing season. For all constructs, seed for the second year consisted of tubers saved from the first year field test plots.

Many of the transgenic clones which had successfully produced tubers in the greenhouse failed to produce significant seed under field conditions (Table 17.1). The majority of clones which failed to produce tubers in the field were those expressing the GUS transgene. While the GUS clones were introduced into the field a year earlier than the other transgenic lines, the propagation and growing conditions in the two years were similar (data not shown). The data suggests that expression of the GUS transgene negatively affected viability of transgenic lines under field conditions (Dale and McPartlan, 1992).

As indicated in Table 17.1, the number of transgenic clones which retained a typical phenotype was less then 10% of the original number of clones from tissue culture. In general, the performance of the GUS-expressing transgenic lines was considerably poorer than that of either the other transgenic lines or the controls. These data are again consistent with observations by Dale and McPartlan (1992), who suggest that expression of GUS in vegetative tissue negatively affected a number of agronomic traits.

The agronomic/quality properties of the 38 transgenic clones which produced significant seed in the field are shown in Table 17.2. The data presented in Table 17.2 represent performance in a single year. The majority of the transgenic clones (62%) produced abnormal plants and/or tubers. Abnormalities in vine phenotype (20%) were observed much less frequently than in tubers. The most common vine abnormalities were: smaller, more compact vine with pronounced rolling of apical leaflets; erect vine with smaller rugose leaflets; small erect vine with short stem internodes, twisted petioles and small cupped leaflets. A number of the transformed clones had severely stunted vines with spindly stems and small malformed leaflets, and did not produce sufficient tubers to be maintained under field conditions.

The tuber abnormalities were as follows: small and elongated with deep eyes; very high tuber number and small size; small size with extreme second growth, loss of skin color and russeting.

The yield among the transgenic lines was variable; 37% produced yields which did not differ significantly from those of the controls (Table 17.2). The variability in yield among the CLaSP clones in shown in Fig. 17.2.

Thirty-seven percent of the transgenic clones produced either more small (21%) and/or more malformed (16%) resulting in a lower yield of U.S. No. 1 tubers than the controls (Table 17.2). The performance of the CLaSP clones relative to control US No. 1 (tubers graded according to USDA standards) yield is shown in Fig. 17.3.

As shown in Table 17.2, very few of the transgenic lines demonstrated decreased performance in either specific gravity or fry quality (a three tuber sample from each plot was stored at 7°C for 2 months and fry color was determined by frying and scoring each fry using the USDA standard french fry color chart). This suggests a relative stability of these traits as compared with

Table 17.2. Performance of transgenic cultivars Russet Burbank (RB) and Lemhi Russet (LH) clones grown from field produced seed relative to normal controls.[1]

Clone	Grade[2]	Yield	US. No 1[3]	Mal-form	> 4 oz.	Specific gravity	Fry qual.[4]	GUS act.[5]
RB Control 1991	+	+	+	+	+	+	+	0
RB GUS 11151	-	+	-	+	-	+	-	30
RB GUS 52021	-	-	-	+	-	+	+	210
RB GUS 72703	-	-	+	+	+	+	+	160
RB Control 1992	+	+	+	+	+	+	+	
RB CLaSP 101	-	-	-	+	+	+	+	
RB CLaSP 102	-	+	-	-	+	+	+	
RB CLaSP 105	-	-	-	+	-	+	+	
RB CLaSP 107	+	-	+	+	+	+	+	
RB CLaSP 112	-	-	-	-	+	+	+	
RB Cec1	+	-	+	+	+	+	+	
RB Cec3	+	-	+	+	+	+	+	
RB Cec4	-	+	-	+	+	+	+	
RB Cec5	-	+	-	+	+	+	+	
RB Cec6	+	-	+	+	+	+	+	
RB Cec7	+	-	+	+	+	+	+	
LH Control 1991	+	+	+	+	+	+	+	0
LH GUS 42101[6]	+	+	+	+	+	+	+	20
LH GUS 42104	+	-	+	+	+	+	+	90
LH GUS 42105	-	+	-	-	+	+	-	120
LH GUS 42106	-	-	-	+	-	+	+	120
LH GUS 52001	-	-	-	+	+	+	+	310
LH GUS 52004	-	-	+	+	+	+	-	70
LH GUS 52005	-	-	+	+	+	+	+	120
LH Control 1992	+	+	+	+	+	+	+	
LH CLaSP 7	+	+	+	-	+	+	+	
LH CLaSP 10	+	-	-	+	-	nd	+	
LH CLaSP 12[6]	+	+	+	+	+	+	+	
LH CLaSP 13	+	-	+	+	+	+	+	
LH CLaSP 18	-	-	-	+	+	+	+	
LH CLaSP 27	-	+	+	+	+	+	+	
LH CLaSP 38	-	+	+	+	+	+	+	
LH CLaSP 40[6]	+	+	+	+	+	+	+	
LH CLaSP 43	-	+	-	-	+	+	+	
LH CLaSP 46	+	+	+	-	+	+	+	
LH CLaSP 52[6]	+	+	+	+	+	+	+	
LH CLaSP 53	+	-	+	+	+	+	+	
LH CLaSP 54	+	+	-	+	-	+	+	
LH CLaSP 57	-	-	-	+	-	+	+	
LH CLaSP 58	-	-	+	+	+	+	+	
LH CLaSP 60	-	+	+	+	+	+	-	
LH CLaSP 63	-	-	-	+	-	nd	-	

[1]Trials: Three replications of 20 hill plots. -: significantly lower performance than control. +: no significant change in performance relative to control. Based on standard LSD test at $P = 0.05$. nd: data not collected.

[2]Appearace of vine and tuber.

[3]Grade of tubers according to USDA standards.

[4]Fry quality determined by frying and scoring each fry using the USDA standard french fry color chart.

[5]μmol MUG hydrolyzed/ml sap/hour.

[6] Normal (typical) clones.

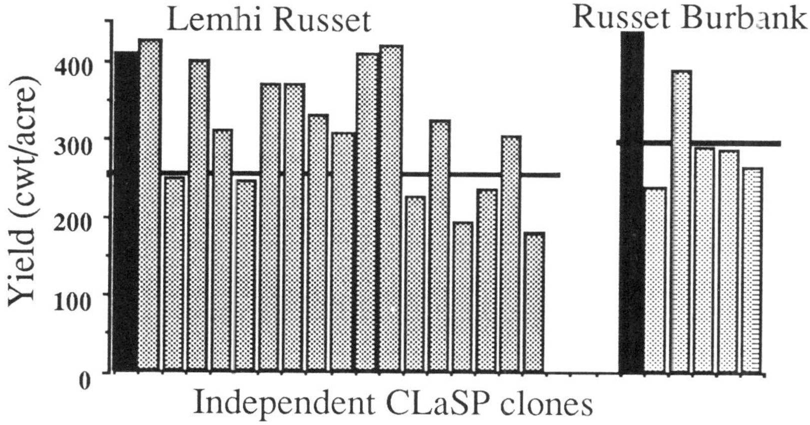

Figure 17.2. Yield of transgenic CLaSP and typical potato clones under field conditions. Yields of cv. Lemhi Russet and cv. Russet Burbank clones containing the CLaSP transgene and normal controls (black columns) grown from field-produced seed pieces. Trials consisted of three replications of 20 hill plots. Independent clones with yields less than values indicated (horizontal bar) show significant change in performance relative to the normal control, based on standard LSD test at $P = 0.05$.

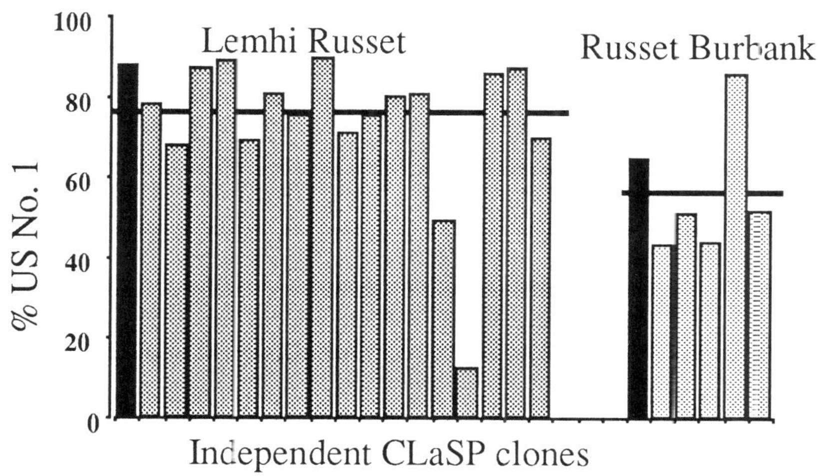

Figure 17.3. % US No. 1 of cv. Lemhi Russet and cv. Russet Burbank clones containing the CLaSP transgene and normal controls (black columns) grown from field-produced seed pieces. Trials consisted of three replications of 20 hill plots. Independent clones with percent yields lower than the values indicated (horizontal bar) show significant change in performance relative to the normal control, based on standard LSD test at $P = 0.05$. Tubers were graded according to USDA standards.

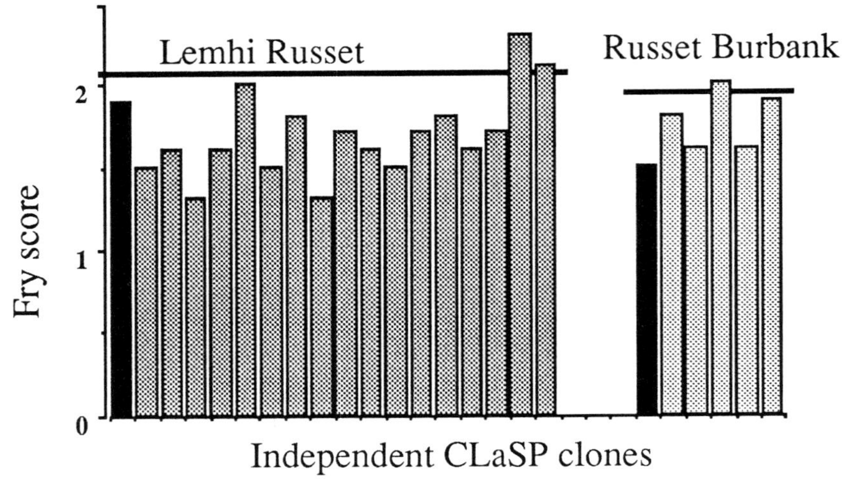

Figure 17.4. Fry scores of cv. Lemhi Russet and cv. Russet Burbank clones containing the CLaSP transgene and normal controls (black columns) grown from field-produced seed pieces. Trials consisted of three replications of 20 hill plots. Independent clones with fry score values greater than the values indicated (horizontal bar) show significant change in performance relative to the normal control, based on standard LSD test at $P = 0.05$. A sample of three tubers from each plot was stored at 7°C for 2 months and fry color was determined by frying for 3.5 minutes at 375°C and scoring each fry using the USDA standard french fry color chart.

yield and grade. The performance of the CLaSP clones relative to control fry score is shown in Fig. 17.4.

The higher variability observed in traits such as yield (Fig. 17.2) and % No. 1 (Fig. 17.3) relative to fry score (Fig. 17.3) and specific gravity (data not shown) has potentially serious implications for identification of abnormal clones. Testing involving relatively large plots over several growing seasons is required in order to obtain reliable data on both yield and grading. These data, and those presented in Table 17.2, suggest that large-scale trials over multiple growing seasons will be required for identification of transgenic clones suitable for the marketplace.

The data presented indicate that the frequency of regenerating transgenic clones with a normal (typical) phenotype was very low for all three transgenes. The frequency of typical phenotypes was much higher in the non-transgenic regenerant controls (Table 17.1). As the low percentage of typical phenotype was observed in regenerants from all three transgenes, this is probably not due to effects of transgene expression. Though all clones express the NPTII selectable marker, there is no evidence that this gene product has a negative effect on plant performance (Dale and McPartlan, 1992).

Of the GUS lines, less than half were capable of producing seed tubers under field conditions. These data are in agreement with those of Dale and McPartlan

(1992), who observed that transgenic plants which produced GUS in both aerial plant parts and tubers performed more poorly than plants which expressed this transgene only in the tuber. The transgenic GUS clones used in this study would be expected to express GUS at a significantly higher level than those described by Dale and McPartlan (1992), who utilized truncated Class I patatin promoters rather than the CaMV 35S promoter employed here. Inactivation of GUS expression was not observed (Ottaviani *et al.*, 1993) and was stable through several vegetative cycles and over two years in the field (Table 17.2).

The relatively high proportion of normal phenotype among the control plants suggests that many of these regenerants did not originate from single somatic cells, but rather from meristematic tissues within the tuber disks used as explant material in the transformation. As the transgenic plantlets were selected for regenerants from a single transformed somatic cell, it is reasonable that the overall frequency of typical plants would be lower in this population.

In contrast to the cv. Lemhi Russet clones, none of the transgenic cv. Russet Burbank clones evaluated here retained a typical phenotype. This may reflect a relatively higher potential for somaclonal variation, or less genetic buffering, in cv. Russet Burbank.

Implications for Commercialization of Transgenic Potatoes

Transgenic potato lines suitable for the marketplace must have transgene-dependent improved characteristics without compromising key agronomic and quality properties. Previous studies (Dale and McPartlan, 1992) indicate that a population of transgenic clones would be expected to contain a significant number of individuals with variations severe enough to limit commercial viability. The purpose of the experiments presented here was to determine the frequency of normal (typical) clones within a transgenic population.

The clones evaluated were transformed with the same protocol and contained three different transgenes encoding an enzymatically active genetic marker (GUS), a modified storage protein (CLaSP) or an antibacterial lytic peptide (CA38). The highest potential for deviation from typical performance was observed in yield and grading, phenotypic parameters which required extensive field evaluation to characterize. With the exception of the transgenic lines expressing GUS, most of the clones were capable of producing seed under both greenhouse and field conditions (Table 17.1). However, the low frequency of typical phenotype in both cultivars suggests that experiments designed to produce transgenic potatoes with commercial properties be initiated with several hundred individual clones in culture.

Acknowledgements

The authors express their gratitude to J.M. Jaynes for the CA38 transgene, to K. Kumaran and M.J. Chrispeels for larval serum protein and signal peptide sequences respectively, to Calgene Inc. for providing pCGN1547, and to J. Steffen-Campbell for expert technical assistance. Reference to a company and/or product by the USDA is only for purposes of information and does not imply approval or recommendation of the product to the exclusion of others that may also be suitable.

References

Bevan, M., Barker, R., Goldsbrough, A., Jarvis, M., Kavanagh, T. and Iturriaga, G. (1986) The structure and transcription start site of a major tuber protein gene. *Nucleic Acids Research* 14:4625-4638.

Dale, P.J. and McPartlan, H.C. (1992) Field performance of transgenic potato plants compared with controls regenerated from tuber discs and shoot cuttings. *Theoretical and Applied Genetics* 84:585-591.

Destefano-Beltran, L., Nagpala, P.G., Cetiner, S.M., Dodds, J.H. and Jaynes, J.M. (1990) Enhancing bacterial and fungal disease resistance in plants: application to potato, In: Vayda, M.E. and Park, W.D. (eds), *Molecular and Cellular Biology of the Potato*, CAB International, Wallingford, pp. 89-95.

Dorel, C., Voelker, T.A., Herman, E.M. and Chrispeels, M.J. (1989) Transport of proteins to the plant vacuole is not by bulk flow through the secretory system, and requires positive sorting information. *Journal of Cell Biology* 108:327-337.

Draper, J., Scott, R. and Hamil, J. (1988) Transformation of dicotyledonous plant cells using the Ti plasmid of *Agrobacterium tumefaciens* and the Ri plasmid of *A. rhizogenes*. In Draper, J., Scott, R., Armitage, P. and Walden, R. (eds), *Plant Genetic Transformation and Gene Expression: A Laboratory Manual*. Blackwell Scientific Publications, Oxford, pp. 161-198.

Feldman, K.A. (1991) T-DNA insertion mutagenesis in *Arabidopsis*: mutational spectrum. *Plant Journal* 1:71-82.

Hoekema, A., Hirsch, P.R., Hooykaas, P.J.J. and Schilperoort, R.A. (1983) A binary plant vector strategy based on separation of the *vir*- and T-region of *Agrobacterium tumefaciens* Ti plasmid. *Nature* 303:179-180.

Jefferson, R.A., Kavanagh, T.A. and Bevan, M.W. (1987) GUS fusions: ß-glucuronidase as a sensitive and versatile gene fusion marker in higher plants. *EMBO Journal* 6:3901-3907.

Karp, A., Jones, M.J.K., Foulger, D., Fish, N. and Bright, S.W.J. (1989) Variability in potato tissue culture. *American Potato Journal* 66:669-684.

Lawson, C., Kaniewski, W., Haley, L., Rozman, R., Newell, C., Sanders, P. and Tumer, N.E. (1990) Engineering resistance to mixed virus infection in a commercial potato cultivar: resistance to potato virus X and potato virus Y in transgenic Russet Burbank. *Bio/Technology* 8:127-134.

McBride, K.E. and Summerfelt, K. (1990) Improved binary vectors for *Agrobacterium*-mediated plant transformation. *Plant Molecular Biology* 14:269-276.

Nordeen, R.O., Sinden, S.L., Jaynes, J.M. and Owens, L.D. (1992) Activity of cecropin SB37 against protoplasts from several plant species and their bacterial pathogens. *Plant Science* 82:101-107.

Odell, J.T., Nagy, F. and Chua, N-H. (1987) Variability in 35S promoter expression between independent transformants. In: Key, J.L. and McIntosh, L. (eds), *Plant Gene Systems and Their Biology*. Alan R. Liss, New York, pp. 321-330.

Ottaviani, M.-P., Smits, T. and Hanisch ten Cate, C.H. (1993) Differential methylation and expression of the ß-glucuronidase and neomycin phosphotransferase genes in transgenic plants of potato cv. Bintje. *Plant Science* 88:73-81

Pavek, J.J., Corsini, D.L., Garner, J.G., Michener, S., Sparks, W.C., Cranahan, G.F., Stanger, C.E., Mosley, A.R., Johnson, M.J., Carter, G.E., Voss, R.E., Martin, M.W. and Johansen, R.H. (1981) Lemhi Russet: a new high yielding potato variety with wide adaptation, attractive tubers, and high internal quality. *American Potato Journal* 58:619-625.

Ray, A., Memmel, N.A., Orchekowski, R.P. and Kumaran, K.A. (1987) Isolation of two cDNA clones coding for larval hemolymph proteins of *Galleria mellonella*. *Insect Biochemistry* 17:603-617.

Rietveld, R.C., Hasegawa, P.M. and Bressan, R.A. (1991) Somaclonal variation in tuber disc-derived populations of potato. I. Evidence of genetic stability across tuber generations and diverse locations. *Theoretical and Applied Genetics* 82:430-440.

Secor, G.A. and Shephard, J.F. (1981) Variability of protoplast derived potato clones. *Crop Science* 21:102-105.

Snyder, G.W. and Belknap, W.R. (1993) An improved method for *Agrobacterium*-mediated transformation of *in vitro*-grown potato microtubers. *Plant Cell Reports* 12:324-327.

Tefler, W.H., Keim, S.P. and Law, J.H. (1983) Arylphorin, a new protein from *Hyalophora cecropia*: comparisons with calliphorin and manducin. *Insect Biochemistry* 13:601-613.

Vayda, M.E. and Belknap, W.R. (1992) The emergence of transgenic potatoes as commercial products and tools for basic science. *Transgenic Research* 1:149-163.

Chapter Eighteen:

Analysis of Containment and Food Safety Issues Associated with the Release of Transgenic Potatoes

A.J. Conner
New Zealand Institute for Crop and Food Research Ltd, Christchurch, New Zealand.

The cultivated potato (*Solanum tuberosum* L.) is at the forefront of the development of genetic engineering in crop plants, with many public research institutes and private companies targeting potato improvement via transformation of existing cultivars with specific genes. This is a consequence of the importance of the potato crop throughout the world, the relative ease with which the crop can be transformed, and genetic limitations associated with traditional potato breeding (Vayda and Belknap, 1992). Many of the initial transgenic plant products to be commercialized in the next decade are anticipated to be potatoes with enhanced characteristics.

Transgenic potatoes were first field tested in the United Kingdom in 1987. In the following 4 years, there were over 70 regulatory approved field trials of transgenic potatoes throughout the world, excluding multi-site trials of the same line in one year (Chasseray and Duesing, 1992). These field tests have involved evaluation of potato plants transformed with genes conferring a wide range of traits (OECD, 1990), including selectable marker genes and reporter genes, insect resistance, virus resistance, herbicide resistance, bacterial and fungal resistance, production of thaumatin, and metabolic changes such as altered sugar/starch content and response to bruising. It is to be expected that potatoes transgenic for many further genes conferring similar or additional traits will be field tested within the next few years. Many of the transgenic potato lines used in the earlier small-scale field trials are now being evaluated in scale-up field trials at multiple sites, with the specific goal of eventually making a commercial release. However, the release and agricultural use of genetically engineered plants has raised public concerns throughout the world. Debates on genetic engineering in plants have involved environmental issues relating to the growth of transgenic crops in the field (Tiedje *et al.*, 1989), and the safety of food produced from these crops (Conner, 1993a). Since public perception can be just as influential scientific deduction, it is important that sound data are collected to

The Molecular and Cellular Biology of the Potato, Second Edition
Edited by W.R. Belknap, M.E. Vayda and W.D. Park

provide a scientific basis for the assessment of the perceived risks, in order to gain public acceptance of genetic engineering technology.

This chapter assesses the issues relating to the environmental containment and food safety evaluation of transgenic potatoes. Aspects of the biology of potato crops relevant to the safety evaluation of transgenic plants are briefly summarized. This is followed by an outline of the nature of potato transformation with respect to the integration and expression of transgenes, and the phenotypic appearance of transgenic potato plants. This information is then used as a basis to discuss the issues of concern associated with the release and commercial use of transgenic potatoes in agriculture. As a component of field testing programs, several research laboratories have generated valuable data relevant to the safety assessment of transgenic potatoes. This information is summarized to provide base-line data on which informed decisions can be made concerning the safety transgenic potatoes.

Aspects of Potato Biology Relevant to Safety Evaluation

The cultivated potato originated from the Peru/Bolivia region of South America (Simmonds, 1976). It is a self-compatible, autotetraploid ($2n = 4x = 48$), and although it suffers from severe inbreeding depression, the majority of true seeds arise from self-pollination (Plaisted, 1980). Potato cultivars are vegetatively propagated clones, and often produce few or no flowers. Many cultivars are male sterile and do not produce true seeds, unless interplanted with a male fertile cultivar. Some cultivars also have reduced female fertility. However, many of the recently developed cultivars, as a consequence of breeding programs based on repeated intercrossing, have been inadvertently selected for high fertility. Potato flowers do not possess a nectary, so are therefore not attractive to honeybees. Bumblebees are the main insect visiting potato flowers, which actively gather the abundant pollen from anthers and effect pollination in the process (Plaisted, 1980).

The number of true seeds produced in fertile cultivars is highly dependent upon the cultivar grown, environmental conditions, and insect activity, but can be as high as 150-250 million per hectare (Accatino, 1980; Lawson, 1983). Seeds may remain dormant in soil for up to 2 years and have been reported to retain viability over a 7 year rotation to the next potato crop. Consequently, true potato seeds can contribute to a significant weed problem in subsequent crops (Lawson, 1983). Potato plants as weeds can also arise from the inability to remove all tubers from a field during harvest, resulting in the appearance of volunteer (ground keeper or self-set) plants the following year. Up to 367,000 tubers per hectare have been estimated to remain in the field after mechanical harvest (Lutman, 1977). Depending on the severity of the winter, many of these tubers will remain viable until the next spring. Potato tubers require 50 frost-hour equivalents at or below - 2°C to kill tubers, e.g. 25 h at -2°C, or 5 h at -10°C etc. (Lumkes and Sijtsma, 1972). In the temperate British climate, up to

80% of the tubers left after harvest die in even mild winters (Lutman, 1977). The remaining viable volunteer tubers do not remain dormant and will always sprout the next season. In addition to competing with the subsequent crop, the appearance of volunteer potatoes from either true seed or tubers can carry through disease and pest inocula that can affect the health of following potato crops. The survival of true seed in soil for many years can also cause problems with maintaining genetic purity during the production of seed tubers.

Wild tuberous *Solanum* species are very bitter in taste due to the presence of potentially toxic amounts of glycoalkaloids. One of the first steps in the domestication of the potato crop was the emergence of plants with non-bitter tubers (Simmonds, 1976). Modern day potato cultivars still have high glycoalkaloid content in their green shoot tissue and tubers upon exposure to light. In some cultivars significant levels of glycoalkaloids can also accumulate in the tubers not exposed to light. The variation in glycoalkaloid content of tubers can be attributed to both genetic effects and the environmental conditions under which the plants are grown (IFBC, 1990). The level of glycoalkaloids in potato tubers is routinely determined for advanced lines in modern breeding programs, and is nowadays effectively controlled by statutory or voluntary guidelines (Morris and Lee, 1984). A classic example involves the cultivar Lenape, bred by the USDA in the 1970s with unusually high solids content and therefore desirable processing attributes. This cultivar also had late blight resistance derived from a wild *Solanum demissum* accession. The glycoalkaloid content in Lenape tubers was found to be several times higher than normal levels during routine monitoring of potato tubers by a food company. The cultivar was quickly withdrawn from release (IFBC, 1990).

Nature of Transformation in Potatoes

Potato transformation is routinely accomplished by co-cultivation of either leaf segments (e.g. De Block, 1988; Wenzler *et al.*, 1989; Conner *et al.*, 1991), anodal stem segments (e.g. Visser *et al.*, 1989a; Newell *et al.*, 1991), or minituber segments (e.g. Sheerman and Bevan, 1988; Stiekema *et al.*, 1988; Ishida *et al.*, 1989) with *Agrobacterium* strains carrying binary vectors. Following co-cultivation, only a small proportion of the cells are the recipients of a gene insertion event. A method is therefore required to identify the transformed cells. The usual approach is the co-introduction of selectable marker genes that confer resistance to phytotoxic chemicals such as antibiotics or herbicides. Only plant cells receiving such selectable marker genes can grow, divide and develop into complete plants on culture medium supplemented with appropriate selective agent. The neomycin phosphotransferase II gene, from the bacterial transposon Tn5, is the most commonly used selectable marker gene for plant transformation. The expression of this gene in plant cells confers resistance to kanamycin and related antibiotics. This kanamycin resistance selectable marker system has been routinely used for the recovery of virtually all transgenic potato plants developed to date.

The complete nucleotide sequence of the modified T-DNA of binary vectors containing the genes targeted for transfer, is either known or can be deduced from the component DNA fragments. The modified T-DNA is integrated into the potato genome in a complete, truncated or rearranged manner, and may occur as single copies or tandem repeats at one or more integration sites (e.g. Ooms *et al.*, 1987; Stockhaus *et al.*, 1987; Sheerman and Bevan, 1988; Visser *et al.* 1989b; Conner *et al.*, 1991). This is consistent with results in other plant species. The amount of DNA integrated is usually only 5-20 kbp, which is very minute relative to the 1597-1862 Mbp for a 1*C* nucleus in potato cultivars (Arumuganathan and Earle, 1991). The initial transgenic plants regenerated following transformation are hemizygous for the transgene insertion. Consequently, the integration events are essentially equivalent to a small 'addition/deletion mutation'.

The integration of transgenes into plant genomes is generally recognized as occurring in a 'random' manner throughout the genome of plants (Walden *et al.*, 1991). Results in potato are consistent with this conclusion. The mapping of the DNA sequences flanking T-DNA insertion sites in potato has established that integration occurs throughout the genome along the length of all chromosomes (Chapter 2). Furthermore, about half these T-DNAs inserted into high-copy/repetitive DNA, and the other half into low/single copy DNA. Since approximately 75% of the potato genome is expected to be repetitive DNA (Meyerowitz and Pruitt, 1985), this suggests a marked bias toward 'random' integration of transgenes into single copy DNA. This is supported by studies on T-DNA tagging of genetic elements in plants. The high frequency of transgenic plants expressing genes transferred without promoters or with truncated promoters suggests that T-DNA preferentially inserts into regions of the genome that are transcribed, i.e. single copy DNA (Walden *et al.*, 1991). In potatoes, varying patterns of organ-specific, ß-glucuronidase (GUS) expression were detected in 53% (n = 374) of independently selected potato lines transformed with a promoterless GUS gene (Conner *et al.*, 1993). Other studies, based on a small number of transgenic potato lines, have reported similar frequencies of activated GUS expression following transfer of this gene with a severely truncated cauliflower mosaic virus 35S promoter (Goldsbrough and Bevan, 1991) or without a promoter (Lindsey *et al.*, 1993). Together, the results from gene mapping of T-DNA inserts in potato to low/single copy DNA (Chapter 2) and the similar high frequency of promoter trapping via T-DNA insertion in potatoes (Conner *et al.*, 1993), suggest that approximately half of the T-DNA insertions into the potato genome involve integration into existing genes. Such integration events raise the possibility of insertional mutagenesis being a relatively common event (see below).

The magnitude of transgene expression, and sometimes the developmental pattern of expression, can vary markedly among a population of plants independently transformed with the same transgene. This general phenomenon is referred to as the 'position effect', and is believed to be a consequence of the random integration of the transgene into different sites of the plant chromatin for each transformation event (Peach and Velten, 1991; Nap *et al.*, 1993). The variable nature of transgene expression is commonly reported among populations of transgenic potatoes (e.g. Stiekema *et al.*, 1988; Wenzler *et al.*, 1989; van der Leij *et al.*, 1991). In general, only about 10-20% of transgenic lines usually have the desired magnitude and specificity of transgene expression. When a transgenic line with high expression of the transgene is identified, the continuity and stability of expression can occasionally be influenced by a variety of allelic and non-allelic interactions. These interactions usually involve a loss-of-function of the transgene, and are usually attributed to such factors as *trans*-inactivation, *cis*-inactivation, methylation and/or co-suppression (Matzke and Matzke, 1993); they may also be a consequence of the integration site of the transgene in the plant genome.

Field testing of transgenic potato lines has revealed dramatic phenotypic changes in plant appearance in many transgenic lines (e.g. Dale and McPartlan, 1992; Conner *et al.*, 1994). This often involves weakly growing transgenic lines with altered shoot morphology, reduced incidence of flowering, and/or significantly reduced tuber yield usually involving a low number of small tubers. These phenotypic changes have been attributed to somaclonal variation, epigenetic or genetic events occurring during the cell culture and plant regeneration phase of transformation. Such effects are well known among potato plants regenerated from cell culture (e.g. Shepard *et al.*, 1980; Potter and Jones, 1991). Among transgenic lines the frequency of these off-types has been recorded as 15-80% depending on the potato cultivar (Jongedijk *et al.*, 1992); these off types often do not become apparent until plants are grown in the field (Conner *et al.*, 1994). The elimination of these off-types is therefore one of the main objectives of initial small-scale field trials with a new series of transgenic potato lines.

Most field evaluations of transgenic potatoes have only involved small-scale trials, which do not allow effective analysis of transgenic plant performance under appropriate agronomic conditions. It is likely that many transgenic potato lines with only slight reductions in yield performance will only be eliminated following scale-up field trials. For example, three transgenic chlorsulfuron-resistant potato lines that did not differ from the parental cultivar in small scale field trials, showed reduced yield performance relative to the parental cultivar when trialled with the more rigorous experimental plot design and statistical analysis possible with larger scale field trials (Moses *et al.*, 1993).

In conclusion, potato transformation is highly unpredictable with respect to the integration and expression of transgenes, and the frequency of somaclonal variation among transgenic lines. It is therefore necessary to select and regenerate large numbers of independently transformed potato lines, in order to recover several lines with appropriate expression of the transgene, while maintaining the phenotype and yield performance of the parental cultivar. This is especially important for clonal crops such as potato, since further breeding to eliminate somaclonal variation will result in the loss of the genetic integrity of the parental cultivar.

Safety Issues Relating to the Release of Transgenic Potatoes

The emphasis for the safety assessment of transgenic potatoes should not be whether the genetically engineered plants possess any environmental or food safety risks. Instead it should focus on whether transgenic potatoes pose any new risks over those that might be expected from existing potato crops or the release of new potato cultivars resulting from traditional plant breeding. In this discussion of the safety concerns, the focus is on the generic issues associated with potatoes as a transgenic crop, rather than on the specific transgenic phenotypes. Safety considerations for specific traits in transgenic crop plants have been previously summarized for herbicide resistance (Duke *et al.*, 1991), virus resistance (Tepfer, 1993; Timmerman, 1994), and insect resistance (Field *et al.*, 1993; Wearing and Hokkanen, 1994).

Insertional mutagenesis

The high frequency at which T-DNA insertion occurs into native genes in the potato genome (see above), raises the issue of insertional mutagenesis being a relatively common event. Transgene integration may result in the disruption of a native gene at the insertion locus, the effects of which are expected to be inherited in a recessive manner. Inbreeding is therefore necessary to recover the homozygous individuals in which gene disruption phenotypes become apparent. This has been well documented during T-DNA mutagenesis of *Arabidopsis* (Feldmann, 1991). Insertional mutagenic events are not expected to be observed among the plants originating as the primary transformants regenerated from tissue culture. Since transgenic lines of clonal crops are propagated vegetatively from the original primary transformant, phenotypic effects resulting from the disruption of native genes are not expected. This is especially so in potato, which is an autotetraploid and has three other potential alleles to complement an insertional mutagenic event resulting from a functional deletion. However, potato cultivars are highly heterozygous, with their allelic configuration often being in the simplex state (*Aaaa*) (Howard, 1970). Consequently, transgene insertion at such loci may occasionally result in the phenotypic effects of insertional mutagenesis being observed in the primary transformants.

In rare instances, insertional mutagenesis may also involve gene activation resulting from transcriptional gene fusions ('read-through' from T-DNA promoters) or even the formation of fusion proteins via translational gene fusions. Such events are likely to be expressed in a dominant manner and therefore observed in the primary transformant.

Unpredictable nature of transformed plants

One of the main difficulties associated with the biosafety assessment of transgenic plants such as potatoes relates to the unpredictable nature of transformation, especially the variability associated with position effects, transgene inactivation and somaclonal effects. Such unpredictability raises concerns that transgenic plants will behave in an inconsistent manner when grown in the field. However, it must be remembered that traditional plant breeding also results in unpredictable behavior of plant performance and phenotypic expression of genes recombined into new genetic backgrounds, especially following wide hybridizations. Furthermore, although the nature of DNA integration cannot be predicted for any specific transgenic plant prior to transformation, the manner in which the modified T-DNA has been integrated, and any resulting position effects on transgene expression, can be accurately determined following transformation.

A key point relevant to the development and use of transgenic plants is that molecular biology allows the exact components of the modified T-DNA to be controlled in a precise manner with respect to the intended coding regions and their controling regulatory elements. This is in contrast to traditional plant breeding, where it is common practice for undefined genes to be transferred from wild related species into crop plants. Such breeding involves the transfer of large blocks of undefined chromatin linked to the gene(s) of interest, which often encode characters such as pest and disease resistance and may be responsible for the accumulation of unknown 'toxins'. New cultivars resulting from such breeding have been widely grown in agriculture and used in food production for many years. As a consequence, it has been argued that genetic engineering offers a greater confidence for producing the desired outcome compared with traditional genetics and breeding (Conner, 1993a).

Selectable marker genes

The use of selectable marker genes is an essential component for the development of transgenic plants (see above). The use of antibiotic resistance genes as markers for the selection of transformed plant cells has raised a number of concerns regarding their safety. These concerns have involved both the gene itself and its protein expression products, and include issues such as: increased fitness of transgenic plants; horizontal gene transfer to microbe populations; toxic, immunologic and allergenic responses to protein expression products; and compromise of the oral therapeutic use of antibiotics from the consumption of protein expression products in the diet. The safety of antibiotic resistance genes as selectable markers for potato transformation will not be considered further here, since the issues relate to plants in general and not specifically potatoes. Compelling arguments that the commonly used marker genes conferring kanamycin resistance do not pose any additional bio-safety concerns over their already common occurrence in nature can be found elsewhere (Flavell *et al.*, 1992; Nap *et al.*, 1992; Redenbaugh *et al.*, 1992, 1993).

Sophisticated methods utilizing the Cre/Lox site-specific recombination system have been proposed for the elimination of selectable marker genes from transgenic plants (Dale and Ow, 1991). However, this approach is not appropriate for clonal crops such as potato, since it requires the outcrossing of transgenic lines to other specially developed transgenic lines. Such sexual crosses and true seed production would result in the loss of the genetic integrity of the parental cultivar. Since potato cultivars are highly heterozygous with considerable inbreeding depression, exceptionally large populations of seedlings have to be evaluated in order in recover new clones that approach the agronomic performance of the original elite line (Howard, 1970).

Environmental Concerns Regarding Transgenic Potatoes

The 'escape' of viable plant material carrying foreign genes is considered to be a potential problem associated with the initial field testing of transgenic plants (Tiedje *et al.*, 1989). In crops such as potatoes, this may result from gene transfer via the dispersal of pollen to other potato crops, hybridization to related weedy species, and the appearance of volunteer plants from tubers and seeds left in the ground following crop harvest and establishing weedy populations. These issues have been well discussed by Helgeson and Davies (1992).

An important consideration for risk assessment of gene flow via sexual reproduction is the fertility status of the parental potato cultivar. Many potato cultivars are male sterile and therefore naturally contained from gene flow through pollen dispersal. Some cultivars only flower very rarely or have greatly reduced female fertility. In addition to gene transfer to other potato crops, introgression of transgenes into related species may occur following interspecific hybridization. Most tuberous *Solanum* species can be readily intercrossed, with natural hybridization being well documented in the South American center of potato genetic diversity (Hawkes, 1962). Natural populations of these tuberous species only occur to a limited extent outside this area, and do not occur in the major potato-growing regions of the world. However, non-tuberous *Solanum* species commonly occur in temperate regions, with *Solanum nigrum* (black nightshade) frequently occurring as a weed in potato crops. The chances of interspecific hybridization between cultivated potato and non-tuberous *Solanum* species are exceedingly remote (see below). Furthermore, there are many natural barriers to gene introgression even if a rare hybridization event occurred (Hadley and Openshaw, 1980).

Volunteer potato plants are well known to cause significant weed problems in subsequent crops (Lutman, 1977; Lawson, 1983). A logical concern is whether transgenes will contribute to increased weediness in these populations. It has been suggested that genetic engineering for pest and disease resistance in potatoes may give slightly increased fitness to volunteer populations establishing from volunteer tubers or true seed production (Dale, 1993). However, such effects are considered to be insignificant in an agricultural setting, and would be no different from traditional potato breeding for such traits. Many potato cultivars with resistance to pests and diseases have been successfully commercialized in the past (Howard, 1970), without concerns over increased fitness of volunteers. It could be argued that the appearance of pest- and disease-resistance volunteer potatoes would be in preference to volunteer populations that carry over pest and disease inocula from previous crops. A concern over the development of transgenic herbicide-resistant potatoes is the possible compromise of the chemical control of potato volunteers (Dale, 1993). However, since the same potato cultivar can be independently transformed with resistance to two different herbicides, novel management options can be design to assist in the elimination of volunteers (Conner *et al.*, 1990). For example, when such isogenic lines are alternated each growing season and the appropriate herbicide is applied, herbicide-resistant volunteers arising from the previous season's non-harvested tubers can be eliminated.

The easiest way to examine the issue of increased weediness of transgenic potatoes is to assess their performance during initial small-scale field trials, with careful post-trial site monitoring. Any consideration of possible increased weediness during such initial trials must also recognize that many of the transgenic lines will be off-types due to somaclonal variation. These plants will have considerably lower fitness than traditional potato crops, due to their poor growth, lack of flower development, decreased number of tubers and/or the reduced longevity of smaller tubers (see above).

Food Safety Concerns Regarding Transgenic Potatoes

The potential and perceived hazards associated with the food safety of transgenic crops have been extensively investigated by the International Food Biotechnology Council (IFBC, 1990). These concerns can be attributed to three mechanisms (Conner, 1993a): protein products of the transgenes, pleiotropic and secondary effects resulting from expression of the transgenes, and insertional mutagenesis as a consequence of transgene integration.

When evaluating the risks of transgene expression products in potatoes, the first factor to investigate is whether the protein is present in the harvested tubers. Concerns to consider include: toxicity responses involving both acute effects and long term growth inhibitions; immunological responses; and allergenic responses. From a positive perspective, the nature of the expression products is known, and the specificity and magnitude of transgene expression can be determined. As a consequence, testing and evaluation of any negative concerns are possible. Even when protein expression products do occur in tubers, cooking may eliminate any potential hazard, since many proteins are heat-labile.

Many genes transferred to plants code for the production of enzymes responsible for catalyzing biochemical reactions. This may result in a range of pleiotropic or secondary effects such as substrate depletion, product accumulation and altered metabolic flow-through in downstream biochemical pathways. A good example of secondary effects involves the overexpression of a yeast ornithine decarboxylase gene in tobacco, which led to an elevation of metabolic flux through the biochemical pathway to result in the accumulation of nicotine (Hamill *et al.*, 1990). However, such effects will be dependent upon the key regulatory points and rate-limiting steps along biochemical pathways, and may vary between plant species and genotypes of the same species. Possible biochemical consequences of pleiotropic and secondary effects of transgene expression are predictable, and appropriate analyses can be performed if possible hazardous situations are envisaged.

The issue of insertional mutagenesis is discussed above. With respect to transgenic potatoes, activation of the gene(s) responsible for the accumulation of glycoalkaloids in potato tubers is of concern. It will therefore be important to assess glycoalkaloid concentrations in transgenic potato tubers, something which is routinely determined in advanced potato lines of traditional breeding programs. Secondary effects of transgene expression and insertional mutagenesis do not present new safety issues with respect to potato production. Identical effects may unwittingly occur during traditional potato breeding programs, especially those involving wide hybridizations. They may arise as a consequence of gene complementation, gene recombination and chromosomal rearrangements such as translocations and inversions. Nevertheless, it is important that the content of significant nutritional components and natural toxicants (glycoalkaloids) found in tuberous *Solanum* species, are determined as a component of safety evaluation.

Safety Assessment Data on Transgenic Potatoes

Advocates of genetic engineering have argued that transgenic crop plants do not pose any new environmental or food safety hazards or risks over those associated with traditional plant breeding and generally accepted agricultural and food industry practices. However, to ensure public acceptance of this technology, it is important that data are collected to provide a scientific basis for the assumption that transgenic potatoes are no different from traditional potatoes, other than for the specific trait that is altered. As a contribution towards establishing whether transgenic potato plants pose any greater environmental and food safety risks than traditional potato plants, we have conducted a series of investigations on relevant issues in conjunction with our field testing program on transgenic potato plants. The most extensive information has involved the monitoring of herbicide-resistant potato lines (SCI1 to SCI5) (Conner *et al.*, 1989). These transgenic lines express three foreign genes: a selectable marker gene conferring resistance to kanamycin, a reporter gene conferring GUS activity, and a gene encoding a chlorsulfuron-insensitive form of acetolactate synthase conferring protection against applications of the herbicide chlorsulfuron. The results of these studies are summarized below, along with relevant data collected by other research teams during the monitoring of their field trials on transgenic potatoes.

Pollen dispersal from transgenic potatoes

We have performed two extensive studies that monitor pollen dispersal from field trials of self-fertile potatoes transgenic for chlorsulfuron resistance. During a small-scale field trial (22 m x 13.5 m) in the 1988/89 summer (Tynan *et al.*, 1990). the frequency of transgenic seedlings from wild-type potato plants growing within the trial was about 1% (*n* = 4476), whereas the frequency of transgenic progeny up to 4.5 m from the trial was only five in 10.000 (*n* = 40,871). Transgenic progeny were not recovered from wild-type plants growing 4.5-10 m from the trial (*n* = 16,034). The same transgenic lines were used in a more comprehensive study of pollen dispersal during a scale-up trial (88 m x 24 m) over the summer of 1989/90 (Conner, 1993b). In this study, only five in 10,000 of the progeny from wild-type plants within the trial were transgenic (*n* = 54,213). From over 250,000 progeny of wild-type plants growing around the trial only 9 transgenic seedlings were recovered, all from rows 3.75 m or 6.0 m west of the trial. It is possible that these transgenic progeny originated from single berries in each row, thereby representing only two pollination events.

During five other field trials of transgenic potatoes in 1988/89 and 1989/90 we only detected very minimal transgenic pollen dispersal (Conner, 1993b). The frequency of transgenic progeny from wild-type potato plants growing within one of these trials was four in 10,000 (*n* = 41,977). No other examples of transgenic pollen being transferred onto wild-type plants either within or surrounding these trials was detected. However, there was only limited seed set in these trials due to late planting dates which hindered flower and berry formation and/or somaclonal variation in some of the transgenic lines resulting in non-flowering phenotypes. Similar results to these latter findings have also been observed in the United Kingdom (J. Scheffler, unpublished results - see Helgeson and Davies, 1992).

Gene transfer to weeds

A recent thorough study has established that potato is naturally contained from gene flow via pollen to the temperate weeds *S. dulcamara* and *S. nigrum* (Eijlander and Stiekema, 1994). These authors attempted to circumvent the interspecific hybridization barriers between potatoes and these weedy species by using genotypes with varying ploidy levels and performing crosses under a range of environmental conditions. Although rare potato pollen grains germinated on *S. dulcamara* stigmata, all crosses between *S. dulcamara* and potatoes resulted in flower abscission, despite using pollen mixtures, bud pollination and style cutting procedures. Crosses between potato and *S. nigrum* showed unilateral incompatibility. In rare cases *S. nigrum* pollen germinated on potato stigmata, but failed to penetrate more than the top third of the style. In contrast, potato pollen tubes reached *S. nigrum* ovaries and resulted in some berry formation, although no hybrid seeds developed. Emasculation of *S. nigrum* was essential for this cross, otherwise potato pollen was outcompeted by *S. nigrum* pollen. Hybrid plants between potato and *S. nigrum* could only be obtained using embryo rescue by *in vitro* ovary culture 10-23 days after pollination. However, all *S. nigrum* x potato hybrid plants were infertile and failed to develop tubers.

Unsuccessful attempts to hybridize potatoes with other Solanaceous plants such as *Nicotiana, Petunia, Datura, S. nigrum* and *S. dulcamara* have been reported (Dale, 1993). We have made similar attempts using New Zealand populations of *S. nigrum.* When pollen from *S. nigrum* was placed on potato stigmata, it germinated and initiated pollen tube growth. All pollen tube growth ceased within the top third of the style, with the growing points showing swellings typical of incompatible reactions (J.E. Grant, unpublished results). However, when potato pollen was placed on *S. nigrum* stigmata, pollen tube growth continued right down the style toward the ovary. These results of unilateral incompatibility between potato and *S. nigrum* confirm the findings of Eijlander and Stiekema (1994). In contrast, we failed to get any berry formation on *S. nigrum* plants following pollination with potato pollen (A.J. Conner, unpublished results). All flowers (n = 143) aborted within one week of pollinating emasculated *S. nigrum* flowers with potato pollen from cultivars Iwa, Rua or Cliffs Kidney (when using *S. nigrum* pollen as a control, 95% of emasculated flowers formed berries and set viable seed).

The progeny of Solanaceous weeds growing among transgenic potatoes have been monitored in some field trials for the occurrence of individuals carrying the transgenes. No evidence for such gene transfer from potato to *S. nigrum* was observed in New Zealand, despite screening of over 53,000 seedlings for the presence of kanamycin-resistant selectable marker gene and a GUS reporter gene (Conner, 1993b). Similar screening of 7600 *S. nigrum* progeny originating from plants growing among transgenic potatoes in Australia failed to recover any examples of interspecific transfer of transgenes (S.F. Chandler, personal communication).

Establishment of volunteers

The trial sites of transgenic potatoes are generally monitored for the appearance of volunteer potato plants for several years following the trial (Dale, 1993; Schechtman, 1993). Any observed volunteer plants are usually required to be removed by hand digging or destroyed by herbicide treatment. As a component of our field trials on transgenic potatoes, the appearance of volunteer potato plants was continually monitored and recorded at the trial sites in subsequent years (Conner, 1993b). The density of volunteers ranged from 1.3/m^2 to 2.9/m^2 in the year immediately following the field trial, which reduced to 0.02/m^2 in year 2, with no volunteers observed in year 3.

The proportion of these volunteers that were transgenic ranged from 3 to 31%, which was considerably lower than the proportion of transgenic potato plants established in the original field trial (Conner, 1993b). This may have resulted from greater care being taken over the harvest of plots within the trial compared with the buffer rows, and/or the marked somaclonal variation observed in many of the transgenic lines which resulted in very weak field-grown plants yielding very few tubers or very small tubers (Conner *et al.*, 1994).

A substantial reduction in the proportion of transgenic volunteers was also observed following hand digging rather than mechanical harvest (Conner, 1993b). Hand harvest would have allowed the recovery of many of the smaller tubers produced by some transgenic lines, which would have otherwise fallen through a mechanical harvester. These results establish that although volunteer transgenic potato plants do appear at trial sites the year following the initial field trial, they can be easily managed and eliminated. In many potato growing areas, the natural course of winter freezing and thawing is sufficiently severe to kill volunteer tubers.

Nutritional Evaluation

As a component of our field testing program on transgenic potatoes we have performed a comparative nutritional analysis of tubers harvested from a transgenic chlorsulfuron-resistant line (SCI4) and the parental cultivar (Iwa) (Monro *et al.*, 1993). The transgenic line also expressed a kanamycin-resistant selectable marker gene and a GUS reporter gene. SCI4 and Iwa differed only slightly in their content of protein, fat, available carbohydrate, soluble and insoluble dietary fiber, ash, dry matter and energy (kJ/g). The content of all amino acids was also virtually identical in these two potato genotypes.

Although no nutritional changes were expected to be observed in this study, the finding of no change in the content of the branched chained amino acids (valine, leucine and isoleucine) is of interest. The basis for the resistance to chlorsulfuron in the transgenic line involved the expression of a mutant acetolactate synthase gene, which encodes an important step in the pathway toward branched chain amino acids (Haughn *et al.*, 1988). The expression of this transgene would supplement the expression of the endogenous potato gene, resulting in an overexpression of acetolactate synthase in the transgenic line. A suggested secondary effect of over-expressing enzymes that catalyze a step in biochemical pathways is flow-through effects to downstream metabolites (Conner *et al.*, 1990). However, such secondary effects will be dependent upon the key regulatory points and rate-limiting steps along the pathway and were not evident with respect to the accumulation of branched chain amino acids in Iwa potato transgenic for chlorsulfuron resistance.

Rat feeding studies

We have evaluated tubers harvested from a transgenic chlorsulfuron-resistant line (SCI4) and the parental cultivar (Iwa) in comparative rat feeding trials. The samples used for nutritional evaluations (see above) have also been used in standard feed evaluations with male Sprague-Dawley rats (Monro *et al.*, 1993). These trials did not detect any difference between the two sources of potato tubers with respect to energy digestibility, apparent metabolizable energy, and relative protein value. A slight difference was detected in the protein digestibility between the two samples. At the end of a long-term growth trial (55 days) in which potato was the sole source of protein and the major source of energy, rats fed the transgenic line did not differ in their mean weight gain, feed intake or feed conversion efficiency from those fed the parental cultivar. All rats appeared in good health at the end of the trial. Post mortem examination did not reveal any pathological changes or any obvious differences between the two groups in the superficial appearance of their main organs.

Cooking and sensory evaluation

We have compared tubers harvested from transgenic potato lines and their parental cultivars using the standard cooking tests and sensory evaluations employed in our traditional potato breeding programs. These tests have involved: measurement of specific gravity; scoring steamed tubers for mealiness, sloughing and after-cooking darkening; measurement of crisp (potato chip) color using reflectance of red light (640 nm) by an Agtron M40 reflectometer; and sensory evaluation using a standard triangle test with a trained taste panel selected for their sensory acuity. For all of these parameters no differences between transgenic lines and their parental cultivars were detected. This has included the evaluation of potato tubers transgenic for selectable marker genes conferring kanamycin resistance (Williams *et al.*, 1988) and our chlorsulfuron-resistant lines (SCI1 to SCI5) (W.J. Harvey and J.P. Lammerink, unpublished results). In other work on the sensory evaluation of transgenic potatoes, trained panelists were able to detect the sweet taste of thaumatin in blind taste tests on potato tissue transgenic for the thaumatin II gene from *Thaumatococcus daniellii* (Witty and Harvey, 1990).

Glycoalkaloid analysis

To date we have not performed any chemical analyses of glycoalkaloids in our transgenic potatoes. However, indirect evidence that there are no significant changes in glycoalkaloid concentrations comes from 'bioassay' assessments described above. When comparing tubers from a transgenic line with the parental cultivar, rat feeding trials did not detect any change in feed intake or long-term growth assessment and human sensory evaluations did not detect any changes in taste parameters such as bitterness.

Conclusions

Any perceived environmental or food safety risks associated with the introduction of transgenic potatoes must be evaluated against the risks associated with the release of new potato cultivars resulting from traditional plant breeding. The long and impressive record of traditional potato breeding for the safe introduction of cultivars with new gene combinations must serve as a positive guide. It should not be assumed that transgenic potatoes are dangerous and have to be proved to be 'safe'. Instead risk assessment should focus on whether transgenic potatoes pose any new risks over those that might be expected from the generally accepted agricultural and food industry practices associated with existing potato crops, or the genetic manipulation of potatoes via traditional plant breeding.

Data collected from the monitoring field trial sites of transgenic potatoes for 'escapes' of transgenic plant material has clearly established negligible environmental risks from such preliminary field trials. The dispersal of pollen carrying foreign genes was limited to within 6 m of the trials, and then only occurred at exceptionally low frequencies. Consequently, isolation distances between field trials of transgenic potatoes and other potato crops do not have to be great to mitigate gene transfer via pollen. Gene transfer to weedy species such as *S. nigrum* has not been detected, despite careful pollination studies and the screening of a large seed population originating from *S. nigrum* plants growing among the transgenic potatoes. Transgenic potatoes are therefore naturally contained from non-tuberous weedy *Solanum* species. Volunteer potato plants have subsequently appeared at the field trial sites; however, these were managed and eliminated within 2 years. Nutritional evaluations, rat feeding studies, and cooking and sensory assessments have not detected any substantial differences between a transgenic potato line and the parental cultivar. These data provide a first step toward the establishment of the biosafety of kanamycin resistance as a selectable marker gene for potato transformation.

At the generic level, genetic engineering itself does not pose any new environmental or food safety hazards or risks over those associated with traditional means of genetic manipulation. Therefore, it should only be necessary for regulatory bodies to assess transgenic potatoes at the phenotypic level with respect to the expression of the targeted genetic change. However, to ensure that the general public accept the commercialization of transgenic potatoes, it is important that data are accumulated to provide a scientific basis for the assumption that transgenic plants are no different from traditional plants, other than for the specific trait that is altered. The data summarized in this chapter are only a small start. When conducting initial field trials on transgenic potatoes, it is important that both research institutes and companies take the opportunity to collect data relating to the perceived environmental and food safety risks of these plants. It will be necessary to document such information for different potato cultivars, genetically engineered for a variety of transgenes, and field-grown in different climatic regions from around the globe. There is an urgent need to establish an international database to collate and disseminate this information. The repeated documentation of the data summarized in this chapter will facilitate the general release of transgenic potatoes into agricultural use.

Acknowledgements

I thank Jan Grant, Winna Harvey and Peter Lammerink (Crop & Food Research, New Zealand), and Steve Chandler (Calgene Pacific Australia) for permission to cite their unpublished results. I also thank Bill Belknap for his encouragement to assemble this information, his patience while it was slowly achieved, and his feedback on the content of the original manuscript. This manuscript was completed while the author was the recipient of a Claude McCarthy Fellowship in the Department of Plant Science, Lincoln University, Canterbury, New Zealand.

References

Accatino, P. (1980) Agronomic management in the utilisation of true potato seed: preliminary results. In: *Production of Potatoes From True Seed-Report of a Planning Conference.* International Potato Centre, Lima, Peru, pp. 61-99.

Arumuganathan, K. and Earle, E.D. (1991) Nuclear DNA content of some important plant species. *Plant Molecular Biology Reporter* 9:208-218.

Chasseray, E. and Duesing, J. (1992) Field trials of transgenic plants-an overview. *Agro-Food-Industry Hi-Tech* 3(4):5-10.

Conner, A.J. (1993a) Food safety issues relating to genetic engineering of crop plants. *Agricultural Science* 6(3):36-41.

Conner, A.J. (1993b) Monitoring 'escapes' from field trials of transgenic potatoes: a basis for assessing environmental risks. In: *Seminar on Scientific Approaches for the Assessment of Research Trials with Genetically Modified Plants, Jouy-en-Josas, France, April 1992.* Organisation for Economic Co-operation and Development, Paris, pp. 34-40.

Conner, A.J., Williams, M.K., Abernethy, D.J. and Genet R.A. (1989) Chlorsulfuron-resistant potatoes via *Agrobacterium*-mediated transformation. *Journal of Cellular Biochemistry* 13D:333.

Conner, A.J., Bezar, H.J. and Ashby, J.W. (1990) Genetic engineering for weed, disease and pest control: science versus politics? *Proceedings of the New Zealand Weed and Pest Control Conference* 43:200-208.

Conner, A.J., Williams, M.K., Gardner, R.C., Deroles, S.C., Shaw, M.L. and Lancaster, J.E. (1991) *Agrobacterium*-mediated transformation of New Zealand potato cultivars. *New Zealand Journal of Crop and Horticultural Science* 19:1-8.

Conner, A.J., Shum, T.S., Tynan, J.L., Williams, M.K. and Christey, M.C. (1993) Frequency of transcriptional gene fusion during *Agrobacterium-*

mediated transformation in potato. In: *Abstracts of the Third International Symposium on the Molecular Biology of the Potato*. Santa Cruz, CA, p. 78.

Conner, A.J., Williams, M.K., Abernethy, D.J., Fletcher, P.J. and Genet, R.A. (1994) Field testing of transgenic potatoes. *New Zealand Journal of Crop and Horticultural Science* 22:in press.

Dale, E.C. and Ow, D.W. (1991) Gene transfer with subsequent removal of the selection gene from the host genome. *Proceedings of the National Academy of Sciences USA* 88:10558-10562.

Dale, P.J. (1993) Potato-general presentation. In: *Seminar on Scientific Approaches for the Assessment of Research Trials with Genetically Modified Plants, Jouy-en-Josas, France, April 1992*. Organisation for Economic Co-operation and Development, Paris, pp. 12-20.

Dale, P.J. and McPartlan, H.C. (1992) Field performance of transgenic potato plants compared with controls regenerated from tuber discs and shoot cuttings. *Theoretical and Applied Genetics* 84:585-591.

De Block, M. (1988) Genotype-independent leaf disc transformation of potato (*Solanum tuberosum*) using *Agrobacterium tumefaciens*. *Theoretical and Applied Genetics* 76:767-774.

Duke, S.O., Christy, A.L., Hess, F.D. and Holt, J.S. (1991) *Herbicide-resistant Crops*. Council for Agricultural Science and Technology, Ames, Iowa, 24 pp.

Eijlander, R. and Stiekema, W.J. (1994) Biological containment of potato (*Solanum tuberosum*)-outcrossing to its related wild species black nightshade (*Solanum nigrum*) and bittersweet (*Solanum dulcamara*). *Sexual Plant Reproduction* 7:29-40.

Feldmann, K.A. (1991) T-DNA insertion mutagenesis in *Arabidopsis*: mutational spectrum. *The Plant Journal* 1:71-82.

Field, R.J., Conner, A.J. and Foreman, M.H. (1993) The impact of developing herbicide-resistant crop plants. In: *Proceedings 1: 10th Australian Weeds Conference and 14th Asian Pacific Weed Science Society Conference*. Weed Society of Queensland, Brisbane, pp.315-318.

Flavell, R.B., Dart, E., Fuchs, R.L. and Fraley, R.T. (1992) Selectable marker genes: safe for plants? *Bio/Technology* 10:141-144.

Goldsbrough, A. and Bevan, M. (1991) New patterns of gene activity in plants detected using an *Agrobacterium* vector. *Plant Molecular Biology* 16:263-269.

Hadley, H.H. and Openshaw, S.J. (1980) Interspecific and intergeneric hybridization. In: Fehr, W.R. and Hadley, H.H. (eds), *Hybridization of Crop Plants*. American Society of Agronomy and Crop Science Society of America, Madison, WI, pp. 133-159.

Hamill, J.D., Robins R.J., Parr, A.J., Evans, D.M., Furze, J.M. and Rhodes, M.J.C. (1990) Over-expressing a yeast ornithine decarboxylase gene in transgenic roots of *Nicotiana rustica* can lead to enhanced nicotine accumulation. *Plant Molecular Biology* 15:27-38.

Haughn, G.W., Smith, J., Mazur, B. and Somerville, C. (1988) Transformation with a mutant *Arabidopsis* acetolactate synthase gene renders tobacco resistant to sulfonyl urea herbicides. *Molecular and General Genetics* 204:266-271.

Hawkes, J.G. (1962) Introgression in certain wild potato species. *Euphytica* 11:26-35.
Helgeson, J.P. and Davies, H.V. (eds) (1992) *Workshop on Safeguards for Planned Introduction of Transgenic Potatoes*. USDA APHIS, Hyattsville, MD, 11 pp.
Howard, H.W. (1970) *Genetics of the Potato*-Solanum tuberosum. Logos Press, London, 126 pp.
IFBC (1990) Biotechnologies and food-assuring the safety of foods produced by genetic manipulation. *Regulatory Toxicology and Pharmacology* 12:S1-S196.
Ishida, B.K., Snyder, G.W. and Belknap, W.R. (1989) The use of *in vitro*-grown microtuber discs in *Agrobacterium*-mediated transformation of Russet Burbank and Lemhi Russet potatoes. *Plant Cell Reports* 8:325-328.
Jongedijk, E., de Schutter, A.A.J.M., Stolte, T., van den Elzen, P.J.M. and Cornelissen, B.J.C. (1992) Increased resistance to potato virus X and preservation of cultivar properties in transgenic potato under field conditions. *Bio/Technology* 10:422-429.
Lawson, H.M. (1983) True potato seed as arable weeds. *Potato Research* 26:237-246.
Lindsey, K., Wei, W., Clarke, M.C., McArdle, H.F., Rooke, L.M. and Topping, J.F. (1993) Tagging genomic sequences that direct transgene expression by activation of a promoter trap in plants. *Transgenic Research* 2:33-47.
Lumkes, L.M. and Sijtsma, R. (1972) Mogelijkheden aardappelen als onkruid in volgewassen voorkomen en/of te bestrijden. *Landbouw en Plantenziekten* 1:17-36.
Lutman, P.J.W. (1977) Investigations into some aspects of the biology of potatoes as weeds. *Weed Research* 17:123-132.
Matzke, M. and Matzke, A.J.M. (1993) Genomic imprinting in plants - parental effects and trans-inactivation phenomena. *Annual Review of Plant Physiology and Plant Molecular Biology* 44:53-76.
Meyerowitz, E.M. and Pruitt, R.E. (1985) *Arabidopsis thaliana* and plant molecular genetics. *Science* 229:1214-1218.
Monro, J.A., James, K.A.C. and Conner, A.J. (1993) Comparative nutritional evaluation of a transgenic herbicide-resistant potato and the parent cultivar. *FoodInfo Report*, New Zealand Institute for Crop and Food Research, Christchurch, No. 6:47 pp.
Morris, S.C. and Lee, T.H. (1984) The toxicity and teratogenicity of Solanaceae glycoalkaloids, particularly those of potato (*Solanum tuberosum*); a review. *Food Technology in Australia* 36:118-123.
Moses, T.J., Field R.J. and Conner, A.J. (1993) Field testing of potato lines genetically modified for chlorsulfuron resistance. In: *Proceedings 1: 10th Australian Weeds Conference and 14th Asian Pacific Weed Science Society Conference*. Weed Society of Queensland, Brisbane, pp. 319-322.
Nap, J.P., Bijvoet, J. and Stiekema W.J. (1992) Biosafety of kanamycin-resistant transgenic plants. *Transgenic Research* 1:239-249.
Nap, J.P., Keizer, P. and Janson, R. (1993) First-generation transgenic plants and statistics. *Plant Molecular Biology Reporter* 11:156-164.
Newell, C.A., Rozman, R., Hinchee, M.A., Lawson, E.C., Haley, L., Sanders, P., Kaniewski, W., Tumer, N.E., Horsch, R.B. and Fraley, R.T. (1991)

Agrobacterium-mediated transformation of *Solanum tuberosum L. cv.* 'Russet Burbank'. *Plant Cell Reports* 10:30-34.

OECD (1990) *Database File-Field Releases of Genetically Modified Organisms.* Organisation for Economic Co-operation and Development, Paris, 94 pp.

Ooms, G., Burrell, M.M., Karp, A., Bevan, M. and Hille, J. (1987) Genetic transformation in two potato cultivars with T-DNA from disarmed *Agrobacterium. Theoretical and Applied Genetics* 73:744-750.

Peach, C. and Velten, J. (1991) Transgene expression variability (position effect) of CAT and GUS reporter genes driven by linked divergent T-DNA promoters. *Plant Molecular Biology* 17:49-61.

Plaisted, R.L. (1980) Potato. In: Fehr, W.R. and Hadley, H.H. (eds), *Hybridization of Crop Plants.* American Society of Agronomy and Crop Science Society of America, Madison, WI, pp. 483-494.

Potter, R. and Jones, M.G.K. (1991) An assessment of genetic stability of potato *in vitro* by molecular and phenotypic analysis. *Plant Science* 76:239-248.

Redenbaugh, K., Hiatt, W., Martineau, B., Kramer, M., Sheehy, R., Sanders, R., Houck, C. and Emlay, D. (1992) *Safety Assessment of Genetically Engineered Fruits and Vegetables - A Case Study of the Flavr SavrTM Tomato.* CRC Press, Boca Raton, Florida, 267 pp.

Redenbaugh, K., Berner, T., Emlay, D., Frankos, B., Hiatt, W., Houck, C., Kramer, M., Malyj, L., Martineau, B., Rachman, N., Rudenko, L., Sanders, R., Sheehy, R. and Wixtrom, R. (1993) Regulatory issues for commercialization of tomatoes with an antisense polygalacturonase gene. *In Vitro Cellular and Developmental Biology* 29P:17-26.

Schechtman, M. (1993) US evaluations of field tests of transgenic potatoes. In: *Seminar on Scientific Approaches for the Assessment of Research Trials with Genetically Modified Plants, Jouy-en-Josas, France, April (1992).* Organisation for Economic Co-operation and Development, Paris, pp. 26-32.

Sheerman, S. and Bevan, M.W. (1988) A rapid transformation method for *Solanum tuberosum* using binary *Agrobacterium tumefaciens* vectors. *Plant Cell Reports* 7:13-16.

Shepard, J.F., Bidney, D. and Shahin E. (1980) Potato protoplasts in crop improvement. *Science* 208:17-24.

Simmonds, N.W. (1976) Potatoes - *Solanum tuberosum* (*Solanaceae*). In: Simmonds, N.W. (ed.), *Evolution of Crop Plants.* Longman, London, pp. 279-283.

Stiekema, W.J., Heidekamp, F., Louwerse, J.D., Verhoeven, H.A. and Dijkhuis, P. (1988) Introduction of foreign genes into potato cultivars Bintje and Desiree using an *Agrobacterium tumefaciens* binary vector. *Plant Cell Reports* 7:47-50.

Stockhaus, J., Eckes, P., Blau, A., Schell, J. and Willmitzer, L. (1987) Organ-specific and dosage-dependent expression of a leaf/stem specific gene from potato after tagging and transfer into potato and tobacco plants. *Nucleic Acids Research* 15:3479-3491.

Tepfer, M. (1993) Viral genes and transgenic plants - what are the potential environmental risks? *Bio/Technology* 11:1125-1132.

Tiedje, J.M., Colwell, R.K., Grossman, R.L., Hodson, R.E., Lenski, R.E., Mack, R.N. and Regal, P.J. (1989) The planned introduction of genetically engineered organisms - ecological considerations and recommendations. *Ecology* 70:298-315.

Timmerman, G.M. (1994) The safety of releasing genetically engineered virus-resistant crops - risk assessment. In: Khetarpal, R.K. (ed.), *Plant Virus Disease Control.* Lewis Publishers, Michigan, in press.

Tynan, J.L., Williams, M.K. and Conner, A.J. (1990) Low frequency of pollen dispersal from a field trial of transgenic potatoes. *Journal of Genetics and Breeding* 44:303-305.

van der Leij, F.R., Visser, R.G.F., Oosterhaven, K., van de Kop, D.A.M., Jacobsen, E. and Feenstra, W.J. (1991) Complementation of the amylose-free starch mutant of potato (*Solanum tuberosum*) by the gene encoding granule-bound starch synthase. *Theoretical and Applied Genetics* 82:289-295.

Vayda, M.E. and Belknap, W.R. (1992) The emergence of transgenic potatoes as commercial products and tools for basic research. *Transgenic Research* 1:149-163.

Visser, R.G.F., Jacobsen, E., Witholt, B. and Feenstra, W.J. (1989a) Efficient transformation of potato (*Solanum tuberosum* L.) using a binary vector in *Agrobacterium rhizogenes*. *Theoretical and Applied Genetics* 78:594-600.

Visser, R.G.F., Hesseling-Meinders, A., Jacobsen, E., Nijdam, H., Witholt, B. and Feenstra, W.J. (1989b) Expression and inheritance of inserted markers in binary vector carrying *Agrobacterium rhizogenes*-transformed potato (*Solanum tuberosum* L.). *Theoretical and Applied Genetics* 78:705-714.

Walden, R., Hayashi, H. and Schell, J. (1991) T-DNA as a gene tag. *The Plant Journal* 1:281-288.

Wearing, C.H. and Hokkanen, H.M.T. (1994) Pest resistance to *Bacillus thuringiensis* - ecological crop assessment for *Bt* gene incorporation and management strategies. In: Hokkanen, H.M.T. and Lynch, J.M. (eds), *Biological Control - Benefits and Risks.* Cambridge University Press, Cambridge, in press.

Wenzler, H., Mignery, G., May, G. and Park, W. (1989) A rapid and efficient transformation method for the production of large numbers of transgenic potato plants. *Plant Science* 63:79-85.

Williams, M.K., Conner, A.J., Gardner, R.C., Deroles, S.C. and Lammerink, J.P. (1988) Gene transfer to potato using *Agrobacterium.* In: McWhirter, K.S., Downes, R.W. and Read, B.J. (eds), *Ninth Australian Plant Breeding Conference, Proceedings.* Agricultural Research Institute, Wagga Wagga, pp. 71-72.

Witty, M. and Harvey, W.J. (1990) Sensory evaluation of transgenic *Solanum tuberosum* producing r-thaumatin II. *New Zealand Journal of Crop and Horticultural Science* 18:77-80.

Index